M000277154

# Statistical Physics, Optimization, Inference, and Message-Passing Algorithms

# École de Physique des Houches
Special Issue, 30 September–11 October 2013

# Statistical Physics, Optimization, Inference, and Message-Passing Algorithms

Edited by
## Florent Krzakala
*École Normale Supérieure and Sorbonne Universités,*
*Université Pierre et Marie Cuire*

## Federico Ricci-Tersenghi
*Università di Roma La Sapienza*

## Lenka Zdeborová
*Centre National de la Recherche Scientique and*
*Commissariat à l'énergie atomique Saclay*

## Riccardo Zecchina
*Politecnico di Torino*

## Eric W. Tramel
*École Normale Supérieure*

## Leticia F. Cugliandolo
*Sorbonne Universités, Université Pierre et Marie Curie*

OXFORD
UNIVERSITY PRESS

# OXFORD
## UNIVERSITY PRESS

Great Clarendon Street, Oxford, OX2 6DP,
United Kingdom

Oxford University Press is a department of the University of Oxford.
It furthers the University's objective of excellence in research, scholarship,
and education by publishing worldwide. Oxford is a registered trade mark of
Oxford University Press in the UK and in certain other countries

© Oxford University Press 2016

The moral rights of the authors have been asserted

First Edition published in 2016

Impression: 1

All rights reserved. No part of this publication may be reproduced, stored in
a retrieval system, or transmitted, in any form or by any means, without the
prior permission in writing of Oxford University Press, or as expressly permitted
by law, by licence or under terms agreed with the appropriate reprographics
rights organization. Enquiries concerning reproduction outside the scope of the
above should be sent to the Rights Department, Oxford University Press, at the
address above

You must not circulate this work in any other form
and you must impose this same condition on any acquirer

Published in the United States of America by Oxford University Press
198 Madison Avenue, New York, NY 10016, United States of America

British Library Cataloguing in Publication Data
Data available

Library of Congress Control Number: 2015944567

ISBN 978–0–19–874373–6

Printed and bound by
CPI Group (UK) Ltd, Croydon, CR0 4YY

Links to third party websites are provided by Oxford in good faith and
for information only. Oxford disclaims any responsibility for the materials
contained in any third party website referenced in this work.

*To Alena and Julie,*
*who did their best not to disturb the organization of the school.*

École de Physique des Houches

Service inter-universitaire commun
à l'Université Joseph Fourier de Grenoble
et à l'Institut National Polytechnique de Grenoble

Subventionné par l'Université Joseph Fourier de Grenoble,
le Centre National de la Recherche Scientifique,
le Commissariat à l'Énergie Atomique

*Directeur:*

Leticia F. Cugliandolo, Sorbonne Universités, Université Pierre et Marie Curie
Laboratoire de Physique Théorique et Hautes Energies, Paris, France

*Directeurs scientifiques de la session:*

Florent Krzakala, École Normale Supérieure and Sorbonne Universités, Université
Pierre et Marie Cuire, Paris, France
Federico Ricci-Tersenghi, Università La Sapienza di Roma, Italy
Lenka Zdeborovà, Centre National de la Recherche Scientique and Commissariat à
l'énergie atomique Saclay, France
Riccardo Zecchina, Politecnico di Torino, Italy
Eric W. Tramel, École Normale Supérieure, Paris, France
Leticia F. Cugliandolo, Sorbonne Universités, Université Pierre et Marie Curie
Laboratoire de Physicque Théorique et Hautes Energies, Paris, France

# Previous sessions

## Publishers

- Session VIII: Dunod, Wiley, Methuen
- Sessions IX and X: Herman, Wiley
- Session XI: Gordon and Breach, Presses Universitaires
- Sessions XII–XXV: Gordon and Breach
- Sessions XXVI–LXVIII: North Holland
- Session LXIX–LXXVIII: EDP Sciences, Springer
- Session LXXIX–LXXXVIII: Elsevier
- Session LXXXIX– : Oxford University Press

# Preface

This book brings together a collection of chapters based on notes from lectures given at the autumn school *Statistical Physics, Optimization, Inference, and Message-Passing Algorithms* held at the Les Houches School of Physics, France, from Monday 30 September 2013 to Friday 11 October 2013. The school was organized by Florent Krzakala from UPMC & ENS Paris, Federico Ricci-Tersenghi from l'Università di Roma I "La Sapienza", Lenka Zdeborová from CEA Saclay & CNRS, and Riccardo Zecchina from Politecnico di Torino.

## Scope and motivation

The connection between statistical physics, combinatorial optimization, and statistical inference has been known for over thirty years. Perhaps the most fruitful cross-fertilization between these fields has been between message-passing algorithms in computer science and the cavity method and Bethe approximation in physics.

In physics, the use of message-passing algorithms can be traced back to Hans Bethe, who introduced a mean-field-like approximation to describe the behavior of many-body systems as early as 1935. Giorgio Parisi, Marc Mézard, and Miguel Angel

Virasoro later proposed the cavity method as an alternative understanding of results obtained by using the famous, but sometimes controversial, replica trick. Their original motivation came from the field of disordered magnets called spin glasses. However, they soon realized that these techniques were broadly useful beyond this field and published the now-classic book *Spin Glass Theory and Beyond*. In information theory, message-passing algorithms were introduced by Robert G. Gallager in 1963 in his work on low-density parity-check error correcting codes, while, in machine learning, Judea Pearl independently invented belief propagation in 1982 while working on Bayesian inference. These works were originally developed independently, yet they both utilize exactly the same concept under different names.

Over the last decade, the use of techniques related to message passing has spread with increasing speed into new fields and problems. This has led to groundbreaking work in information theory, machine learning, probability, optimization, statistics, signal processing, and other fields. Many high-level conferences and journals accept contributions related to these topics. Every year, there are several international research conferences and workshops devoted solely to this subject, with the aim of strengthening the community of researchers coming from different backgrounds and facilitating the exchange of expertise. In 2012, these events included *Mathematical Challenges in Graphical Models and Message-Passing Algorithms* at IPAM UCLA (USA), *Bridging Statistical Physics and Optimization, Inference and Learning* at the Les Houches Physics School (France), *Statistical Mechanics of Unsatisfiability and Glasses* at Mariehamn (Finland), and *Disorder, Algorithms and Complexity* at the Aspen Center for Physics, Colorado (USA).

However, a school aimed at teaching the foundations and applications of these techniques to students from varied backgrounds was sorely missed. The aim, then, of the Les Houches School *Statistical Physics, Optimization, Inference, and Message-Passing Algorithms* was to present the background necessary for entry into this fast-developing field. The lectures explained state-of-the-art developments in message-passing algorithms and their applications. It was our intention that the same concepts be explained in several different lectures from slightly different perspectives, perhaps with different notations and vocabulary.

The notes on which the chapters in this book were based were written by two or three volunteer students from the audience and revised by the corresponding lecturer. All the lecturers are active leading experts in their fields, each having made groundbreaking contributions. We hope that students and researchers from physics, computer science, information theory, machine learning, mathematics, and other related fields will find these notes instructive and thought-provoking.

## Lectures

The lectures by Devavrat Shah (*MIT*) covered the basics of inference and learning. He explained how inference problems are represented within structures known as graphical models. The theoretical basis of the belief propagation algorithm was then explained and derived. This lecture set the stage for generalizations and applications of message-passing algorithms.

Cristopher Moore (*Santa Fe Institute*) introduced the notion of computational complexity and unveiled the basics of random graph theory and random constraint satisfaction problems and their associated phase transitions. Finally, for the example of community detection problems, using belief propagation and related analysis, he illuminated the connections between phase transitions and computational complexity.

Giorgio Parisi (*La Sapienza Rome*) introduced the now famous replica method. He presented the celebrated solution of the Sherrington–Kirkpatrick model, and the theory of replica symmetry breaking and its physical interpretation. He discussed where these results stand nowadays from a rigorous point of view.

Marc Mézard (*ENS Paris*) introduced the cavity method and derived belief propagation, this time from the perspective of statistical physics. He showed how this approach can be used to obtain the replica-symmetric solution of the Sherrington–Kirkpatrick model, thus illuminating the connection between the cavity and replica methods. He applied replica symmetry breaking to random constraint satisfaction problems and derived striking predictions for the space of solutions and related survey propagation algorithms.

Andrea Montanari (*Stanford University*) presented the basics of statistical estimation for linear models. He connected the theory of denoising to compressed sensing, a recent revolutionary concept from signal processing. He derived a version of belief propagation adapted to linear estimation—so-called approximate message passing—which is closely related to the Thouless–Anderson–Palmer equations from physics. He also discussed other application-focused problems (hidden clique discovery and reconstruction of low-rank matrices) where message-passing algorithms and the analysis of phase transitions lead to remarkable, and otherwise unobtainable, results.

Laurent Daudet (*ESPCI Paris*) gave a seminar about the application of compressed sensing in an optical imaging experiments.

Rüdiger Urbanke (*EPFL Lausanne*) discussed recent developments in error correcting codes. He described and compared two known classes of codes that are able to achieve the Shannon capacity: polar codes and spatially coupled low-density parity-check codes. One of the lectures on spatial coupling and its properties was given by Henry Pfister (*Texas A&M University*).

Amin Coja-Oghlan (*University of Frankfurt*) lectured about recent proofs of some of the predictions made in statistical physics for the random regular $K$-satisfiability problem. Remarkably, the proof techniques closely follow the path set by the cavity method. Such a development makes it plausible that within several years all the predictions of the cavity method will be made mathematically rigorous.

Mike Molloy (*University of Toronto*) gave a lecture about mathematical progress in the understanding of the so-called freezing of variables that has been introduced in physics as a property of random constraint satisfaction problems.

David Gamarnik (*MIT*) used statistical physics tools to show that local algorithms provide some precise estimations for the independent set problem. He also discussed how the limits of such algorithms are linked to the clustering of solutions—a prediction of the cavity method.

Manfred Opper (*Technical University of Berlin*) presented belief propagation as a special case in a more general framework of approximate variational inference and expectation propagation.

Riccardo Zecchina (*Politecnico di Torino*) described an important example of how to include global constraints in belief propagation and how to write consistent message-passing algorithms for a class of dynamical problems.

# Acknowledgements

The organization of the school would not have been possible without generous funding from the CNRS, the European Research Council under the European Union's 7th Framework Programme (FP/2007–2013)/ERC Grant Agreement 307087-SPARCS and 267915-OPTINF, the Japan Society for the Promotion of Science, the Institute Henri Poincaré in Paris, the European Network NETADIS, the Swiss National Foundation under grant 200020-140388, and the Italian Research Minister through the FIRB Project No. RBFR086NN1.

We also acknowledge gratefully the work of Flora Gheno, Murielle Gardette, and Isabel Lelievre, the secretaries of the Les Houches School, who were instrumental in the smooth organization and operation of the school. Special thanks go to Božena Zdeborová for taming the dragon during the lectures.

# Contents

# List of participants

The participants are listed with affiliations and positions that they held when the school took place.

## ORGANIZERS

KRZAKALA FLORENT
UPMC & ENS Paris, France

RICCI-TERSENGHI FEDERICO
La Sapienza Roma, Italy

ZDEBOROVÁ LENKA
CEA Saclay & CNRS, France

ZECCHINA RICCARDO
Politecnico Torino, Italy

## LECTURERS

COJA-OGHLAN AMIN
University Frankfurt, Germany

GAMARNIK DAVID
MIT, USA

MÉZARD MARC
ENS Paris, France

MONTANARI ANDREA
Stanford, USA

MOORE CRISTOPHER
Santa Fe Institute, USA

OPPER MANFRED
TU Berlin, Germany

PARISI GIORGIO
La Sapienza Roma, Italy

SHAH DEVAVRAT
MIT, USA

URBANKE RÜDIGER
EPFL, Switzerland

# ACADEMIC PARTICIPANTS

DAUDET LAURENT
ESPCI Paris, France

LU YUE
Harvard University, USA

MOFFATT IAIN
University of London, UK

MOLLOY MIKE
University of Toronto, Canada

PFISTER HENRY
Texas A&M University, USA

SCHABANEL NICOLAS
Paris Diderot, France

ZUK OR
University of Chicago, USA

# POSTDOCS

ANGELINI MARIA CHIARA
CEA Saclay & CNRS, France

BENNA MARCUS
Columbia University, USA

BENNETT PATRICK
University of Toronto, Canada

CALTAGIRONE FRANCESCO
CEA Saclay & CNRS, France

DECELLE AURÉLIEN
La Sapienza Roma, Italy

DE LUCA ANDREA
ENS Paris, France

DREMEAU ANGELIQUE
ESPCI Paris, France

HASSANI SEYED HAMED
EPFL, Switzerland

LAGE CASTELLANOS ALEJANDRO
Politecnico Torino, Italy

MARTI DANIEL
ENS Paris, France

MOZEIKA ALEXANDER
Aalto University, Finland

RAYMOND JACK
La Sapienza Roma, Italy

SAKATA AYAKA
Tokyo Institute of Technology, Japan

TRAMEL ERIC
ENS Paris, France

VEHKAPERA MIKKO
KTH, Sweden

ZHANG PAN
Santa Fe Institute, USA

# STUDENTS

BARBIER JEAN
ENS Paris, France

COHEN OR
Weizmann Institute, Israel

DE BACCO CATERINA
Université Paris-Sud, France

DEL FERRARO GINO
KTH, Sweden

DITTMANN JONAS
Universität Wurzburg, Germany

DUBREUIL ALEXIS
University of Chicago, USA

EL-KHATIB RAFAH
EPFL, Switzerland

FALKNER STEFAN
Emory University, USA

GIRNYK MAKSYM
KTH, Sweden

GIURGIU ANDREI
EPFL, Switzerland

GUGGIOLA ALBERTO
ENS Paris, France

HARRISON ELIZABETH
Aston University, UK

HEMERY MATHIEU
UPMC & Grenoble, France

HETTERICH SAMUEL
Universität Frankfurt, Germany

HUTTEL JANINA
Universität Frankfurt, Germany

JACOBS ABIGAIL
University of Colorado, USA

KAMILOV ULUGBEK
EPFL, Switzerland

LEMOY RÉMI
Aalto University, Finland

LOKHOV ANDREY
Université Paris-Sud, France

MANOEL ANDRE
University of Sao Paolo, Brazil

MONDELLI MARCO
EPFL, Switzerland

MORONE FLAVIANO
La Sapienza Roma, Italy

NG TONY
Brandeis University, USA

PAGA PIERRE
King's College London, UK

PERUGINI GABRIELE
La Sapienza Roma, Italy

RASSMANN FELICIA
Universität Frankfurt, Germany

SAADE ALAA
ENS Paris, France

SAKAI YUJI
University of Tokyo, Japan

SCHÜLKE CHRISTOPHE
CEA Saclay & CNRS, France

SHRESTHA MUNIK
Santa Fe Institute, USA

SINGH VIJAY
Emory University, USA

TAKABE SATOSHI
University of Tokyo, Japan

UEDA MASAHIKO
Kyoto University, Japan

VANAPARTHY SANTHOSH KUMAR
Texas A&M University, USA

WANG CHUANG
Institute for Theoretical Physics, Chinese Academy of Sciences, China

WATANABE SHUNSUKE
Tokyo Institute of Technology, Japan

XU YINGYING
Tokyo Institute of Technology, Japan

# PRESENTED POSTERS

1. Marco Mondelli, *Scaling exponent of list decoders with applications to polar codes.*
2. Andrei Giurgiu, *A proof of the exactness of the replica symmetric formula for LDPC codes above the MAP threshold.*
3. El-Khatib Rafah, Displacement convexity—a framework for the analysis of spatially coupled codes.
4. Santhosh Kumar Vanaparthy, *Threshold saturation for spatially-coupled LDPC and LDGM codes on BMS channels.*
5. Hamed Hassani, *New lower bounds for CSPs using spatial coupling.*
6. Jean Barbier, *Robust error correction for real-valued signals via message-passing decoding and spatial coupling.*
7. Francesco Caltagirone, *Dynamics and termination cost of spatially coupled mean-field models.*
8. Cohen Or, *Low temperature expansion of steady-state measures of non-equilibrium Markov chains via graphical representation.*
9. Aurelien Decelle, *Belief propagation inspired Monte Carlo.*
10. Yuji Sakai, *Markov chain Monte Carlo method with skew detailed balance conditions.*
11. Satoshi Takabe, *Typical behavior of the linear programming method for combinatorial optimization problems.*

12. Masahiko Ueda, *Calculation of 1RSB transition temperature of spin glass models on regular random graphs under replica symmetric ansatz.*
13. Wang Chuang, *Tensor renormalization group method for spin glass.*
14. Gino Del Ferraro, *Mean field spin glasses treated with PDE techniques.*
15. Flaviano Morone, *Large deviations of correlation functions in random magnets.*
16. Andre Manoel, *Statistical mechanics for the analysis and design of information hiding algorithms.*
17. Ayaka Sakata, *Time evolution of autocorrelation function in dynamical replica theory.*
18. Alexander Mozeika, *Projected generalized free energies for non-equilibrium states.*
19. Stefan Falkner, *A renormalization group approach to quantum walks.*
20. Moffatt Iain, *The Potts–Tutte connection in an external field.*
21. Jonas Dittmann, *TV regularized tomographic reconstruction from few (X-ray) projections based on CS.*
22. Yingying Xu, *Statistical mechanics approach to 1-bit compressed sensing.*
23. Christophe Schülke, *Blind calibration in compressed sensing using message passing algorithms.*
24. Wang Chuang, *Partition function expansion for generalized belief propagation.*
25. Harrison Elizabeth, *Probabilistic control in smart-grids.*
26. Maksym Girnyk, *A statistical mechanical approach to MIMO channels.*
27. Ulugbek Kamilov, *Wavelet-domain approximate message passing for Bayesian image deconvolution.*
28. Rémi Lemoy, *Variable-focused local search on random 3-SAT.*
29. Alberto Guggiola, *Mapping between sequence and response space in the olfactory system in* Drosophila.
30. Marcus Benna, *Long-term memory with bounded synaptic weights.*
31. Jack Raymond, *Utilizing the Hessian to improve variational inference.*
32. Andrey Lokhov, *Dynamic message-passing equations and application to inference of epidemic origin.*
33. Aurelien Decelle, *Decimation based method to improve inference using the pseudo-likelihood method.*
34. Munik Shrestha, *Spread of reinforcement driven trends in networks with message-passing approach.*
35. Dani Martí, *Scalability properties of multimodular networks with dynamic gating.*
36. Abigail Zoe Jacobs, *Latent space models for network structure: an application to ecology.*
37. Shunsuke Watanabe, *The analysis of degree-correlated networks based on cavity method.*
38. Caterina De Bacco, *Shortest non-overlapping routes on random graphs.*

# 1
# Statistical inference
# with probabilistic graphical models

## Devavrat SHAH

Laboratory for Information and Decisions Systems, Operations Research Center,
Department of Electrical Engineering and Computer Science,
Massachusetts Institute of Technology,
Cambridge, MA 02139

*Lecture notes taken by*
Angélique Drémeau, École Normale Supérieure, France
Christophe Schülke, Université Paris Diderot, France
Yingying Xu, Tokyo Institute of Technology, Japan

*Statistical Physics, Optimization, Inference, and Message-Passing Algorithms.* First Edition.
F. Krzakala et al. © Oxford University Press 2016. Published in 2016 by Oxford University Press.

# Chapter Contents

## 1.1 Introduction to graphical models

### 1.1.1 Inference

Consider two random variables $A$ and $B$ with a joint probability distribution $P_{A,B}$. From the observation of the realization of one of those variables, say $B = b$, we want to infer the one that we did not observe. To that end, we compute the conditional probability distribution $P_{A|B}$, and use it to obtain an estimate $\hat{a}(b)$ of $a$.

To quantify how good this estimate is, we introduce the **error probability**:

$$P_{error} \triangleq P(A \neq \hat{a}(b)|B = b) \tag{1.1}$$
$$= 1 - P(A = \hat{a}(b)|B = b),$$

and we can see from the second equality that minimizing this error probability is equivalent to the following maximization problem, called the maximum a posteriori (**MAP**) problem:

$$\hat{a}(b) = \operatorname*{argmax}_{a} P_{A|B}(a|b). \tag{1.2}$$

The problem of computing $P_{A|B}(a|b)$ for all $a$ given $b$ is called the marginal (**MARG**) problem. When the number of random variables increases, the MARG problem becomes difficult, because an exponential number of combinations has to be calculated.

**Fano's inequality** provides us with an information-theoretic way of gaining insight into how much information about $a$ the knowledge of $b$ can give us:

$$P_{error} \geq \frac{H(A|B) - 1}{\log|A|}, \tag{1.3}$$

with

$$H(A|B) = \sum_{b} P_B(b) H(A|B = b),$$

$$H(A|B = b) = \sum_{a} P_{A|B}(a|b) \log \left( \frac{1}{P_{A|B}(a|b)} \right).$$

Fano's inequality formalizes only a theoretical bound that does not tell us how to actually make an estimation. From a practical point of view, graphical models (GMs) constitute here a powerful tool allowing us to write algorithms that solve inference problems.

### 1.1.2 Graphical models

#### *Directed GMs*

Consider $N$ random variables $X_1, \ldots, X_N$ on a discrete alphabet $\mathcal{X}$, and their joint probability distribution $P_{X_1 \cdots X_N}$. We can always factorize this joint distribution in the following way:

$$P_{X_1 \cdots X_N} = P_{X_1} P_{X_2|X_1} \cdots P_{X_N|X_1 \cdots X_{N-1}} \tag{1.4}$$

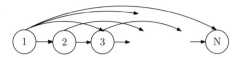

**Fig. 1.1**  A directed graphical model representing the factorized form (1.4).

**Fig. 1.2**  A simpler graphical model representing the factorized form (1.5).

and represent this factorized form by the directed graphical model shown in Fig. 1.1. In this graphical model, each node is associated with a random variable, and each directed edge represents a conditioning. Because of the way that we factorized the distribution, we obtain a complicated graphical model, in the sense that it has many edges. A much simpler graphical model would be the one shown in Fig. 1.2. The latter graphical model corresponds to a factorization in which each of the probability distributions in the product is conditioned on only one variable:

$$P_{X_1 \cdots X_N} = P_{X_1} P_{X_2 | X_1} \cdots P_{X_N | X_{N-1}}. \tag{1.5}$$

In the most general case, we can write a distribution represented by a directed graphical model in the factorized form

$$P_{X_1 \cdots X_N} = \prod_i P_{X_i | X_{\Pi_i}}, \tag{1.6}$$

where $X_{\Pi_i}$ is the set containing the parents of $X_i$ (the vertices from which an edge points to $i$).

The following **notation** will be used in the rest of this chapter:

- random variables are capitalized: $X_i$;
- realizations of random variables are lower case: $x_i$;
- a set of random variables $\{X_1, \ldots, X_N\}$ is denoted by $\underline{X}$;
- a set of realizations of $\underline{X}$ is denoted by $\underline{x}$;
- the subset of random variables of indices in $S$ is denoted by $X_S$.

### Undirected GMs

Another type of graphical model is the undirected graphical model. In that case, we define the graphical model not through the **factorization**, but through **independence**.

Let

$$\mathcal{G}(\mathcal{V}, \mathcal{E}) \quad \text{be an undirected graphical model, where}$$
$$\mathcal{V} = \{1, \ldots, N\} \quad \text{is the set of vertices, and}$$
$$\mathcal{E} \subseteq \mathcal{V} \times \mathcal{V} \quad \text{is the set of edges.}$$

Each vertex $i \in \mathcal{V}$ of this GM represents one random variables $X_i$, and each edge $(i,j) \in \mathcal{E}$ represents a conditional dependence. As the GM is undirected, we have $(i,j) \equiv (j,i)$.

We define

$$N(i) \triangleq \{j \in \mathcal{V} | (i,j) \in \mathcal{E}\} \quad \text{the set containing the neighbours of } i. \qquad (1.7)$$

The undirected graphical model captures the following dependence:

$$P_{X_i|X_{\mathcal{V}\backslash\{i\}}} \equiv P_{X_i|X_{N(i)}}, \qquad (1.8)$$

meaning that only variables connected by edges have a conditional dependence.

Let $A \subset \mathcal{V}$, $B \subset \mathcal{V}$, $C \subset \mathcal{V}$. We write $X_A \perp X_B \mid X_C$ if $A$ and $B$ are disjoint and if all paths leading from one element of $A$ to one element of $B$ lead over an element of $C$, as illustrated in Fig. 1.3. In other words, if we remove $C$, then $A$ and $B$ are unconnected (Fig. 1.4).

Undirected GMs are also called **Markov random fields** (MRFs).

### *Cliques*

(**Definition**) A clique is a subgraph of a graph in which all possible pairs of vertices are linked by an edge. A maximal clique is a clique that is contained by no other clique (see, e.g., Fig. 1.5).

**Theorem 1.1 (Hammersley and Clifford, 1971)** *Given an MRF $\mathcal{G}$ and a probability distribution $P_{\underline{X}}(\underline{x}) > 0$, then*

$$P_{\underline{X}}(\underline{x}) \propto \prod_{C \in \mathcal{C}} \phi_C(x_C), \qquad (1.9)$$

*where $\mathcal{C}$ is the set of cliques of $\mathcal{G}$.*

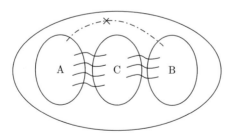

**Fig. 1.3** Schematic view of a graphical model in which $X_A \perp X_B \mid X_C$. All paths leading from $A$ to $B$ go through $C$.

**Fig. 1.4** Simple view showing the independence of $A$ and $B$ conditioned on $C$.

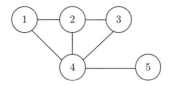

**Fig. 1.5** In this graphical model, the maximal cliques are $\{1, 2, 4\}$, $\{2, 3, 4\}$, and $\{4, 5\}$.

**Proof** (Grimmet, 1973) for $\mathcal{X} = \{0, 1\}$.
We will show the equivalent formulation

$$P_{\underline{X}}(\underline{x}) \propto e^{\sum_{C \in \mathcal{C}} V_C(x_C)} \tag{1.10}$$

by exhibiting the solution

$$V_C(x_C) = \begin{cases} Q(C) & \text{if } x_C = \mathbb{1}_C, \\ 0 & \text{otherwise}, \end{cases} \tag{1.11}$$

with

$$Q(C) = \sum_{A \subseteq C} (-1)^{|C-A|} \underbrace{\ln P_{\underline{X}}\left(x_A = \mathbb{1}_A, x_{V \setminus A} = 0\right)}_{\triangleq G(A)}. \tag{1.12}$$

Suppose we have an assignment $\underline{X} \mapsto N(\underline{X}) = \{i | x_i = 1\}$. We want to prove that

$$G(N(\underline{X})) \triangleq \ln P_{\underline{X}}(\underline{x})$$
$$= \sum_{C \in \mathcal{C}} V_C(x_C)$$
$$= \sum_{C \subseteq N(\underline{x})} Q(C). \tag{1.13}$$

This is equivalent to proving the following two claims:

C1: $\quad \forall S \subset \mathcal{C}, \quad G(S) = \sum_{A \subseteq S} Q(A);$

C2: $\quad$ if $A$ is not a clique, $\quad Q(A) = 0.$

Let us begin by proving C1:

$$\sum_{A \subseteq S} Q(A) = \sum_{A \subseteq S} \sum_{B \subseteq A} (-1)^{|A-B|} G(B)$$
$$= \sum_{B \subseteq S} G(B) \left( \sum_{B \subseteq A \subseteq S} (-1)^{|A-B|} \right), \tag{1.14}$$

where we note that the term in large parentheses is zero except when $B = S$, because we can rewrite it as

$$\sum_{0 \le l \le k} (-1)^l \binom{l}{k} = (-1+1)^k = 0. \tag{1.15}$$

Therefore, $G(S) = \sum_{A \subseteq S} Q(A)$.

For C2, suppose that $A$ is not a clique, which allows us to choose $(i, j) \in A$ with $(i, j) \notin \mathcal{E}$. Then

$$Q(A) = \sum_{B \subseteq A \setminus \{i,j\}} (-1)^{|A-B|} \left[ G(B) - G(B+i) + G(B+i+j) - G(B+j) \right].$$

Let us show that the term in square brackets is zero by showing

$$G(B+i+j) - G(B+j) = G(B+i) - G(B),$$

or equivalently

$$\ln \frac{P_X \left( x_B = \mathbb{1}_B, x_i = 1, x_j = 1, x_{\mathcal{V} \setminus \{i,j,B\}} = \mathbb{0} \right)}{P_X \left( x_B = \mathbb{1}_B, x_i = 0, x_j = 1, x_{\mathcal{V} \setminus \{i,j,B\}} = \mathbb{0} \right)}$$

$$= \ln \frac{P_X \left( x_B = \mathbb{1}_B, x_i = 1, x_j = 0, x_{\mathcal{V} \setminus \{i,j,B\}} = \mathbb{0} \right)}{P_X \left( x_B = \mathbb{1}_B, x_i = 0, x_j = 0, x_{\mathcal{V} \setminus \{i,j,B\}} = \mathbb{0} \right)},$$

where $\mathcal{V} \setminus \{i, j, B\}$ stands for the set of all vertices except $i$, $j$, and those in $B$. We see that the only difference between the left- and right-hand sides is the value taken by $x_j$. Using Bayes' rule, we can rewrite both sides in the form

$$\ln \frac{P_X \left( X_i = 1 | X_j = \pm 1, X_B = \mathbb{1}_B, X_{\mathcal{V} \setminus \{i,j,B\}} = \mathbb{0} \right)}{P_X \left( X_i = 0 | X_j = \pm 1, X_B = \mathbb{1}_B, X_{\mathcal{V} \setminus \{i,j,B\}} = \mathbb{0} \right)}.$$

As $(i, j) \notin \mathcal{E}$, the conditional probabilities on $X_i$ do not depend on the value taken by $X_j$, and therefore the right- and left-hand sides are equal, $Q(A) = 0$, and C2 is proved.  □

### 1.1.3   Factor graphs

Thanks to the Hammersley–Clifford theorem, we know that we can write a probability distribution corresponding to an MRF $\mathcal{G}$ in the following way:

$$P_X (\underline{x}) \propto \prod_{C \in \mathcal{C}^*} \phi_C(x_C), \tag{1.16}$$

where $\mathcal{C}^*$ is the set of maximal cliques of $G$. In a general definition, we can also write

$$P_X (\underline{x}) \propto \prod_{F \in \mathcal{F}} \phi_F(x_F), \tag{1.17}$$

where the partition $\mathcal{F} \subseteq 2^{\mathcal{V}}$ has nothing to do with any underlying graph.

In what follows, we give two examples in which the introduction of factor graphs is a natural approach to an inference problem.

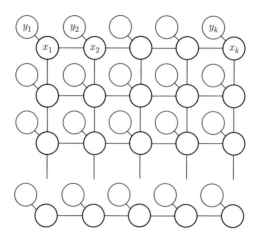

**Fig. 1.6**  Graphical model representing a two-dimensional image. The thick circles correspond to the pixels of the image $x_k$, and each one is linked to a noisy measurement $y_k$. Adjacent pixels are linked by edges that allow modeling the assumed smoothness of the image.

### Image processing

We consider an image with binary pixels ($\mathcal{X} = \{-1, 1\}$) and a probability distribution

$$p(\underline{x}) \propto e^{\sum_{i \in V} \theta_i x_i + \sum_{(i,j) \in E} \theta_{ij} x_i x_j}. \tag{1.18}$$

For each pixel $x_k$, we record a noisy version $y_k$ (Fig. 1.6). We consider natural images, in which big jumps in intensity between two neighboring pixels are unlikely. This can be modeled with

$$a \sum_i x_i y_i + b \sum_{(i,j) \in \mathcal{E}} x_i x_j. \tag{1.19}$$

This way, the first term pushes $x_k$ to match the measured value $y_k$, while the second term favors piecewise-constant images. We can identify $\theta_i \equiv a y_i$ and $\theta_{ij} \equiv b$.

### Crowd-sourcing

Crowd-sourcing is used for tasks that are easy for humans but difficult for machines and that are as hard to verify as to evaluate. Crowd-sourcing consists in assigning to each of $M$ human "workers" a subset of $N$ tasks to evaluate and to collect their answers $A$. Each worker has a different error probability $p_i \in \{\frac{1}{2}, 1\}$: either he gives random answers or he is fully reliable. The goal is to infer both the correct value of each task $t_j$ and the $p_i$ of each worker. The factor graph corresponding to that problem is represented in Fig 1.7.

The conditional probability distribution of $\underline{t}$ and $\underline{p}$ knowing the answers $A$ reads

$$P_{\underline{t},\underline{p}|\underline{A}} \propto P_{\underline{A}|\underline{t},\underline{p}} P_{\underline{t},\underline{p}}$$

$$\propto P_{\underline{A}|\underline{t},\underline{p}}, \tag{1.20}$$

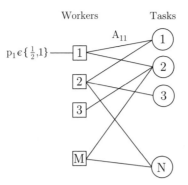

**Fig. 1.7** Graphical model illustrating crowd-sourcing. Each worker $i$ is assigned a subset of the tasks for evaluation, and for each of those tasks $a$, his answer $A_{ia}$ is collected.

where we have assumed a uniform distribution on the joint probability $P_{\underline{t},\underline{p}}$. Then

$$P_{\underline{A}|\underline{t},\underline{p}} = \prod_e P_{A_e|t_e,p_e}, \tag{1.21}$$

with

$$P_{A_e|t_e,p_e} = \left(\left(\frac{p_e}{1-p_e}\right)^{A_e t_e}(1-p_e)p_e\right)^{\frac{1}{2}}. \tag{1.22}$$

### 1.1.4 MAP and MARG

*MAP*

The MAP problem consists in solving

$$\max_{\underline{x}\in\{0,1\}^N} \sum_i \theta_i x_i + \sum_{(i,j)\in\mathcal{E}} \theta_{ij} x_i x_j. \tag{1.23}$$

When $\theta_{ij} \to -\infty$, neighboring nodes can no longer be in the same state. This is the hard-core model, which is very hard to solve.

*MARG*

The MARG focuses on the evaluation of marginal probabilities, depending on only one random variable, for instance,

$$P_{X_1}(0) = \frac{Z(X_1 = 0)}{Z}, \tag{1.24}$$

as well as conditional marginal probabilities,

$$P_{X_2|X_1}(X_2 = 0|X_1 = 0) = \frac{Z(X_1 = 0, X_2 = 0)}{Z(X_1 = 0)}, \tag{1.25}$$

$$P_{X_N|X_1 \cdots X_{N(1)}}(X_N = 0|X_1 \cdots X_{N-1} = 0) = \frac{Z(\text{all } 0)}{Z(\text{all but } X_N \text{ are } 0)}, \tag{1.26}$$

and

$$P_{X_1}(0) \times \cdots \times P_{X_N|X_1 \cdots X_{N-1}}(0) = \frac{1}{Z}. \tag{1.27}$$

Both of these problems are computationally hard. Can we design efficient algorithms to solve them?

## 1.2   Inference algorithms: elimination, junction tree, and belief propagation

In the MAP and MARG problems described previously, the hardness comes from the fact that with growing instance size, the number of combinations of variables over which to maximize or marginalize becomes quickly intractable. But when dealing with GMs, one can exploit the structure of the GM in order to reduce the number of combinations that have to be taken into account. Intuitively, the smaller the connectivity of the variables in the GM, the smaller this number of combinations becomes. We will formalize this by introducing the elimination algorithm, which gives us a systematic way of making fewer maximizations/marginalizations on a given graph. We will see how substantially the number of operations is reduced on a graph that is not completely connected.

### 1.2.1   The elimination algorithm

We consider the GM in Fig. 1.8, which is not fully connected. The subgraphs drawn with different line styles represent the maximal cliques. Using the decomposition (1.16), we can write

$$P_{\underline{X}}(\underline{x}) \propto \phi_{123}(x_1, x_2, x_3) \cdot \phi_{234}(x_2, x_3, x_4) \cdot \phi_{245}(x_2, x_4, x_5). \tag{1.28}$$

We want to solve the MARG problem on this GM, for example, for calculating the marginal probability of $x_1$,

$$P_{X_1}(x_1) = \sum_{x_2, x_3, x_4, x_5} P_{\underline{X}}(\underline{x}). \tag{1.29}$$

A priori, this requires us to evaluate $|\mathcal{X}|^4$ terms, each taking $|\mathcal{X}|$ different values. In the end, $3|\mathcal{X}||\mathcal{X}|^4$ operations are needed for calculating this marginal naively. But if we

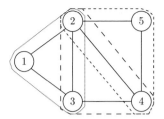

**Fig. 1.8** A GM and its maximal cliques.

take advantage of the factorized form (1.28), we can eliminate some of the variables. The elimination process goes along the following lines:

$$P_{X_1}(x_1) \propto \sum_{x_2,x_3,x_4,x_5} \phi_{123}(x_1,x_2,x_3) \cdot \phi_{234}(x_2,x_3,x_4) \cdot \phi_{245}(x_2,x_4,x_5), \quad (1.30)$$

$$\propto \sum_{x_2,x_3,x_4} \phi_{123}(x_1,x_2,x_3) \cdot \phi_{234}(x_2,x_3,x_4) \cdot \sum_{x_5} \phi_{245}(x_2,x_4,x_5) \quad (1.31)$$

$$\propto \sum_{x_2,x_3,x_4} \phi_{123}(x_1,x_2,x_3) \cdot \phi_{234}(x_2,x_3,x_4) \cdot m_5(x_2,x_4) \quad (1.32)$$

$$\propto \sum_{x_2,x_3} \phi_{123}(x_1,x_2,x_3) \cdot \sum_{x_4} \phi_{234}(x_2,x_3,x_4) \cdot m_5(x_2,x_4) \quad (1.33)$$

$$\propto \sum_{x_2,x_3} \phi_{123}(x_1,x_2,x_3) \cdot m_4(x_2,x_3) \quad (1.34)$$

$$\propto \sum_{x_2} \left( \sum_{x_3} \phi_{123}(x_1,x_2,x_3) \, m_4(x_2,x_3) \right) \quad (1.35)$$

$$\propto \sum_{x_2} m_3(x_1,x_2) \quad (1.36)$$

$$\propto m_2(x_1). \quad (1.37)$$

With this elimination process completed, the number of operations necessary to compute the marginal scales as $|\mathcal{X}|^3$ instead of $|\mathcal{X}|^5$, thereby greatly reducing the complexity of the problem by using the structure of the GM. Similarly, we can rewrite the MAP problem as follows:

$$\max_{x_1,x_2,x_3,x_4,x_5} \phi_{123}(x_1,x_2,x_3) \cdot \phi_{234}(x_2,x_3,x_4) \cdot \phi_{245}(x_2,x_4,x_5) \quad (1.38)$$

$$= \max_{x_1,x_2,x_3,x_4} \phi_{123}(x_1,x_2,x_3) \cdot \phi_{234}(x_2,x_3,x_4) \cdot \max_{x_5} \phi_{245}(x_2,x_4,x_5) \quad (1.39)$$

$$= \max_{x_1,x_2,x_3,x_4} \phi_{123}(x_1,x_2,x_3) \cdot \phi_{234}(x_2,x_3,x_4) \cdot m_5^{\star}(x_2,x_4) \quad (1.40)$$

$$= \max_{x_1,x_2,x_3} \phi_{123}(x_1,x_2,x_3) \cdot \max_{x_4} \phi_{234}(x_2,x_3,x_4) \cdot m_5^{\star}(x_2,x_4) \quad (1.41)$$

$$= \max_{x_1, x_2, x_3} \phi_{123}(x_1, x_2, x_3) \cdot m_4^\star(x_2, x_3) \qquad (1.42)$$

$$= \max_{x_1, x_2} \left( \max_{x_3} \phi_{123}(x_1, x_2, x_3) \, m_4^\star(x_2, x_3) \right) \qquad (1.43)$$

$$= \max_{x_1, x_2} m_3^\star(x_1, x_2) \qquad (1.44)$$

$$= \max_{x_1} \left( \max_{x_2} m_3^\star(x_1, x_2) \right), \qquad (1.45)$$

leading to

$$x_1^\star = \operatorname*{argmax}_{x_1} m_2^\star(x_1). \qquad (1.46)$$

Just as for the MARG problem, the complexity is reduced from $|\mathcal{X}|^5$ (a priori) to $|\mathcal{X}|^3$. We would like to further reduce the complexity of the marginalizations (in $|\mathcal{X}|^3$). One simple idea would be to reduce the GM to a linear graph as in Fig. 1.9.

By grouping variables in the GM (Fig. 1.8), it is in fact possible to obtain a linear graph, as shown in Fig. 1.10, with the associated potentials $\phi_{123}(Y_{123})$, $\phi_{234}(Y_{234})$, and $\phi_{245}(Y_{245})$, and the consistency constraints $Y_{123}|_{23} \equiv Y_{234}|_{23}$ and $Y_{234}|_{24} \equiv Y_{245}|_{24}$. For other GMs, the simplest graph achievable by grouping variables might be a tree instead of a simple chain. But not all groupings of variables will lead to a tree graph that correctly represents the problem. In order for the grouping of variables to be correct, we need to build the tree attached to the maximal cliques, and we have to resort to the junction tree property.

### 1.2.2 Junction tree property and chordal graphs

The junction tree property allows us to find groupings of variables under which the GM becomes a tree (if such groupings exist). On this tree, the elimination algorithm will need fewer maximizations/marginalizations than on the initial GM. However, there is a remaining problem: running the algorithm on the junction tree does not give a straightforward solution to the initial problem, since the variables on the junction tree

**Fig. 1.9** A linear graph. Each marginalization is computed in $|\mathcal{X}|^2$ operations.

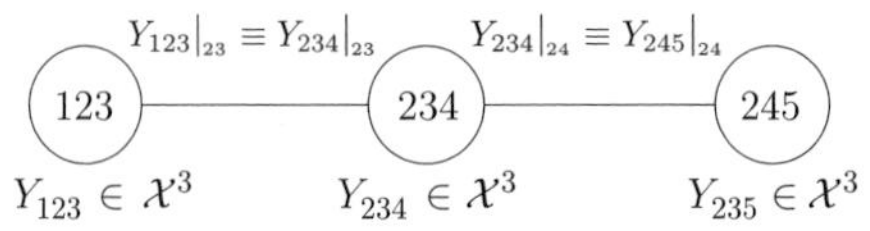

$$Y_{123}|_{23} \equiv Y_{234}|_{23} \qquad Y_{234}|_{24} \equiv Y_{245}|_{24}$$

$$\boxed{123} \quad\text{—}\quad \boxed{234} \quad\text{—}\quad \boxed{245}$$

$$Y_{123} \in \mathcal{X}^3 \qquad Y_{234} \in \mathcal{X}^3 \qquad Y_{235} \in \mathcal{X}^3$$

**Fig. 1.10** Linear GM obtained by grouping variables.

are groupings of variables of the original problem. This means that further maximizations/marginalizations are then required to have a solution in terms of the variables of the initial problem.

### *Junction tree (JCT) property* (Definition)

A graph $\mathcal{G} = (\mathcal{V}, \mathcal{E})$ is said to possess the JCT property if it has a junction tree $\mathcal{T}$, which is defined as follows: it is a tree graph such that

- its nodes are maximal cliques of $\mathcal{G}$;
- an edge between nodes of $\mathcal{T}$ is allowed only if the corresponding cliques share at least one vertex;
- for any vertex $v$ of $\mathcal{G}$, let $\mathcal{C}_v$ denote the set of all cliques containing $v$. Then $\mathcal{C}_v$ forms a connected subtree of $\mathcal{T}$.

Two questions then arise:

- Do all graphs have a JCT?
- If a graph has a JCT, how can we find it?

### *Chordal graph* (Definition)

A graph is chordal if all of its loops have chords. Figure 1.11 illustrates the concept.

**Proposition 1.2**  *$\mathcal{G}$ has a junction tree $\Leftrightarrow$ $\mathcal{G}$ is a chordal graph.*

**Proof** of the implication $\Leftarrow$. Let us take a chordal graph $\mathcal{G} = (\mathcal{V}, \mathcal{E})$ that is not complete, as represented in Fig. 1.12.

We will use the two following lemmas that can be shown to be true:

1. If $\mathcal{G}$ is chordal, has at least three nodes and is not fully connected, then $\mathcal{V} = \mathcal{A} \cup \mathcal{B} \cup \mathcal{S}$, where all three sets are disjoint and $\mathcal{S}$ is a fully connected subgraph that separates $\mathcal{A}$ from $\mathcal{B}$.
2. If $\mathcal{G}$ is chordal and has at least two nodes, then $\mathcal{G}$ has at least two nodes each with all neighbors connected. Furthermore, if $\mathcal{G}$ is not fully connected, then there exist two nonadjacent nodes each with all its neighbors connected.

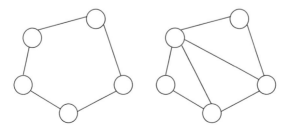

**Fig. 1.11**   The graph on the left is not chordal, the one on the right is.

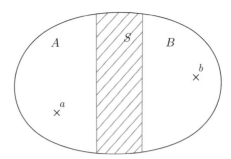

**Fig. 1.12**  On a chordal graph that is not complete, two vertices $a$ and $b$ that are not connected, separated by a subgraph $S$ that is fully connected.

The property *"If $\mathcal{G}$ is a chordal graph with $N$ vertices, then it has a junction tree"* can be shown by induction on $N$. For $N = 2$, the property is trivial. Now, suppose that the property is true for all integers up to $N$. Consider a chordal graph with $N + 1$ nodes. By the second lemma, $\mathcal{G}$ has a node $a$ with all its neighbors connected. Removing it creates a graph $\mathcal{G}'$ that is chordal and therefore has a JCT, $\mathcal{T}'$. Let $C$ be the maximal clique that $a$ participates in. Either $C \setminus a$ is a maximal-clique node in $\mathcal{T}'$, and in this case adding $a$ to this clique node results in a junction tree $\mathcal{T}$ for $\mathcal{G}$, or $C \setminus a$ is not a maximal-clique node in $\mathcal{T}'$. Then, $C \setminus a$ must be a subset of a maximal-clique node $D$ in $\mathcal{T}'$. Then, we add $C$ as a new maximal-clique node in $\mathcal{T}'$, which we connect to $D$ to obtain a junction tree $\mathcal{T}$ for $\mathcal{G}$. □

### Procedure to find a JCT

Let $G$ be the initial GM, and $\mathcal{G}(\mathcal{V}, \mathcal{E})$ be the GM in which $\mathcal{V}$ is the set of maximal cliques of $G$ and $(c_1, c_2) \in \mathcal{E}$ if the maximal cliques $c_1$ and $c_2$ share a vertex. Let us take $e = (c_1, c_2)$ with $c_1, c_2 \in \mathcal{V}$ and define the weight of $e$ as $w_e = |c_1 \cap c_2|$. Then, finding a junction tree of $G$ is equivalent to finding the max-cut spanning tree of $\mathcal{G}$. Denoting by $T$ the set of edges in a tree, we define the weight of the tree as

$$W(T) = \sum_{e \in T} w_e \tag{1.47}$$

$$= \sum_{e \in T} |c_1 \cap c_2|$$

$$= \sum_{v \in V} \sum_{e \in T} \mathbb{1}_{\{v \in e\}},$$

and we claim that $W(T)$ is maximal when $T$ is a JCT.

**Procedure** to obtain the maximal weight spanning tree:

- list all edges in decreasing order;
- include $e_i$ in $\mathcal{E}_{i-1}$ if possible.

At the end of the algorithm we are left with the maximal weight spanning tree.

### *Tree width* (Definition)

The width of a tree decomposition is the size of its maximal clique minus one.

### *Toy examples*

Figure 1.13 shows examples for two different values of the tree width.

### 1.2.3 Belief propagation (BP) algorithms

Up to now, everything we have done has been exact. The elimination algorithm is an exact algorithm. But as we are interested in **efficient algorithms**, as opposed to exact algorithms with too great a complexity to end in a reasonable time, we will from now on **introduce approximations**.

Coming back to the elimination algorithm (1.30)–(1.37), we can generalize the notation used as follows:

$$m_i(x_j) \propto \sum_{x_i} \phi_i(x_i) \cdot \phi_{i,j}(x_i, x_j) \cdot \prod_k m_k(x_i). \tag{1.48}$$

Considering now the same but oriented GM (arrows on Fig. 1.14), we get

$$m_{i \to j}(x_j) \propto \sum_{x_i} \phi_i(x_i) \cdot \phi_{i,j}(x_i, x_j) \cdot \prod_{k \in N(i) \backslash j} m_{k \to i}(x_i), \tag{1.49}$$

where $N(i)$ is the neighborhood of $x_i$.

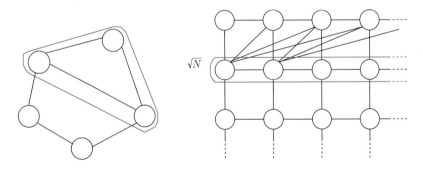

**Fig. 1.13** Left: tree width $= 2$. Right: tree width $= \sqrt{N}$.

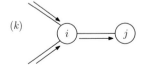

**Fig. 1.14** Message passing on a graph.

The MARG problem can then be solved using the **sum–product** procedure:

---

**Sum–product BP**

- $t = 0$,

$$\forall (i,j) \in E, \forall(x_i, x_j) \in \mathcal{X}^2 : \quad m_{i \to j}^0(x_j) = m_{j \to i}^0(x_i) = 1. \qquad (1.50)$$

- $t > 0$,

$$m_{i \to j}^{t+1}(x_j) \propto \sum_{x_i} \phi_i(x_i) \cdot \phi_{ij}(x_i, x_j) \cdot \prod_{k \in N(i) \backslash j} m_{k \to i}^t(x_i), \qquad (1.51)$$

$$P_{X_i}^{t+1}(x_i) = \prod_{k \in N(i)} m_{k \to i}^{t+1}(x_i). \qquad (1.52)$$

---

For the MAP problem, the **max–sum** procedure is considered:

---

**Max–sum BP**

- $t = 0$,

$$m_{i \to j}^0(x_j) = m_{j \to i}^0(x_i) = 1. \qquad (1.53)$$

- $t > 0$,

$$m_{i \to j}^{t+1}(x_j) \propto \max_{x_i} \phi_i(x_i) \cdot \phi_{ij}(x_i, x_j) \cdot \prod_{k \in N(i) \backslash j} m_{k \to i}^t(x_i), \qquad (1.54)$$

$$x_i^{t+1} = \underset{x_i}{\mathrm{argmax}}\, \phi_i(x_i) \cdot \prod_{k \in N(i)} m_{k \to i}^{t+1}(x_i). \qquad (1.55)$$

---

Note that here we use only the potentials of pairs. However, in the case of cliques, we have to consider the JCT and iterate on it. To understand this point, let us apply the sum–product algorithm to factor graphs.

### *Factor graphs*

Considering the general notation in Fig. 1.15, the sum–product BP algorithm is particularized such that

$$m_{i \to f}^{t+1}(x_i) = \prod_{f' \in N(i) \backslash f} m_{f' \to i}^t(x_i), \qquad (1.56)$$

$$m_{f \to i}^{t+1}(x_i) = \sum_{x_j, j \in N(f) \backslash i} f(x_i, x_j) \prod_{j \in N(f) \backslash i} m_{j \to f}^t(x_j). \qquad (1.57)$$

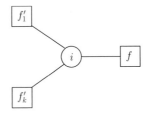

**Fig. 1.15**  A simple factor graph.

*On a tree*, the leaves are sending the right messages at time 1 already, and after a number of time steps proportional to the tree diameter,[1] all messages are correct: the steady point is reached and the algorithm is exact. Therefore, **BP is exact on trees**. The JCT property discussed before is therefore useful, and can in certain cases allow us to construct graphs on which we know that BP is exact. However, the problem mentioned before remains: if BP is run on the JCT of a GM, subsequent maximizations/marginalizations will be necessary to recover the solution in terms of the initial problem's variables.

## 1.3    Understanding belief propagation

We have seen how to use the (exact) elimination algorithm in order to design the BP algorithms max–product and sum–product, which are exact only on trees. The JCT property has taught us how to group variables of an initial loopy GM such that the resulting GM is a tree (when it is possible), on which we can then run BP with a guarantee of an exact result. However, the subsequent operations that are necessary to obtain the solution in terms of the initial problem's variables can be a new source of intractability. Therefore, we would like to know what happens if we use BP on the initial (loopy) graph anyway. The advantage is that BP remains tractable because of the low number of operations per iteration. The danger is that BP is no longer exact and therefore we need to ask ourselves the following three questions:

1. Does the algorithm have fixed points?
2. What are those fixed points?
3. Are they reached?

The analysis will be performed with the sum–product BP algorithm, but could be carried out similarily for the max–product version.

### 1.3.1    Existence of a fixed point

The algorithm is of the type

$$m^{t+1} = F\left(\underline{m}^t\right) \quad \text{with} \quad \underline{m}^t \in [0,1]^{2|\mathcal{E}||\mathcal{X}|} \tag{1.58}$$

and **the existence of a fixed point is guaranteed** by a theorem.

---

[1] The eccentricity of a vertex $v$ in a graph is the maximum distance from $v$ to any other vertex. The diameter of a graph is the maximum eccentricity over all vertices in a graph.

### 1.3.2   Nature of the fixed points

Let us recall that we factorized $P_{\underline{X}}(\underline{x})$ in the following way:

$$P_{\underline{X}}(\underline{x}) \propto \prod_{i \in \mathcal{V}} \phi_i(x_i) \prod_{(i,j) \in \mathcal{E}} \psi_{ij}(x_i, x_j)$$

$$= \frac{1}{Z} e^{Q(\underline{x})}. \tag{1.59}$$

The fixed points are a solution of the following problem:

$$P_{\underline{X}} \in \operatorname*{argmax}_{\mu \in M(\mathcal{X}^N)} \mathbb{E}_{\mu}[Q(X)] + H(\mu), \tag{1.60}$$

with

$$\mathbb{E}_{\mu}[Q(X)] + H(\mu) = \sum_{\underline{x}} \mu(\underline{x}) Q(\underline{x}) - \sum_{\underline{x}} \mu(\underline{x}) \log \mu(\underline{x}) = F(\mu). \tag{1.61}$$

Let us find a bound for this quantity. From (1.59), we obtain $Q(\underline{x}) = \log P_{\underline{X}}(\underline{x}) + \log Z$. Then

$$F(\mu) = \left( \sum_{\underline{x}} \mu(\underline{x}) \log Z \right) + \left( \sum_{\underline{x}} \mu(\underline{x}) \log \frac{P_{\underline{X}}(\underline{x})}{\mu(\underline{x})} \right) \tag{1.62}$$

$$= \log Z + \mathbb{E}_{\mu} \left[ \log \frac{P_{\underline{X}}}{\mu(\underline{x})} \right]$$

$$\leq \log Z + \log \mathbb{E}_{\mu} \left[ \frac{P_{\underline{X}}}{\mu} \right] \quad \text{using Jensen's inequality}$$

$$\leq \log Z,$$

and equality is reached when the distributions $\mu$ and $P$ are equal.

The maximization in (1.60) is performed over the space of all possible distributions, which is far too large a search space. But if we restrict ourselves to trees, we know that $\mu$ has the form

$$\mu \propto \prod_i \mu_i \prod_{(i,j)} \frac{\mu_{ij}}{\mu_i \mu_j}. \tag{1.63}$$

BP has taught us that

$$\mu_i \propto \phi_i \prod_{k \in N(i)} m_{k \to i}, \tag{1.64}$$

$$\mu_{ij} \propto \prod_{k \in N(i) \backslash j} m_{k \to i} \, \phi_i \, \psi_{ij} \, \phi_j \prod_{l \in N(j) \backslash i} m_{l \to j}. \tag{1.65}$$

If we marginalize $\mu_{ij}$ with respect to $x_j$, we should obtain $\mu_i$: $\sum_{x_j} \mu_{ij}(x_i, x_j) = \mu_i(x_i)$. Writing this out, we obtain

$$\prod_{k \in N(i)\setminus j} m_{k \to i}\phi_i \left( \sum_{x_j} \psi_{ij}\phi_j \prod_{l \in N(i)\setminus j} m_{l \to j} \right) = \phi_i \prod_{k \in N(i)} m_{k \to i}, \qquad (1.66)$$

and this should lead us to what we believe from the fixed points of BP. Let us recharacterize this in terms of the fixed points. In order to simplify the notation, we will write $\phi$ instead of $\log \phi$ and $\psi$ instead of $\log \psi$:

$$F_{\text{Bethe}}(\mu) = \mathbb{E}_\mu \left[ \sum_i \phi_i + \sum_{i,j} \psi_{ij} \right] - \mathbb{E}_\mu [\log \mu] \qquad (1.67)$$

We now use the factorization

$$\mathbb{E}_\mu [\log \mu] = -\sum_i \mathbb{E}_{\mu_i} [\log \mu_i] - \sum_{ij} \left( \mathbb{E}_{\mu_{ij}} [\log \mu_{ij}] - \mathbb{E}_{\mu_i} [\log \mu_i] - \mathbb{E}_{\mu_j} [\log \mu_j] \right) \quad (1.68)$$

and obtain a new expression for the Bethe free energy

$$F_{\text{Bethe}} = \sum_i (1 - d_i) \left( H_{\mu_i} + \mathbb{E}_{\mu_i} [\phi_i] \right) + \sum_{ij} \left( H(\mu_{ij}) + \mathbb{E}_{\mu_{ij}} [\psi_{ij} + \phi_i + \phi_j] \right), \quad (1.69)$$

where $d_i$ is the degree of node $i$.

### *Background on nonlinear optimization*

The problem

$$\max_q \quad G(q) \quad \text{s.t.} \quad Aq = b \qquad (1.70)$$

can be expressed in a different form by using Lagrange multipliers $\lambda$,

$$L(q, \lambda) = G(q) + \lambda^T (Aq - b), \qquad (1.71)$$

and maximizing:

$$\max_q L(q, \lambda) = M(\lambda) \leq G(q^*),$$

$$\inf_\lambda M(\lambda) \leq G(q^*).$$

Let us look at all $\lambda$ such that $\nabla_q L(q) = 0$. In a sense, BP is finding stationary points of this Lagrangian.

### *Belief propagation as a variational problem*

In our case, the conditions we will enforce with Lagrange multipliers are as follows:

$$\mu_{ij}(x_i, x_j) \geq 0, \qquad (1.72)$$

$$\sum_{x_i} \mu_i(x_i) = 1 \qquad\qquad \to \lambda_i, \qquad (1.73)$$

$$\sum_{x_j} \mu_{ij}(x_i, x_j) = \mu_i(x_i) \qquad\qquad \to \lambda_{j \to i}(x_i), \qquad (1.74)$$

$$\sum_{x_i} \mu_{ij}(x_i, x_j) = \mu_j(x_j) \qquad\qquad \to \lambda_{i \to j}(x_j). \qquad (1.75)$$

The complete Lagrangian reads

$$
\mathcal{L} = F_{\text{Bethe}}(\mu) + \sum_i \lambda_i \left( \sum_{x_i} \mu_i(x_i) - 1 \right)
$$

$$
+ \sum_{ij} \left[ \left( \sum_{x_j} \mu_{ij}(x_i, x_j) - \mu_i(x_i) \right) \lambda_{j \to i}(x_i) \right.
$$

$$
\left. + \left( \sum_{x_i} \mu_{ij}(x_i, x_j) - \mu_j(x_j) \right) \lambda_{i \to j}(x_j) \right]. \tag{1.76}
$$

We need to minimize this Lagrangian with respect to all possible variables, which we do by setting the partial derivatives to zero:

$$
\frac{\partial \mathcal{L}}{\partial \mu_i(x_i)} = 0 \tag{1.77}
$$

$$
= -(1 - d_i)(1 + \log \mu_i(x_i)) + (1 - d_i)\phi_i(x_i) + \lambda_i - \sum_{j \in N(i)} \lambda_{j \to i}(x_i).
$$

This imposes the following equality for the distribution $\mu_i$:

$$
\boxed{\mu_i(x_i) \propto e^{\phi_i(x_i) + (d_i - 1)^{-1} \sum_{j \in N(i)} \lambda_{j \to i}(x_i)}.} \tag{1.78}
$$

We now use the transformation $\lambda_{j \to i}(x_i) = \sum_{k \in N(i) \setminus j} \log m_{k \to i}(x_i)$ and obtain

$$
\sum_{j \in N(i)} \lambda_{j \to i}(x_i) \equiv (d_i - 1) \sum_{j \in N(i)} \log m_{j \to i}(x_i). \tag{1.79}
$$

In the same way, we can show that

$$
\frac{\partial \mathcal{L}}{\partial \mu_{ij}(x_i, x_j)} = 0 \quad \Rightarrow \quad \boxed{\mu_{ij}(x_i, x_j) \propto e^{\phi_i(x_i) + \phi_j(x_j) + \psi_{ij}(x_i, x_j) + \lambda_{j \to i}(x_i) + \lambda_{i \to j}(x_j)}}
$$

In this way, we find the distributions $\mu_i$ and $\mu_{ij}$ that are the fixed points of BP.

### 1.3.3    Can the fixed points be reached?

We will now try to determine whether the algorithm can actually reach the fixed points that we have exhibited in the previous section. Let us look at the simple (but loopy) graph in Fig. 1.16.

At time $t = 1$, we have

$$
m_{2 \to 1}^1(x_1) \propto \sum_{x_2} \phi_2(x_2)\phi_{12}(x_1, x_2) \underbrace{m_{3 \to 2}^0(x_2)}_{=1} \tag{1.80}
$$

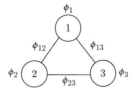

**Fig. 1.16**

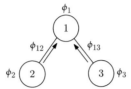

**Fig. 1.17**

and

$$m^1_{3\to1} \propto \sum_{x_3} \phi_3\phi_{13}, \tag{1.81}$$

which also corresponds to the messages of the modified graph in Fig. 1.17.

At time $t = 2$, the messages will be

$$m^2_{2\to1} \propto \sum_{x_2} \phi_2\phi_{12}m^1_{3\to2}(x_2), \tag{1.82}$$

corresponding to the messages on the modified graph in Fig. 1.18.

If we increase $t$, the corresponding non-loopy graph becomes longer at each time step. Another way of seing this is by looking at the recursion equation:

$$F_{ij}(m^*) = m^*_{ij}, \tag{1.83}$$

$$m^{t+1}_{ij} = F_{ij}(m^t),$$

$$|m^{t+1}_{ij} - m^*_{ij}| = |F_{ij}(m^t) - F_{ij}(m^*)|$$

$$= |\nabla F_{ij}(\theta)^T(m^t - m^*)| \quad \text{(mean value theorem)},$$

$$|m^{t+1} - m^*|_\infty \le |\nabla F_{ij}(\theta)|_1|m^t - m^*|_\infty. \tag{1.84}$$

From this last inequality, it is clear that if we can prove that $|F_{ij}|_1$ is bounded by some constant $\rho < 1$, then convergence is proved. Unfortunately, it is not often easy to do this.

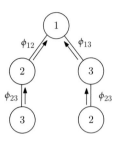

**Fig. 1.18**

### The hardcore model

In the hardcore model, we have

$$\phi_i(x_i) = 1 \quad \text{for all } x_i \in \{0, 1\}, \tag{1.85}$$

$$\psi_{ij}(x_i, x_j) = 1 - x_i x_j. \tag{1.86}$$

Instead of using BP, let us use the following gradient-descent-like algorithm:

$$y(t+1) = \left[ y(t) + \alpha(t) \frac{\partial F}{\partial y_i} \bigg|_{y(t)} \right], \tag{1.87}$$

where the operator $[\cdot]$ is a clipping function that ensures that the result stays in the interval $(0, 1)$. This is a projected version of a gradient algorithm with variable step size $\alpha(t)$. Choosing this step size with the rule

$$\alpha(t) = \frac{1}{\sqrt{t}} \frac{1}{2^d}, \tag{1.88}$$

we can show that in a time $T \sim n^2 2^d / \epsilon^4$, we will find $F_b$ up to $\epsilon$, and convergence is proved.

## 1.4    Learning graphical models

In this final section, we focus on the learning problem. In particular, we consider three different cases:

- Parameter learning
  Given a graph, the parameters are learned from the observation of the entire set of realizations of all random variables.
- Graphical model learning
  Both the parameters and the graph are learned from the observations of the entire set of realizations of all random variables.
- Latent graphical model learning
  The parameters and the graph are learned from partial observations: some of the random variables are assumed to be hidden.

### 1.4.1　Parameter learning

*Single-parameter learning*

We consider the following simple setting where $x_i$ is a Bernoulli random variable with parameter $\theta$:

$$P_X(x_i, \theta) = \begin{cases} \theta & \text{if } x_i = 1, \\ 1 - \theta & \text{if } x_i = 0. \end{cases} \tag{1.89}$$

Given observations $\{x_1, \ldots, x_S\}$, we are interested in the MAP estimation of the parameter $\theta$:

$$\hat{\theta}^{MAP} = \underset{\theta \in [0,1]}{\operatorname{argmax}} P(\theta | x_1, \ldots x_S)$$

$$= \underset{\theta \in [0,1]}{\operatorname{argmax}} P(x_1, \ldots x_S | \theta) \, p(\theta), \tag{1.90}$$

where maximizing $P(x_1, \ldots x_S | \theta)$ leads to the maximum likelihood (ML) estimator $\hat{\theta}^{ML}$ of $\theta$.

Denoting $\mathcal{D} \triangleq \{x_1, \ldots x_S\}$, the observed set of realizations, we define the empirical likelihood as follows:

$$\ell(\mathcal{D}; \theta) = \frac{1}{S} \log P(x_1, \ldots x_S | \theta)$$

$$= \frac{1}{S} \sum_i \log P(x_i | \theta)$$

$$= \hat{P}(1) \log \theta + \hat{P}(0) \log(1 - \theta), \tag{1.91}$$

with $\hat{P}(1) = S^{-1} \sum_i^S \mathbb{1}_{\{x_i=1\}}$. Taking the derivative of (1.91) and setting the result to zero, we obtain the **maximum likelihood estimator** $\hat{\theta}^{ML}$:

$$\frac{\partial}{\partial \theta} \ell(\mathcal{D}; \theta) = \frac{\hat{P}(1)}{\theta} - \frac{\hat{P}(0)}{1 - \theta} = 0,$$

$$\Rightarrow \boxed{\hat{\theta}^{ML} = \hat{P}(1).} \tag{1.92}$$

What is the number of samples $S$ needed to achieve $\hat{\theta}^{ML}(S) \approx (1 \pm \epsilon)\theta$? Considering the binomial variable $B(S, \theta)$ (which is the sum of $S$ independently drawn Bernoulli variables from (1.89)), we can write

$$P(|B(S, \theta) - S\theta| > \epsilon S\theta) \sim \exp(-\epsilon^2 S\theta) \leq \delta,$$

$$\Rightarrow \boxed{S \geq \frac{1}{\theta} \frac{1}{\epsilon^2} \log \frac{1}{\delta}.} \tag{1.93}$$

*Directed graphs*

We consider the following setting in which we have not one but many random variables to learn on a directed graph:

$$P_{\underline{X}}(\underline{x}) \propto \prod_i P_{X_i | X_{\Pi_i}}(x_i | x_{\Pi_i}), \tag{1.94}$$

where $\Pi_i$ stands for the parents of node $i$, and $P_{X_i|X_{\Pi_i}}(x_i|x_{\Pi_i}) \triangleq \theta_{x_i,x_{\Pi_i}}$. Again, we look at the empirical likelihood

$$\ell(\mathcal{D};\underline{\theta}) = \sum_i \sum_{x_i,x_{\Pi_i}} \hat{P}(x_i,x_{\Pi_i}) \log \theta_{x_i,x_{\Pi_i}}$$

$$= \sum_i \sum_{x_i,x_{\Pi_i}} \hat{P}(x_i|x_{\Pi_i})\hat{P}(x_{\Pi_i}) \left[ \log \frac{\theta_{x_i,x_{\Pi_i}}}{\hat{P}(x_i|x_{\Pi_i})} + \log \hat{P}(x_i|x_{\Pi_i}) \right]$$

$$= \sum_i \sum_{x_i,x_{\Pi_i}} \hat{P}(x_i|x_{\Pi_i})\hat{P}(x_{\Pi_i}) \log \frac{\theta_{x_i,x_{\Pi_i}}}{\hat{P}(x_i|x_{\Pi_i})}, \qquad (1.95)$$

and set the derivative to zero in order to obtain the ML estimation of $\underline{\theta}$, resulting in

$$\sum_{x_i} \hat{P}(x_i|x_{\Pi_i}) \log \frac{\theta_{x_i,x_{\Pi_i}}}{\hat{P}(x_i|x_{\Pi_i})} = \mathbb{E}_{\hat{P}} \left[ \log \frac{\theta_{x_i,x_{\Pi_i}}}{\hat{P}(x_i|x_{\Pi_i})} \right],$$

$$\Rightarrow \qquad \boxed{\hat{\theta}^{ML}_{x_i,x_{\Pi_i}} = \hat{P}(x_i|x_{\Pi_i}).} \qquad (1.96)$$

### Undirected graphs

Let us now consider the case of undirected graphs. To reduce the number of indices, we will write $i$ instead of $x_i$ in the following.

On a tree, $\qquad P_X = \prod_i P_i \prod_{ij} \frac{P_{ij}}{P_i P_j} \rightarrow$ possible estimator: $\hat{P}_i \frac{\hat{P}_{ij}}{\hat{P}_i \hat{P}_j}$.

On a chordal graph, $\qquad P_X \propto \frac{\prod_C \phi_C(x_C)}{\prod_S \phi_S(x_S)} \rightarrow$ possible estimator: $\frac{\hat{P}_C}{\hat{P}_S}$.

On a triangle-free graph, $\qquad P_X \propto \prod_i \phi_i \prod_{ij} \psi_{ij}$.

For the last case, let us use the Hammersley–Clifford theorem. Let $\mathcal{X} = \{0,1\}$. On a triangle-free graph, the maximal clique size is 2, and therefore we can write

$$P_{\underline{X}}(\underline{x}) \propto \exp \left( \sum_i U_i(x_i) + \sum_{ij} V_{ij}(x_i,x_j) \right). \qquad (1.97)$$

Using the fact that we have an MRF, we obtain

$$\frac{P(X_i = 1, X_{\text{rest}} = \underline{0})}{P(X_i = 0, X_{\text{rest}} = \underline{0})} \propto \exp\left(Q(i)\right). \qquad (1.98)$$

Also, because of the fact that on an MRF, a variable conditioned on its neighbours is independent of all the others, we can write

$$\frac{P(X_i = 1, X_{\text{rest}} = \underline{0})}{P(X_i = 0, X_{\text{rest}} = \underline{0})} = \frac{P(X_i = 1, X_{N(i)} = \underline{0})}{P(X_i = 0, X_{N(i)} = \underline{0})}, \qquad (1.99)$$

and therefore this quantity can be calculated with $2^{|N(i)|+1}$ operations.

### 1.4.2   Graphical model learning

What can we learn from a set of realizations of variables when the underlying graph is not known? We focus now on the following maximization:

$$\max_{\mathcal{G},\theta_{\mathcal{G}}} \ell(\mathcal{D};\mathcal{G},\theta_{\mathcal{G}}) = \max_{\mathcal{G}} \underbrace{\max_{\theta_{\mathcal{G}}} \ell(\mathcal{D};\mathcal{G},\theta_{\mathcal{G}})}_{\hat{\ell}(\mathcal{D};\mathcal{G}) \triangleq \ell(\mathcal{D};\mathcal{G},\hat{\theta}_{\mathcal{G}}^{ML})} . \tag{1.100}$$

From the previous subsection, we have $\hat{\theta}_{\mathcal{G}}^{ML}$, and therefore we only need to find a way to evalute the maximization on the possible graphs.

#### *Directed graphs*

On a directed graph $\mathcal{G} \to (i, \Pi_i)$, the empirical likelihood reads

$$\hat{\ell}(\mathcal{D};\mathcal{G}) = \sum_i \sum_{x_i, x_{\Pi_i}} \hat{P}(x_i, x_{\Pi_i}) \log \hat{P}(x_i | x_{\Pi_i})$$

$$= \sum_i \sum_{x_i, x_{\Pi_i}} \hat{P}(x_i, x_{\Pi_i}) \log \left[ \frac{\hat{P}(x_i, x_{\Pi_i})}{\hat{P}(x_i)\hat{P}(x_{\Pi_i})} \hat{P}(x_i) \right]$$

$$= \sum_i \sum_{x_i, x_{\Pi_i}} \hat{P}(x_i, x_{\Pi_i}) \log \frac{\hat{P}(x_i, x_{\Pi_i})}{\hat{P}(x_i)\hat{P}(x_{\Pi_i})} + \sum_{x_i} \hat{P}(x_i) \log \hat{P}(x_i)$$

$$= \sum_i I(\hat{X}_i; \hat{X}_{\Pi_i}) - H(\hat{X}_i). \tag{1.101}$$

The search for the graph maximizing the empirical likelihood thus consists in maximizing the mutual information: $\max_{\mathcal{G}} \sum_i I(\hat{X}_i; \hat{X}_{\Pi_i})$. In a general setting, this is not easy.

**Reducing the search space to trees:** However, some methods exist, such as the Chow–Liu algorithm (Chow and Liu, 1968), which relies on the procedure used to obtain the maximal weighted spanning tree (cf. Section 1.2).

#### *Undirected graphs*

What can we do in the case of undirected graphs? Let us restrict ourselves to the binary case $\underline{x} \in \{0,1\}^N$ and to exponential families:

$$P_{\underline{X}}(\underline{x}) = \exp\left( \sum_i \theta_i x_i + \sum_{i,j} \theta_{ij} x_i x_j - \log Z(\underline{\theta}) \right). \tag{1.102}$$

Again, we denote by $\mathcal{D} = \{\underline{x}^1, \ldots, \underline{x}^S\}$ the observed dataset, and the log-likelihood can be written as

$$\ell(\mathcal{D};\underline{\theta}) = \underbrace{\sum_i \theta_i \mu_i + \sum_{i,j} \theta_{ij} \mu_{ij}}_{\langle\theta,\mu\rangle} - \log Z(\underline{\theta}). \tag{1.103}$$

As $\ell(\mathcal{D}; \underline{\theta})$ is a concave function of $\underline{\theta}$, it can be efficiently solved using a gradient descent algorithm of the form

$$\boxed{\underline{\theta}^{t+1} = \underline{\theta}^t + \alpha(t)\nabla_{\underline{\theta}}\ell(\mathcal{D}; \underline{\theta})|_{\underline{\theta}=\underline{\theta}^t}.} \tag{1.104}$$

The difficulty in this formula is the evaluation of the gradient

$$\nabla_{\underline{\theta}}\ell(\mathcal{D}; \underline{\theta}) = \mu - \mathbb{E}_{\underline{\theta}}(\underline{X}), \tag{1.105}$$

whose second term is an expectation that has to be calculated, using the sum–product algorithm or with a Markov chain Monte Carlo method for instance.

Another question is whether we will be learning interesting graphs at all. Graph-learning algorithms tend to link variables that are not linked in the real underlying graph. To avoid this, complicated graphs should be penalized by introducing a regularizer. Unfortunately, this is a highly nontrivial problem, and graphical model learning algorithms do not always perform well at present.

### 1.4.3   Latent graphical model learning: the expectation-maximization algorithm

In this last case, we distinguish two different variables:

- $Y$ stands for observed variables;
- $X$ denotes the hidden variables.

The parameter $\theta$ is estimated from the observations, namely,

$$\hat{\theta}^{ML} = \operatorname*{argmax}_{\theta} \log P_Y(y; \theta). \tag{1.106}$$

The log-likelihood is derived by marginalizing on the hidden variables:

$$\ell(y; \theta) = \log P_Y(y; \theta)$$

$$= \log \sum_x P_{X,Y}(x, y; \theta) \tag{1.107}$$

$$= \log \sum_x q(x|y)\frac{P_{X,Y}(x, y; \theta)}{q(x|y)} \tag{1.108}$$

$$= \log \mathbb{E}_q\left[\frac{P}{q}\right] \geq \mathbb{E}_q\left[\frac{P}{q}\right] \triangleq \mathcal{L}(q; \theta). \tag{1.109}$$

This gives raise to the expectation-maximization (EM) algorithm (Dempster *et al.*, 1977):

---

**EM algorithm**
Until convergence, iterate between

- **E-step**: estimation of the distribution $q$
  $\theta^t \rightarrow q^{t+1} = \mathrm{argmax}_q \, \mathcal{L}(q; \theta^t)$.
- **M-step**: estimation of the parameter $\theta$
  $q^{t+1} \rightarrow \theta^{t+1} = \mathrm{argmax}_\theta \, \mathcal{L}(q^{t+1}; \theta)$.

---

# References

Chow, C. K. and Liu, C. N. (1968). Approximating discrete probability distributions with dependence trees. *IEEE Transactions on Information Theory*, **14**(3), 462–467.

Dempster, A. P., Laird, N. M., and Rubin, D. (1977). Maximum likelihood from incomplete data via the EM algorithm. *Journal of the Royal Statistical Society, Series B*, **39**(1), 1–38.

Grimmet, G. R. (1973). A theorem about random fields. *Bulletin of the London Mathematical Society*, **5**(1), 81–84.

Hammersley, J. M. and Clifford, P. (1971). Markov fields on finite graphs and lattices. Available online: http://www.statslab.cam.ac.uk/~grg/books/hammfest/hamm-cliff.pdf.

# 2
# Computational complexity, phase transitions, and message-passing for community detection

## Cristopher MOORE

Santa Fe Institute,
1399 Hyde Park Road,
Santa Fe, NM 87501,
USA

*Lecture notes taken by*
Aurélien Decelle, La Sapienza, Italy
Janina Hüttel, Goethe Universität Frankfurt am Main, Germany
Alaa Saade, École Normale Supérieure, France

*Statistical Physics, Optimization, Inference, and Message-Passing Algorithms.* First Edition.
F. Krzakala et al. © Oxford University Press 2016. Published in 2016 by Oxford University Press.

# Chapter Contents

## 2.1 Computational complexity

Computational complexity is a branch of complex systems that has the advantages of being very well defined and presenting a lot of rigorous results. It allows qualitative distinctions between different kinds of "hardness" in computational problems, such as the difference between polynomial and exponential time. Among historical examples of combinatorial problems, the "Bridges of Köningsberg" problem is a famous one solved by Euler, where the premises of computational complexity can already be sketched. This problem deals with finding a path through all parts of the city of Köningsberg using each bridge only once (Fig. 2.1). It was solved quickly by Euler, who, by using the dual graph, showed that a solution exists only if the degree of each node is even (except for the starting and ending nodes). This perspective is a profound change in how the problem is viewed, since, by Euler's argument, the verification of the existence of such a path can be done very quickly (in polynomial time, since it is enough to check the degree of all edges) as opposed to an exhaustive search through all possible paths.

Hamilton came later with a problem that looks similar. It is about the possibility of finding a path in a graph that visits each vertex exactly once. Although it appears to be similar, the fundamental difference from the previous problem is the impossibility (as far as we know) of verifying the existence of such a path by a simple (and quick) argument. Therefore, finding a Hamiltonian path seems to be possible only by an exhaustive search (with an exponential number of possibilities). On the other hand, checking if a given path is a Hamiltonian cycle or not is an easy problem. One simply goes through the path, node by node, to check whether it is a solution of the problem or not.

To go from these first simple examples to the theory of computational complexity, we first need a few definitions.

**Definition 2.1 (Problem)** *A problem is given by defining an input (or instance), and a question on that input. For example, in the case of the Hamilton cycle, the input is "a graph G," and the question is "Does there exist a Hamiltonian path on this graph? (yes or no)."*

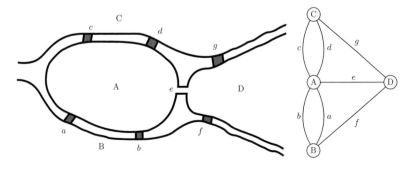

**Fig. 2.1** The "Bridges of Köningsberg" problem. From Moore and Mertens (2011).

**Definition 2.2 (The class P)**  *P is the class of problems that can be "solved" in polynomial time* $\mathrm{poly}(x) = x^c$ *for some constant c and where x is the size of the instance. In the example above, the Euler path belongs to P since, given a graph G, we can answer the question "Does there exist an Eulerian path on G?" in polynomial time.*

**Definition 2.3 (The class NP)**  *NP is a class of decision problems that have a yes/no answer. If the answer is "yes," there is a proof that the answer is "yes" that can be checked in polynomial time.*

Let us look at the Hamiltonian case. Consider the question "Is there a Hamiltonian path?" We do not know any polynomial algorithm answer, but if someone provides us with such a path, it is a proof that the answer is yes, and it can be checked in polynomial time.

Note that the definition of NP problems is not symmetrical. If we claim "Prove that there is not a Hamiltonian path!", it does not seem at all easy to prove that such a path does not exist: as far as we know, in general, all possible paths have to be examined.

**Definition 2.4 (The class NP-complete)**  *A problem is called NP-complete if it "somehow" encompasses any problem in NP.*

To be more precise, we need to introduce the notion of "reduction" of one problem into another. A problem (A) can be reduced to (B) if, given an instance $x$ of (A), there exists a function $f$ such that $f(x)$ is an instance of (B) and $f$ is in P. In addition, $f(x)$ is a "yes"-instance of (B) if and only if $x$ is a "yes"-instance of (A). Such a reduction is written as $A \leq B$. In practice, this means that if I have an algorithm to solve (B), I can use it in some way to solve (A). It also implies that if (B) is in P, then (A) is in P, since $f$ is a polynomial function. And, as can be understood by the definition, if any problem of the NP-complete class is in P, then P = NP.

Here is an example of a reduction between two problems in P.

**Example 2.5 (Reduction—perfect bipartite matching ≤ max-flow)**  The problem of telling whether a bipartite graph has a perfect matching can be mapped on the max-flow problem by adding the nodes $s$ and $t$ to the bipartite graph as illustrated in the diagram at the top of page 33. The node $s$ is linked to the left part of the bipartite graph by directed edges; at the center, the edges are replaced by directed edges going from left to right; and the right part is linked to the node $t$ by directed edges. Each edge has a weight of one. The intuition is that if there is a max-flow of value $m$, then there is a matching consisting of $m$ edges. By the direction of the arrow and the weight of the edges in the max-flow problem, the principle of the proof can be understood.

It can also be demonstrated that max-flow can be reduced to the weighted min-cut problem and vice versa.

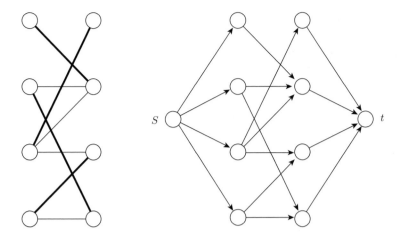

The reduction of one problem into another implies also that if A ≤ B, then, if A is not in P, then B is not in P (which justifies the use of the "less than or equal" symbol). Now we can define more clearly the class of NP-complete problems:

**Definition 2.6 (NP-completeness)** *B is NP-complete if*

- *B ∈ NP;*
- *A ≤ B ∀ A in NP.*

This definition implies that if B is in P, then P = NP, so it is enough to find a polynomial solution to an NP-complete problem to demonstrate the equality. It can be shown that the problem of finding a Hamiltonian path is NP-complete. We now introduce some examples of NP-complete problems.

**Example 2.7 (NP-complete problems)**

- program-SAT: A program Π and a time $t$ given in unary. Does there exist an instance $x$ such that $Π(x) =$"yes" (in $t$ steps or less)?
  This problem is NP-complete because it reproduces the exact structure of any NP problem: a general "yes/no" question that should be checkable in a time at most of size $t$. *Remark:* $t$ is given in unary (i.e., a string of $t$ 1s), instead of in binary. Otherwise $t$ would be exponential as a function of the size of the input, measured in bits.
- circuit-SAT: Boolean circuits are a set of source nodes (bits) connected to logical gates (AND, OR, NOT) on a directed acyclic graph. The output consists of a set of sink nodes. A circuit is called satisfiable if there exists an assignment of the input nodes such that the output is true. The problem can be phrased as follows:
  Input: a circuit C.
  Question: Does there exist an $x$ such that $C(x)$ is true?
  This problem is NP-complete because the program SAT can be reduced to a circuit-SAT instance. This can be understood as any algorithm can be stated

in term of a circuit-SAT form (it is enough to think that a computer is only an ensemble of nodes and logical gates, and that for-loops can be unfolded to produce more layers of these gates). Therefore, any program can be mapped to a circuit-SAT instance and it is an NP-complete problem.

- 3-SAT: the 3-SAT problem is a constraint satisfaction problem (CSP), which is defined by an ensemble of variable nodes (true/false). Then a 3-SAT instance is a formula that consists of a set of clauses all connected by an OR operator. A clause contains three variables linked by an AND operator, and each variable can be negated or not. We can prove the following: circuit-SAT $\leq$ 3-SAT.

This reduction can be proven by first demonstrating that any circuit-SAT can be mapped into a SAT formula, and then showing that any SAT formula can be mapped in a 3-SAT formula.

**Circuit-SAT to SAT formulas:** To map a circuit to a formula, we add variables for the internal values on the wires, and then transform each logical gate into a set of SAT formulas. Then the output of the gate is used as a new variable that is propagated throughout the rest of the circuit.

1. **AND:** $y = x_1 \wedge x_2 \Leftrightarrow (x_1 \vee \bar{y}) \wedge (x_2 \vee \bar{y}) \wedge (\bar{x}_1 \vee \bar{x}_2 \vee y)$.
2. **OR:** $y = x_1 \vee x_2 \Leftrightarrow (\bar{x}_1 \vee y) \wedge (\bar{x}_2 \vee y) \wedge (x_1 \vee x_2 \vee \bar{y})$.
3. **NOT:** $y = \bar{x} \Leftrightarrow (x \vee y) \wedge (\bar{x} \vee \bar{y})$.

Any $k$-SAT clause can be written as a 3-SAT formula. We refer to Moore and Mertens (2011) for the case $k > 3$ and show the cases $k = 1, 2$ here so that the reduction from a circuit-SAT to 3-SAT is complete.

* A single variable $x \Leftrightarrow (x \vee z_1 \vee z_2) \wedge (x \vee z_1 \vee \bar{z}_2) \wedge (x \vee \bar{z}_1 \vee z_2) \wedge (x \vee \bar{z}_1 \vee \bar{z}_2)$. In the 3-SAT formula above, whatever the values of $z_1$ and $z_2$ are, the expression is satisfiable if $x$ is true and unsatisfiable otherwise.
* Two variables $(x \vee y) \Leftrightarrow (x \vee y \vee z) \wedge (x \vee y \vee \bar{z})$. As above, whatever the value $z$ takes, the expression is satisfiable only if $x$ OR $y$ is true.

With these two pieces, it is now easy to map any instance of a circuit SAT to a 3-SAT problem, and therefore circuit-SAT $\leq$ 3-SAT.

- NAE-SAT (NonAllEqual-SAT): This problem is very similar to a SAT problem with the small difference that all the variables in one clause cannot be all true (or false) at the same time. We can observe that any NAE-SAT solution is symmetric (unlike the $k$-SAT case). For this problem, we have the reduction 3-SAT $\leq$ NAE-3-SAT. To prove this reduction, it is easier to prove first that 3-SAT $\leq$ NAE-4-SAT and then to show that NAE-3-SAT $\leq$ NAE-4-SAT. The first point can be proved by adding a variable to each clause of a 3-SAT instance. So any formula $\phi = (x_1 \vee x_2 \vee x_3) \wedge \ldots$ becomes $\phi' = (x_1, x_2, x_3, s) \wedge \ldots$, with the same variable $s$ added to every clause. If $\phi$ is satisfiable, we can satisfy $\phi'$ by setting $s$ to false. Now take a SAT-instance of $\phi'$. Because the problem is symmetric, the symmetric version of that instance will exist, and among those two one where $s$ is false. Therefore, the remaining variables satisfy the 3-SAT formula as well. Then it remains to convert any NAE-4-SAT formula into a NAE-3-SAT formula. This can be done easily by using another variable $z$ and seeing that $(x_i, x_j, x_k, x_l) = (x_i, x_j, z) \wedge (\bar{z}, x_k, x_l)$.

## 2.2 Hardness: P, NP, or EXP?

We have already discussed the hardness of a problem in the previous section. Now we want to discuss counting problems. We already know that problems in NP ask whether an object with a certain property exists. We now define #P, pronounced "sharp P," as the class of problems that ask how many such objects exist. As in NP, we require that this property can be checked in polynomial time.

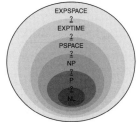

**Definition 2.8** #P *is the class of functions* $A(x)$ *of the form* $A(x) = \#\{w \mid B(x, w)\}$, *where* $B(x, w)$ *is a property that can be checked in polynomial time, and where* $|w| = \text{poly}(|x|)$ *for all* $w$ *such that* $B(x, w)$ *holds.*

From Moore and Mertens (2011)

### 2.2.1 Examples

**Example 2.9 (Perfect Matching)** The problem of deciding if there exists a perfect matching in a given graph $G$ is, as we have already mentioned, in P. On the other hand, the problem of counting these perfect matchings is in #P. Suppose we have a bipartite graph $G$ with $2n$ vertices. We can represent it as an $n \times n$ matrix $A$, where $A_{ij} = 1$ if the $i$th vertex on the left is connected to the $j$th vertex on the right, and $A_{ij} = 0$ otherwise.

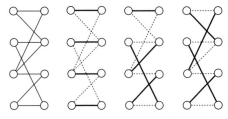

As an example, we can write down the matrix $A$ for the first graph on the left:

$$A = \begin{pmatrix} 1 & 1 & 0 & 0 \\ 0 & 1 & 0 & 1 \\ 1 & 1 & 1 & 0 \\ 0 & 0 & 1 & 1 \end{pmatrix}.$$

From Moore and Mertens (2011)

But how can we express the number of perfect matchings in terms of $A$? Each perfect matching is a permutation that maps the vertices on the left to their partners on the right. Each permutation $\sigma$ corresponds to a matching if and only if for each $i$ there is an edge from $i$ to $\sigma(i)$, i.e., if $A_{i\sigma(i)} = 1$. Therefore, the number of matchings is given by the following quantity, which is called the **permanent of A**:

$$\text{perm}(A) = \sum_{\sigma \in S_n} \prod_{i=1}^{n} A_{i\sigma(i)} \, .$$

Note that this is just like the determinant, except that it does not have the parity $(-1)^\sigma$. Ironically, this makes it harder to compute.

**Example 2.10 (Graph 3-colorability)**   We want to give a hint on how it is possible to reduce graph 3-colorability to planar graph 3-colorability—in other words, how we can transform an arbitrary graph $G$ to a planar graph $G'$, such that $G'$ is 3-colorable if and only if $G$ is. We can easily see that the reverse reduction is trivial. Thus, planar graph 3-colorability is just as hard as graph 3-colorability in general, and the two problems have the same complexity.

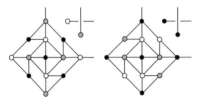

From Moore and Mertens (2011)

This crossover gadget allows us to convert an arbitrary graph into a planar one while preserving the property of 3-colorability. The gadget can be colored in two different ways up to symmetry, depending on whether the transmitted colors are different (left) or the same (right).

**Example 2.11 (Two-dimensional Ising model)**   We want to think about the Ising model as a two-dimensional lattice with $n$ spins that can take the values $\pm 1$. Now we can consider the dual lattice, whose edges cross the edges of our Ising model lattice.

An edge in the dual lattice is indicated by a thick black line if the two spins that are connected through its crossing edge have different values. It is clear that one vertex of the dual lattice is always adjacent to an even number of thick edges. Now we can think about a ground state of the Ising model as a *minimum-weight perfect matching* in a decorated version of the dual graph (see Moore and Mertens (2011) for the gadgets and details). We know that every spin assignment corresponds to a single coloring of the dual lattice.

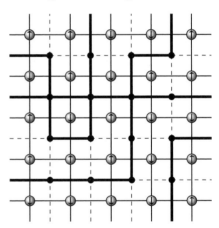

**Example 2.12 (Domino tiling)**

Our next task is to reduce domino tiling (where we want to cover a $n \times n$-chess board $C$ with dominos) to finding a perfect matching in a bipartite graph. We therefore define a graph $G$ where each vertex corresponds to a square in $C$ and is connected to the four vertices corresponding to the neighboring squares. A domino tiling of $C$ is just a perfect matching of $G$. So domino tiling is in P.

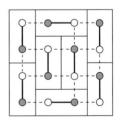

from Moore and Mertens (2011)

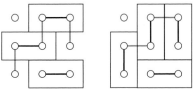

From Moore and Mertens (2011)

We can improve a partial domino tiling, or equivalently a partial matching, by flipping the edges along an alternating path. Now if we want to compute the number of domino tilings for the matrix $C$, we have to compute perm($C$) as above.

**Turning permanents into determinants** Again we have a planar, bipartite graph $G$ with $n$ vertices of each type. Now we color the two types of vertices in black and white (see above). As in Example 2.9, we define an $n \times n$ matrix $A$. We know already that each perfect matching corresponds to a permutation $\sigma \in S_n$, which maps each black vertex to its white partner. We saw in Example 2.9 that perm($A$) counts all of these permutations. But the determinant of $A$ counts them weighted by their parities, $\det(A) = \sum_{\text{matchings } \sigma} (-1)^\sigma$. The idea is to compensate the parity weights of the determinant in order to obtain the correct count of perfect matching. To do this, we place weights $w_{ij} = \pm 1$ on the edges of $G$. This defines a quantity $\tilde{A}_{ij} = w_{ij}$. Now, each matching $\sigma$ of $\tilde{A}$ has a weight

$$w(\sigma) = \prod_i w_{i\sigma(i)},$$

and the determinant of $\tilde{A}$ is given by

$$\det(\tilde{A}) = \sum_{\text{matchings } \sigma} (-1)^\sigma w(\sigma).$$

Now we would like to write the permanent of $A$ as the determinant of $\tilde{A}$. To do this, we should choose the weights $w_{ij}$ such that $(-1)^\sigma w(\sigma)$ has the same sign for all $\sigma$. This means that the matching $\sigma$ should change the weight $w(\sigma)$ by a factor $-1$ when its parity changes. Then we would have $|\det(\tilde{A})| = \text{perm}(A) = $ number of perfect matchings. The trick is to find a proper set of weights. For instance, in the particular case of the chessboard (see the diagram above), one can decide to put a weight $i$ on all vertical edges. Then, by changing two horizontal dimers to two vertical ones, the weight changes by $i^2 = -1$, and so does the parity. It should be emphasized that a set of such weights can always be found for planar graphs. In addition, an analogous approach allows us to solve the two-dimensional Ising model (using a trick to count perfect matchings on planar graphs that are not bipartite). For further details, we refer to Moore and Mertens (2011). We finally notice that some problems (like graph 3-coloring) are just as hard when limited to the planar case, while others (like 4-coloring or counting matchings) get much easier.

We now turn to random graphs, and give some of their basic properties. These properties will be useful when discussing some other problems in more details.

## 2.3 Random graphs

### 2.3.1 Erdős–Rényi random graphs

First we want to give the definition of a random graph.

**Definition 2.13** *A graph $G(n,p)$ with $n$ vertices and edge probability $p$ is one model of a **random graph**. Edge probability $p$ means that between any two vertices $v$ and $w$ there exists an edge with probability $p$. For every pair of vertices, this event is independent of every other pair. An **Erdős–Rényi graph** $G(n,m)$ has $n$ vertices and is uniformly chosen from all graphs with exactly $m$ edges.*

Our first task is to compute the expected degree of one vertex and the expected number of edges in the $G(n,p)$ model. We want to investigate the sparse case $pn = c$, where $c$ is a constant independent of $n$. We use $\mathbb{E}$ to denote the expectation.

$$\mathbb{E}[m] = p\binom{n}{2} \approx \frac{pn^2}{2} = \frac{cn}{2}, \tag{2.1}$$

$$\mathbb{E}[\deg] = p(n-1) \approx pn = c. \tag{2.2}$$

Now we can compute some other interesting quantities in our model. We compute the expected number of triangles and bicycles in $G(n,p)$:

$$\mathbb{E}[\#\Delta] = \binom{n}{3}p^3 \approx \frac{c^3}{6}, \tag{2.3}$$

$$\mathbb{E}[\# \text{ bicycles}] = const \times \binom{n}{k}p^{k+1} \approx n^k \frac{c^{k+1}}{n^{k+1}} \xrightarrow{n \to \infty} 0. \tag{2.4}$$

Bicycles are subgraphs containing $k$ vertices and $k + 1$ edges. Thus, they contain two loops, connected by a path or by an edge belonging to both loops. The constant depends on $k$ but not on $n$. For any fixed $k$, there are probably no bicycles as $n \to \infty$, so most vertices have a treelike neighborhood.

Our next task is to compute the probability that a vertex $v$ has degree $k$:

$$\mathrm{P}[\deg(v) = k] = \binom{n-1}{k}p^k(1-p)^{n-1-k} \approx \frac{n^k}{k!}\frac{c^k}{n^k}\left(1-\frac{c}{n}\right)^n = \frac{e^{-c}c^k}{k!}. \tag{2.5}$$

This is called the **degree distribution**. As $n$ goes to infinity, it becomes a Poisson distribution with mean $c$.

### 2.3.2 The giant component

In this section, we want to describe the most basic phase transition occurring in $G(n,p)$ for $pn = c$. For very small $c$, $G$ consists of small components isolated from each other, and almost all of them are trees. As $c$ grows larger and larger, we add more edges, and these trees grow and connect with each other. Suddenly, at $c = 1$, many of them come together to form a giant component. We want to observe this phase transition later in this chapter. First, we study the expected component size of a vertex $v$. We start at this vertex and go further to its neighbors and to their neighbors and so on. We can think of those vertices as $v$'s children, and the descendants with distance $k$ from our starting vertex $v$ are the so-called **k-th generation** of $v$. By doing that, we can develop the whole component of $v$. For large $n$, we can assume that every child of $v$ again has Poisson($c$) children. That is, we can approximate the process of exploring $v$'s component as a branching process. We only have to count the descendants of $v$ to get the component size:

$$\mathbb{E}[\text{Component size of } v] = 1 + c + c^2 + c^3 + \cdots = \frac{1}{1-c} \text{ for } c < 1. \qquad (2.6)$$

For $c < 1$, this sum converges and the expected component size is finite. For $c > 1$, on the other hand, the sum diverges and $v$ has, in expectation, an infinite number of descendants; in fact, the number of descendants is infinite with positive probability. At this point, the branching process is no longer an accurate model to compute the component size of $v$. For further computations, we call $\gamma$ the fraction of all vertices that lie in the giant component of $G(n,p)$. When $c > 1$, with high probability there is a unique giant component.

We will see two different methods to estimate the value of $\gamma$.

## Method 1

The probability that a vertex $v$ is not part of the giant component is the sum over all $k$ of the probability that $v$ has degree $k$ and none of its children are part of the giant component:

$$1 - \gamma = \sum_k \mathrm{P}[\deg(v) = k](1-\gamma)^k = \sum_k \frac{e^{-c}c^k}{k!}(1-\gamma)^k = e^{-c}e^{c(1-\gamma)} = e^{-c\gamma}. \qquad (2.7)$$

So we get a transcendental equation for the size of the giant component:

$$\gamma = 1 - e^{-c\gamma}. \qquad (2.8)$$

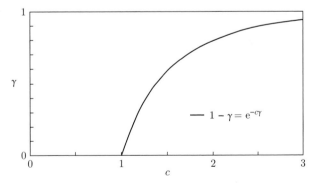

From Moore and Mertens (2011)

## Method 2

We compute $\gamma$ again, but now with a system of differential equations. We start again with a vertex $v$ and explore its connected component, now by using an algorithm. At each point in time, a vertex is labeled **Reached**, **Boundary**, or **Untouched**:

- "Reached" means that a vertex lies in $v$'s component and its neighborhood has been explored.
- "Boundary" means that a vertex lies in $v$'s component but its neighbors have not yet been explored.
- "Untouched" means that it is not yet known if the vertex is in $v$'s component.

---

**Algorithm 1** Cluster expansion

---

**Require:** a vertex $v$
**Ensure:** a connected-graph
1: label $v$ Boundary
2: label all other vertices Untouched
3: **while** there are Boundary vertices **do**
4:     choose a Boundary vertex
5:     label each of $v$'s Untouched neighbors Boundary
6:     label $v$ Reached
7: **end while**

---

This algorithm explores $G$ one vertex at a time, until there are no Boundary vertices left and $v$'s component has been labeled Reached. Let $R$, $B$, and $U$ denote the numbers of vertices of each type at a given point in time. At each step of the algorithm, $R$ increases by 1, and the expected change in $B$ and $U$ is

$$\Delta R = 1, \tag{2.9}$$

$$\mathbb{E}[\Delta B] = -1 + pU = \frac{c}{n}U - 1, \tag{2.10}$$

$$\mathbb{E}[\Delta U] = -pU = -\frac{c}{n}U. \tag{2.11}$$

We have an expected change from $U$ to $B$ of $pU$, since each Untouched vertex is connected to $v$ with probability $p$. By changing the chosen vertex $v$ from Boundary to Reached, we get the term $-1$. When $n$ is large, we can rescale these stochastic difference equations so that they become a system of differential equations:

$$\frac{dr}{dt} = 1, \tag{2.12}$$

$$\frac{db}{dt} = cu - 1, \tag{2.13}$$

$$\frac{du}{dt} = -cu. \tag{2.14}$$

We solve these differential equations with the initial conditions $b(0) = 0$ and $u(0) = 1$:

$$u(t) = e^{-ct} \text{ and } b(t) = 1 - t - e^{-ct}. \tag{2.15}$$

The fraction $\gamma$ of vertices in the giant component is the value of $t$ at which no Boundary vertices remain and the algorithm stops. This gives the same equation as before:

$$b(\gamma) = 1 - \gamma - e^{-c\gamma} = 0. \qquad (2.16)$$

*Remark:* The differential equation approach to the size of the giant component in random graphs is similar to the dynamics for the SIR model, where $p$ is the transmission rate and $\gamma$ is the fraction of the population that eventually becomes infected.

Another interesting quantity we can consider is the expected number of components/trees of $G$ with $k$ vertices. We approximate this with the number of trees:

$$\mathbb{E}[\text{Components/trees of size } k] = \binom{n}{k} p^{k-1}(1-p)k^{k-2}(1-p)^{kn-k}. \qquad (2.17)$$

Here $k^{k-2}$ comes from Cayley's formula for the number of labeled trees with $k$ vertices. For $p = c/n$ and $c = 1$, the above expression becomes a power law by applying Stirling's formula:

$$\mathbb{E}[\text{Components/trees of size } k] \approx \frac{1}{\sqrt{2\pi}} \frac{n}{k^{5/2}} \left(1 + O(k^2/n)\right). \qquad (2.18)$$

### 2.3.3   Giant component and configuration model

One problem with random graphs is that the degree distribution converges toward the Poisson distribution, which is unrealistic in many concrete examples (social networks, biological networks, etc.). The configuration model can be used to remedy this problem. This model deals with a sequence of nodes and degrees. The nodes are chosen randomly before being connected by an edge. During this process, each node is chosen with a probability proportional to the number of unmatched edges it has. Equivalently, each vertex with degree $d$ has $d$ "stubs" or half-edges. The total number of stubs is $2m$, where $m$ is the number of edges, and we choose a uniformly random matching of these stubs with each other.

This procedure may create some self-loops or multiple edges, so, strictly speaking, this model produces random multigraphs. However, for reasonable degree distributions (with bounded mean and variance), the resulting graph is simple with constant probability.

---

**Algorithm 2** Configurational model

---

**Require:** degree sequence
**Ensure:** a graph
  1: **while** there remain unmatched edges **do**
  2:    choose two unmatched edges uniformly and randomly
  3:    put an edge between them
  4: **end while**

---

The next question is to determine where, in this new setting, the giant component appears. As long as the graph is sparse (i.e., the average degree is constant), the appearance of the giant component can be analyzed by a branching process as before. When we follow a link to a vertex of degree $k$, there are $k - 1$ "children" consisting of new edges that come out of it. Thus, the average branching ratio is

$$\lambda = \sum_k (k-1) \frac{k a_k}{\sum j a_j} = \frac{\mathbb{E}[k(k-1)]}{\mathbb{E}[k]}, \tag{2.19}$$

where $a_k$ is the fraction of nodes in the degree sequence that have degree $k$. A giant component appears when $\lambda > 1$, or equivalently

$$\mathbb{E}[k^2] > 2\mathbb{E}[k]. \tag{2.20}$$

Thus, both the first and second moments of the degree distribution matter.

### 2.3.4   The $k$-core

**Definition 2.14**   *The **$k$-core** of a graph $G$ is the largest subgraph where each vertex has minimal degree $k$. Equivalently, it is the graph remaining after removing vertices with $d < k$ neighbors (iteratively).*

This object is related to the $k$-colorable property of a graph: if there is no $k$-core, the graph is $k$-colorable (Pittel *et al.*, 1996) (note that the converse is not necessarily true). We will again use a branching process to characterize the $k$-core: at which $\alpha$ it appears and what fraction of the system is in it at that point. Unlike the giant component, the $k$-core appears suddenly for $k \geq 3$: that is, it includes a constant fraction of vertices when it first appears.

### *Branching process*

We consider only $k > 2$, since for any $c > 0$, $G(n, p = c/n)$ typically contains a 2-core as a loop of size $O(\log n)$ or even a triangle is present with constant probability. To describe the problem with a branching process, we need to consider first a root node $v$, again treating its neighbors as its children, their neighbors as its grandchildren, and so on. Let us (recursively) say that a node in this tree is *well connected* if it has at least $k - 1$ well-connected children. Such a node will survive the process that deletes nodes with degree less than $k$, if we think of this process as moving up the tree from the leaves.

If the fraction of well-connected children is $q$, then the number of well-connected children a given node has is Poisson distributed with mean $cq$. The probability that the number of well-connected children is less than $k$ is then given by

$$Q_k = \sum_{j=0}^{k-1} \frac{e^{-cq} (cq)^j}{j!}. \tag{2.21}$$

This gives us the fixed-point equation

$$q = 1 - Q_{k-1}. \tag{2.22}$$

With this equation, we can find the value of $c$ at which the $k$-core appears, i.e., the smallest $c$ such that this equation has a positive root. For $k = 3$, for instance, we have $c_3^{\text{CORE}} = 3.351$.

The probability that a given node is in the core is a little tricky. Here we treat the node as the root. In order to survive the deletion process, it has to have at least $k$ well-connected children. Thus the fraction of nodes in the core is

$$\gamma_k = 1 - Q_k . \tag{2.23}$$

where $q$ is the root of (2.22). For $k = 3$, this gives $\gamma_3 = 0.268$.

## 2.4   Random $k$-SAT

In analogy with random graphs $G(n, m)$ that have $n$ nodes and $m$ edges, we can define random $k$-SAT formulas $F_k(n, m)$ with $n$ variables and $m$ clauses. We choose each clause uniformly and independently from the $2^k \binom{n}{k}$ possible clauses. That is, we choose a $k$-tuple of variables, and then for each one independently negate it with probability $\frac{1}{2}$.

We focus on the sparse case where the number of clauses scales as $m = \alpha n$, where $\alpha$ is a constant. In the following analysis, we will not care about rare events such as two variables appearing in the same clause, or the same clause appearing twice in the formula.

*Exercise:* show that in the sparse case, the probability of either of these happening tends to zero as $n \to \infty$.

**Phase diagram:** It is now commonly accepted that the random $k$-SAT (here $k = 3$) problem undergoes a phase transition at a critical value $\alpha_k$. This phase transition is characterized by the probability of having a satisfying assignment converging to 1 below $\alpha_k$ and to 0 above.

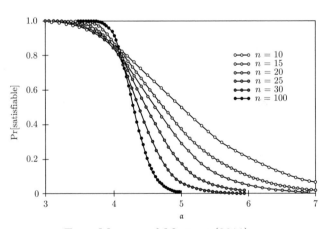

From Moore and Mertens (2011)

Mathematically speaking, this picture is still a conjecture, which is known as:

**Conjecture 2.15 (Threshold conjecture)** $\forall k \geq 3$, $\exists \alpha_k$ *such that* $\forall \epsilon > 0$,

$$\lim_{n \to \infty} \mathrm{P}[F_k(n, \alpha m) \text{ is satisfiable}] = \begin{cases} 1 \ \textit{if} \ \alpha < (1 - \epsilon)\alpha_k , \\ 0 \ \textit{if} \ \alpha > (1 + \epsilon)\alpha_k . \end{cases}$$

The closest rigorous result for this conjecture came from Friedgut, who showed that there is a function $\alpha_k(n)$ for which it is true. But we do not know if $\alpha_k(n)$ converges to a constant as $n \to \infty$, or whether it continues to fluctuate in some way. Physically, this would be like the freezing point of water depending on the number of water molecules, which seems very unlikely, but we still lack a rigorous proof. For the rest of this section, we will assume that the threshold conjecture is true.

### 2.4.1   Easy upper bound

It is easy in a first approach to derive a (not so good) upper bound for the position of the $k$-SAT threshold. We will use the "first-moment" method: for any random variable that takes values $0, 1, 2, \ldots$, such as the number of some object, we have (exercise)

$$P[x > 0] \le E[x].$$

Let us define $x$ as the number of satisfying assignments for a given formula. It is easy to compute the average of $x$, since the clauses are independent. Let write $x$ as the sum over all truth assignments or "configurations," using an indicator function to count only the satisfying ones:

$$x = \sum_{\sigma \in \{0,1\}^n} \mathbb{1}_\sigma, \tag{2.24}$$

$$E[x] = \sum_\sigma E[\mathbb{1}_\sigma] \tag{2.25}$$

$$= \sum_\sigma P(\sigma \text{ satisfies } \phi) = \sum_\sigma \prod_{c \in \phi} P(\sigma \text{ satisfies } c) \tag{2.26}$$

$$= 2^n (1 - 2^{-k})^m. \tag{2.27}$$

For $k = 3$ in particular, this gives

$$E[x] = \left[ 2 \left( \frac{7}{8} \right)^\alpha \right]^n.$$

Therefore, we can compute an upper bound on $\alpha_3$ by finding $\alpha$ such that $2(\frac{7}{8})^\alpha = 1$. This gives $\alpha_3 \le 5.19$. More generally, we have

$$\alpha_k < \frac{\ln 2}{\ln(1 - 2^{-k})} < 2^k \ln 2.$$

An interesting question, which was open until fairly recently, is whether this bound is asymptotically tight. How close to the truth is this simple counting argument? For that matter, why is it not exactly correct? For $k = 3$, in the range $4.27 < \alpha < 5.19$, the expected number of satisfying assignments is exponentially large, and yet most of the time there are none at all. What is going on?

## 2.4.2   Lower bounds from differential equations and the analysis of algorithms

We will go through two proofs for the lower bound of the $k$-SAT threshold. The first (easy) one is based on the analysis of a very simple algorithm called "unit clause propagation" (UC). This algorithm deals with unit clauses, which are clauses containing a single variable ($x$ of $\bar{x}$). How could unit clauses appear in the $k$-SAT problem? Well, if one fixes a variable of the problem, three different options can arise concerning the clauses in which it appears. If it satisfies a clause, then the clause disappears. If the clause remains unsatisfied, then the variable is removed from the clause. In that case, a 3-clause becomes a 2-clause, and a 2-clause becomes a unit clause. If the clause in which the variable appears is a unit clause and does not get satisfied when fixing the variable, then a contradiction appears.

The principle of UC is the following. Whenever there exists a unit clause, the algorithm chooses a unit clause uniformly at random and satisfies it by permanently fixing the variable of the clause. If no more unit clauses are present, the algorithm chooses a variable uniformly at random from the unset variables and fixes it randomly to 0 or 1 with probability $\frac{1}{2}$. When all the clauses have been satisfied, the algorithm returns "satisfiable." If the process encounters a contradiction (unsatisfied clause with all the variables of the clause fixed), it returns "contradiction" and the algorithm stops.

---

**Algorithm 3** Unit clause propagation (UC)

---

**Require:** a $k$-SAT instance
**Ensure:** "satisfiable" or "contradiction"
 1: **while** (there is no contradiction) AND (there exists unsatisfied clauses) **do**
 2:     **if** there exists a unit clause **then**
 3:         (forced step) choose one uniformly at random and satisfy it
 4:     **else**
 5:         (free step) choose $x$ uniformly from all the unset variables $\rightarrow$ set $x = 0, 1$
 6:     **end if**
 7:     **if** there exists an empty clause (or unsat clause) **then**
 8:         **return** "contradiction"
 9:     **end if**
10: **end while**
11: **return** "satisfiable"

---

At all times, the remaining formula is uniformly random conditioned on the numbers $S_3$, $S_2$, and $S_1$ of unsatisfied 3-clauses, 2-clauses, and unit clauses. This comes from the fact that since variables are removed from the formula whenever they are set, we know nothing at all about how the unset variables appear in the remaining clauses. Moreover, both moves effectively give a random variable a random value, since the (free) move is completely random, and the forced one satisfies a random unit clause. This property makes it easier to deal with the algorithm.

Let us imagine now that we are at a time $t$ where there remain $n - T$ variables. When dealing with a 3-clause, the probability that a chosen variable appears in it is $3/(n-T)$. When fixing this variable, there is half a chance that it will satisfy the clause and half a chance that the 3-clause has become a 2-clause. The same calculation can be done for the 2-clause, a variable having $2/(n - T)$ chance to be in it. If we write $S_3 = s_3 n$, $S_2 = s_2 n$, and $T = tn$, and assume that the change in each of these rescaled variables equals its expectation, then we can write a set of differential equations for $s_2$ and $s_3$:

$$\frac{ds_3}{dt} = -\frac{3s_3}{1-t}, \tag{2.28}$$

$$\frac{ds_2}{dt} = -\frac{2s_2}{1-t} + \frac{3}{2}\frac{s_3}{1-t}. \tag{2.29}$$

The solution of these equations is

$$s_3 = \alpha(1 - t)^3, \tag{2.30}$$

$$s_2 = \frac{3}{2}\alpha t(1 - t)^2 \tag{2.31}$$

These equations do not tell us how many unit clauses there are. Since there are $O(1)$ of these as opposed to $O(n)$ of them (indeed, there is a contradiction with high probability as soon as there are $O(\sqrt{n})$ of them), we will model them in a different way, using a branching process rather than a differential equation.

Each time we satisfy a unit clause, we might create some new ones by shortening a 2-clause. Thus, the branching ratio, i.e., the expected number of new unit clauses, is $\lambda = s_2/(1-t)$. If $\lambda > 1$, this branching process explodes, leading to a contradiction. On the other hand, it can be shown that if $\lambda < 1$ throughout the algorithm, then it succeeds in satisfying the entire formula with positive probability. Because of Friedgut's theorem, proving positive probability is enough to prove probability 1. Therefore, we should determine the maximum value $\alpha$ can take such that $\max_t \lambda \leq 1$. We find that

$$\lambda_{\max} = \frac{3}{8}\alpha, \tag{2.32}$$

and therefore

$$\alpha_3 > \frac{3}{8}.$$

This analysis can be generalized for any $k$, giving

$$\alpha_k \gtrsim \frac{2^k}{k}. \tag{2.33}$$

Note that this is a factor of $k$ below the first-moment upper bound. Next, we will see how to close this gap.

### 2.4.3  Lower bounds from the second-moment method

We now use the so-called second moment method to narrow the gap between the two bounds on $\alpha_k$ previously discussed. The second-moment method relies on the inequality

$$P[X > 0] \geq \frac{E[X]^2}{E[X^2]}. \tag{2.34}$$

$X$ denotes the number of solutions to the formula $\phi$. Note that by Friedgut's theorem, we will automatically have $P[X > 0] = 1$ by simply proving that the right-hand side is strictly positive. We will now prove the following lower bound on $\alpha_k$:

$$\alpha_k \geq 2^{k-1} \ln 2 - O(1). \tag{2.35}$$

This will reduce the width of the gap between upper and lower bounds from a factor of $k$ to a factor of 2, independent of $k$. To do so, the new challenge is to compute $E[X^2]$. Using the indicator function $\mathbb{1}_\sigma$ introduced previously—which is equal to 1 if $\sigma$ satisfies $\phi$ and 0 otherwise—we have the following:

$$X^2 = \left( \sum_\sigma \mathbb{1}_\sigma \right)^2 = \sum_{\sigma, \tau \in \{0,1\}^n} \mathbb{1}_\sigma \mathbb{1}_\tau, \tag{2.36}$$

so that

$$E[X^2] = \sum_{\sigma, \tau \in \{0,1\}^n} P[\sigma, \tau \text{ both satisfy } \phi]. \tag{2.37}$$

While using the first-moment method previously, we had to compute $E[X]$, which could be expressed in terms of $P[\sigma \text{ satisfies } \phi]$, which was independent of $\sigma$ because of the definition of $F_k(n, m)$. On the other hand, $P[\sigma, \tau \text{ both satisfy } \phi]$ now depends on the Hamming distance $z$ between the truth assignments $\sigma$ and $\tau$. More precisely, by independence of the clauses, we have

$$E[X^2] = \sum_{\sigma, \tau \in \{0,1\}^n} \prod_{c \in \phi} P[\sigma, \tau \text{ both satisfy } c]$$

$$= \sum_{\sigma, \tau \in \{0,1\}^n} \left( P[\sigma, \tau \text{ both satisfy a random } c] \right)^m.$$

By the inclusion–exclusion principle, we then have

$$P[\sigma, \tau \text{ both satisfy a random } c] = 1 - P[\sigma \text{ doesn't satisfy } c]$$
$$- P[\tau \text{ doesn't satisfy } c] + P[\sigma, \tau \text{ both don't satisfy } c]$$
$$= 1 - 2^{-k} - 2^{-k} + (z/n)^k 2^{-k}$$
$$= f\left( \frac{z}{n} \right).$$

To compute $P[\sigma, \tau$ both don't satisfy $c]$, we used the fact that

$$P[\sigma, \tau \text{ both don't satisfy } c] = P[\tau \text{ doesn't satisfy } c]$$
$$\times P[\sigma \text{ doesn't satisfy } c \mid \tau \text{ doesn't satisfy } c]$$
$$= 2^{-k}(z/n)^k,$$

because $\sigma$ will not satisfy $c$ if and only if none of the variables that differ between $\sigma$ and $\tau$ fall in the set of variables concerned by the clause $c$. This happens with probability $(z/n)^k$.

We therefore have

$$\mathbb{E}[X^2] = \sum_{z=0}^{n} 2^n \binom{n}{z} f\left(\frac{z}{n}\right)^m,$$

where $2^n \binom{n}{z}$ is the number of couples of truth assignments $\sigma$ and $\tau$ such that they differ by $z$ bits. Note that if $z/n = \frac{1}{2}$, which is the most likely overlap between two truth assignments, then we have $f(z/n) = f(\frac{1}{2}) = (1-2^{-k})^2$, as if $\sigma$ and $\tau$ were independent. If this term dominates the sum in $\mathbb{E}[X^2]$, then we have $\mathbb{E}[X^2] = 4^n(1-2^{-k})^{2m} = \mathbb{E}[X]^2$, so that there exists a solution to the $k$-SAT problem. We now need to check for which values of $\alpha$ this approximation holds. To do so, we write in the limit $n \longrightarrow \infty$

$$\mathbb{E}[X^2] = 2^n n \int_0^1 d\zeta \binom{n}{\zeta n} f(\zeta)^m.$$

By Stirling's formula,

$$\binom{n}{\zeta n} \sim \frac{1}{\sqrt{n}} e^{nh(\zeta)},$$

where $h(\zeta) = -\zeta \ln(\zeta) - (1 - \zeta) \ln(1 - \zeta)$, so that

$$\mathbb{E}[X^2] = 2^n \sqrt{n} \int_0^1 d\zeta \, e^{n\phi(\zeta)},$$

with $\phi(\zeta) = h(\zeta) + \alpha \ln f(\zeta)$. Using the Laplace method, we can write

$$\mathbb{E}[X^2] \sim 2^n \sqrt{n} \sqrt{\frac{2\pi}{n|\phi''|}} e^{n\phi^{\max}}.$$

When $\phi^{\max} = \phi(\frac{1}{2})$, corresponding to $e^{\phi(\frac{1}{2})} = 2(1 - 2^{-k})^{2\alpha}$, we recover $\mathbb{E}[X^2] \sim 4^n(1 - 2^{-k})^{2m} = \mathbb{E}[X]^2$, consistently with a previous remark. For values of $\alpha$ where this holds, there exists a solution to the $k$-SAT problem. But we know that this cannot be true for all values of $\alpha$, because we know there is a transition. We therefore need to compare the actual maximum of $\phi$ with $\phi^{\max}$. But because $h$ is symmetric with respect to the $\zeta = \frac{1}{2}$ axis, and $\ln f$ is increasing, it is clear that whenever $\alpha > 0$ also $\phi^{\max} > \phi(\frac{1}{2})$, so that in the limit where $n$ goes to infinity, the second-moment inequality (2.34) just tells us that $P[X > 0] \geq 0$.

The way out is to consider another problem for which the function $f$ is also symmetric with respect to the $\zeta = \frac{1}{2}$ axis. Consider the NAE $k$-SAT problem, where we ask that at least one literal in each clause be false. It is clear that a truth assignment that satisfies an NAE $k$-SAT formula is also a solution to the corresponding $k$-SAT problem, so that a lower bound for $\alpha$ in NAE also holds in $k$-SAT. It turns out the function $f$ in NAE is given by

$$f(\zeta) = 1 - 2^{1-k} + \zeta^k 2^{1-k} + (1-\zeta)^k 2^{1-k},$$

and therefore $\phi$ is symmetric around $\frac{1}{2}$. It is then straightforward to show that the maximum of $\phi$ is in $\frac{1}{2}$, up to some value of $\alpha = \frac{1}{2} 2^k \ln 2 - O(1)$. We therefore have $\alpha_k \geq \alpha_k^{\mathrm{NAE}} \geq \frac{1}{2} 2^k \ln 2 - O(1)$, which is the lower bound we wanted. This lower bound can in fact be improved again to $2^k \ln 2 - O(1)$ by considering another random variable

$$X = \sum_{\sigma \text{ satis. } \phi} \eta^{\# \text{ true literals}}$$

for some carefully chosen $\eta$.

We conclude this section with some physical considerations on what the second-moment inequality (2.34) means. It is interesting to interpret this inequality in terms of the planted ensemble. In the planted ensemble, we start by choosing at random a truth assignment $\sigma$, and then sample formulas $\phi$ such that $\sigma$ satisfies $\phi$. The expectation value of $X^2$ can be expressed in terms of the planted average as follows:

$$\mathbb{E}[X^2] = \sum_{\sigma, \tau} \mathrm{P}[\sigma, \tau \text{ sol.}]$$

$$= \sum_{\sigma} \mathrm{P}[\sigma \text{ sol.}] \sum_{\tau} \mathrm{P}[\tau \text{ sol.} \mid \sigma \text{ sol.}]$$

$$= \mathbb{E}[X]\mathbb{E}[X|\sigma]$$

$$= \mathbb{E}[X]\mathbb{E}_{\text{planted}}[X],$$

so that (2.34) can be rewritten

$$\mathrm{P}[X > 0] \geq \frac{\mathbb{E}[X]}{\mathbb{E}_{\text{planted}}[X]}.$$

By construction, $\mathbb{E}_{\text{planted}}[X]$ is bigger than $\mathbb{E}[X]$, because a formula constructed in the planted ensemble has more solutions on average than a completely random instance. But we see here that if the number of solutions is not too different, i.e., if the planted expectation is not too different from the exact expectation in the large-$n$ limit, then the second-moment inequality tells us something new.

## 2.5   Community detection

### 2.5.1   The stochastic block model

It has been a general trend lately to consider everything as a network, whether in physics, sociology, biology, finance, ... The question that naturally arises is whether this is justified, that is, whether seeing a system as a network actually allows us to answer questions about the system.

In this section, we consider the problem of community detection, i.e., the problem of identifying groups of nodes in a graph that share a common set of features. Community structure is called *assortative* if each group has a larger connectivity between its members than with other communities. The opposite situation is called *disassortative*. While trying to guess communities in a general setting, one does not know a priori if they are associative or disassortative.

To generate instances of this problem, we consider the stochastic block model (SBM). This is a generalization of the Erdős–Renyí ensemble, with $k$ types of vertices. These types can be thought of as colors, groups, spins, and so on. We start by supposing that $k$ is known, and we encode the parameters of the model in a $k \times k$ affinity matrix $p$, where $p_{r,s}$ is the probability that a given node from group $r$ is connected with a given node from group $s$. Denoting the type of node $i$ by $t_i$, we then construct the graph with the rule

$$\mathrm{P}[(i,j) \in E] = p_{t_i,t_j} \, .$$

Our goal is, given the graph $G$, to find the types of the nodes.

Let us begin by rewriting the problem as a statistical physics problem. The probability of a given graph $G$ conditioned on the types of the nodes $t$ and the affinity between communities $p$ is given by

$$\mathrm{P}[G|t,p] = \prod_{(i,j) \in E} p_{t_i,t_j} \prod_{(i,j) \notin E} (1 - p_{t_i,t_j}) \, .$$

In the problem we are considering, $t$ is unknown—it is what we are looking for. But we might know a priori probabilities $q_r$ for $r \in \{1,..k\}$ that a given node has type $r$. We then have

$$\mathrm{P}[G|p,q] = \sum_t \mathrm{P}[G|t,p]\,\mathrm{P}[t|q]$$

$$= \sum_t \prod_{(i,j) \in E} p_{t_i,t_j} \prod_{(i,j) \notin E} (1 - p_{t_i,t_j}) \prod_i q_{t_i}$$

$$= \sum_t e^{-H(t)} \, .$$

This is the partition function of a physical system at inverse temperature $\beta = 1$, with Hamiltonian

$$H(t) = -\sum_{i \sim j} \ln p_{t_i,t_j} - \sum_{i \nsim j} \ln (1 - p_{t_i,t_j}) - \sum_i \ln q_i \tag{2.38}$$

corresponding to a generalized Potts model with coupling constants $J_{ij} = \ln p_{t_i,t_j}$ and external fields $h_i = \ln q_i$. Note that it includes interactions between non-neighboring sites. In the sparse case ($p_{rs} = c_{rs}/n$), the coupling between two non-neighboring sites $i \sim j$ is of order $1/n$. However, we cannot simply get rid of these interactions, since the sum contains $O(n^2)$ such terms.

The Boltzmann distribution over assignments of types $t$ is given, using Bayes' rule, by

$$P[t|G, p, q] = \frac{P[G, t|p, q]}{P[G|p, q]}$$

$$= \frac{P[G|t, p] \, P[t|q]}{P[G|p, q]}$$

$$= \frac{e^{-H(t)}}{P[G|p, q]}.$$

Now suppose we want to find the parameters $p$, $q$ that maximize $\mathcal{Z} = P[G|p, q]$; that is, we want to maximize the total probability of the network, summed over all type assignments. We can relate these optimal values of the parameters to thermodynamic quantities in the following way:

$$\frac{\partial P}{\partial p_{rs}} = \sum_t \frac{\partial}{\partial p_{rs}} e^{-H(t)}$$

$$= -\sum_t e^{-H(t)} \frac{\partial H(t)}{\partial p_{rs}}$$

$$= -\sum_t e^{-H(t)} \left( -\sum_{\substack{i \sim j \\ t_i = r \\ t_j = s}} \frac{1}{p_{rs}} + \sum_{\substack{i \nsim j \\ t_i = r \\ t_j = s}} \frac{1}{1 - p_{rs}} \right),$$

In a particular instance of the problem, we have

$$\frac{\partial P}{\partial p_{rs}} = -\sum_t e^{-H(t)} \left( -\frac{m_{rs}}{p_{rs}} + \frac{n_r n_s - m_{rs}}{1 - p_{rs}} \right)$$

$$\propto +\frac{\langle m_{rs} \rangle}{p_{rs}} - \frac{\langle n_r n_s \rangle - \langle m_{rs} \rangle}{1 - p_{rs}}, \tag{2.39}$$

where $m_{rs}$ denotes the number of edges between group $r$ and group $s$, and $n_r$ denotes the number of nodes in group $r$, and the angle brackets denote the average over the Boltzmann distribution.

We will assume $\langle n_r n_s \rangle = \langle n_r \rangle \langle n_s \rangle$; this holds, in particular, if both $n_r$ and $n_s$ are tightly concentrated with $O(\sqrt{n})$ fluctuations. In that case, (2.39) gives

$$p_{rs} = \frac{\langle m_{rs} \rangle}{\langle n_r \rangle \langle n_s \rangle}.$$

Similarly, by taking the derivative of $P$ with respect to $q_r$, we find

$$q_r = \frac{\langle n_r \rangle}{n}.$$

Thus, if we can estimate the averages with respect to the Boltzmann distribution, we can learn the parameters using the following expectation-maximization algorithm:

- E step: Compute the averages $\langle m_{rs} \rangle$ and $\langle n_r \rangle$ using the current values of $p$ and $q$.
- M step: Update $p$ and $q$ to their most likely values given $\langle m_{rs} \rangle$ and $\langle n_r \rangle$.

We iterate until we reach a fixed point.

The only question that remains is how to estimate averages with respect to the Boltzmann distribution. This can be done by doing a Monte Carlo (MC) simulation, also known as Gibbs sampling. For instance, we can use the heat bath algorithm: at each step, a node in the graph is chosen and "thermalized," which means that its group $t_i$ is sampled from the marginal distribution imposed by its neighbors. More precisely, we set $t_i = s$ with probability proportional to $q_s \prod_{j \sim i} p_{s,t_j}$. It is straightforward that this algorithm verifies detailed balance with respect to the Boltzmann distribution (2.38). After convergence, it will sample configurations with the correct weights. However, to compute averages like $\langle m_{rs} \rangle$, we need many independent samples, which forces us to run the algorithm for a large multiple of the autocorrelation time.

We can do better using belief propagation (BP). The idea of BP is that vertices pass each other's estimates of marginals until consistency (a fixed point) is achieved. More precisely, we write

$$\mu_r^{i \to j} = \mathrm{P}[t_i = r \text{ if } j \text{ were absent}],$$

which is the cavity interpretation of the messages of BP. Our goal is to estimate the one-node and two-node marginals:

$$\mu_r^i = \mathrm{P}[t_i = r],$$
$$\mu_{rs}^{ij} = \mathrm{P}[t_i = r, \ t_j = s].$$

BP estimates of these marginals can be expressed in terms of the messages. In particular,

$$\mu_{rs}^{ij} \propto \mu_r^{i \to j} \mu_s^{j \to i} \begin{cases} p_{rs} & \text{if } i \sim j, \\ 1 - p_{rs} & \text{otherwise}. \end{cases}$$

The BP update rule is given by

$$\mu_r^{i \to j} = \frac{1}{\mathcal{Z}^{i \to j}} q_r \prod_{\substack{k \sim i \\ k \neq j}} \sum_s \mu_s^{k \to i} p_{rs} \prod_{\substack{k \not\sim i \\ k \neq j}} \sum_s \mu_s^{k \to i} (1 - p_{rs}). \tag{2.40}$$

As usual in BP, this expression assumes that nodes other than $i$ are independent conditioned on $t_i$. This is only approximately true if $G$ is not a tree; we believe that it is approximately true when $G$ is locally treelike, as in graphs generated by the sparse stochastic block model. Even in real networks, it works surprisingly well.

Note that (2.40) actually takes place on a fully connected graph, because of the interaction between non-neighboring sites. This yields $O(n^2)$ calculations at each update. To recover sparsity in this formula, we will assume that site $i$ only feels the mean field of its non-neighboring sites, that is, $\mu_r^{k \to i} = \mu_r^k$ for all $k \nsim i$. With this assumption, one iteration of BP requires only $O(m)$ computations, where $m$ is the number of edges: for sparse graphs, we have $m = O(n)$ rather than $O(n^2)$. This makes the algorithm far more scalable.

Once the parameters $p$ and $q$ are set correctly, we can run MC or BP to determine the most likely assignment of group types. But we get a much finer sense of what is really going on by taking a look at the whole distribution of group assignments. Zachary's Karate Club provides a cautionary tale about the risks of the procedure we have just described. Wayne Zachary collected relationship data in a university karate club composed of 34 people in 1977. Because of an argument between the president of the club and the instructor, the club split into two groups. Some followed the president, others followed the instructor. Zachary asked whether it is possible, from the friendship network he collected before the split-up, to retrodict the composition of the two groups. This is easy to do except for one node, who was closer to the president, but ended in the instructor's group because he had a black belt exam three weeks later and did not want to change instructor!

This story tells us that the graph cannot contain all the information relevant to assigning communities to nodes. On the other hand, by looking at the distribution of group assignments $t$, we can get a sense of how tightly each node is linked to its community. For instance, we could have computed that the president, the instructor, and their closest friends had a 99% chance of ending up in their respective groups, while some other nodes only had a 60% chance of ending up in a given group. In this sense, determining the assignment $t$ is a marginalization rather than a maximization problem.

A somewhat similar problem arises at the level of determining the parameters $p$ and $q$. In the network literature, many authors vary these parameters to minimize the ground state energy. That is, they minimize

$$P[t|G, p, q] = \frac{e^{-H(t)}}{\mathcal{Z}}.$$

The problem with this approach is that by varying the parameters $p$ and $q$, one can reach artificially low ground state energies. Zdeborová and Boettcher (2010) showed that in a random 3-regular graph, one can always find a partition of the nodes in two groups such that only 11% of the edges cross from one group to the other. One might think that such a partition shows communities in the graph, although it was generated completely at random. The point is that there are exponentially many ways to partition the nodes in two groups, and it is hardly astonishing that there should exist one with few edges between the two groups: but this is really just overfitting, fitting the random noise in the graph rather than finding statistically significant communities. Instead of looking at the ground state energy, one should focus on the free energy $F = -\ln \mathcal{Z} = -\ln P[G|p, g]$. This is another reason we prefer BP to MC. In MC, computing the entropy is tricky because it requires us to integrate over different temperatures, and running MC for a long time at each temperature.

To see how BP provides us with an estimate of the free energy, we transform the problem of evaluating $F$ into a variational problem. For simplicity, we write $P[\cdot]$ for $P[\cdot \mid p, q]$. Let $Q$ be an arbitrary probability distribution over the possible assignments $t$. The free energy is

$$-F = -\ln \mathcal{Z} = \ln P[G]$$

$$= \ln \sum_t P[G, t]$$

$$= \ln \sum_t Q(t) \frac{P[G, t]}{Q(t)}$$

$$= \ln \mathbb{E}_{t \sim Q(t)} \frac{P[G, t]}{Q(t)} \,.$$

Jensen's inequality (i.e., the concavity of the logarithm) gives $\ln \mathbb{E} X \geq \mathbb{E} \ln X$. Then,

$$-F \geq \mathbb{E}_{t \sim Q(t)} \ln \frac{P[G, t]}{Q(t)}$$

$$\geq \mathbb{E}_{t \sim Q(t)} \ln P(t) - \sum_t Q(t) \ln Q(t) \,,$$

so that

$$F \leq E_Q - S(Q) \,,$$

where $E_Q$ is the average energy if a configuration $t$ has probability $Q(t)$ instead of the Boltzmann probability $P$, and $S(Q)$ is the entropy of the distribution $Q$. Furthermore, this inequality is saturated if and only if $P = Q$.

This allows us to approximate the free energy by minimizing the Gibbs free energy $E_Q - S(Q)$. To make the variational problem tractable, we can constrain $Q$ within a family probability distributions with a small (i.e., polynomial) number of parameters. A popular choice is the mean-field assumption, in which correlations between different sites are neglected. This assumes that $Q$ is simply a product distribution,

$$Q(t) = \prod_i \mu_{t_i}^i \,.$$

BP does better than this rough assumption as it considers two-node correlations, in addition to single-site marginals. In fact, as was shown by Yedidia *et al.* (2000), BP fixed points are the stationary points of the free energy within the family where $Q$ is of the form

$$Q(t) = \frac{\prod_{i \sim j} \mu_{t_i, t_j}^{i, j}}{\prod_i \mu_{t_i}^{d_i - 1}} \,.$$

Note that this form is not even a distribution unless the underlying graph $G$ is a tree. Plugging this form into the Gibbs free energy in general therefore only gives an approximation of the Helmholtz free energy, called the Bethe free energy.

Let us now see how BP performs on a concrete example. Again letting $k$ denote the number of groups, we generate a graph using the following parameters of the stochastic block model:

$$q_r = \frac{1}{k}\,,$$

$$p_{rs} = \frac{c_{rs}}{n}\,,$$

$$c_{rs} = \begin{cases} c_{\text{in}} & \text{if } r = s\,, \\ c_{\text{out}} & \text{if } r \neq s\,. \end{cases}$$

The average connectivity is

$$c = \frac{1}{k}c_{\text{in}} + \frac{k-1}{k}c_{\text{out}}\,.$$

Given the graph, the goal is to infer an assignment of the groups $t$ that maximizes the overlap with the true values. Figure 2.2 shows the running time of BP as a function of the ratio $c_{\text{out}}/c_{\text{in}}$. We notice that it has hardly any dependence on the number of nodes $n$, and that at some particular value $c^*$ a critical slowdown appears, typical of second-order phase transitions.

It is in fact possible to evaluate analytically at which value of the parameters BP begins to fail to detect the communities. All the nodes have the same average degree, so that the uniform distribution over all nodes is a fixed point of the BP equations.

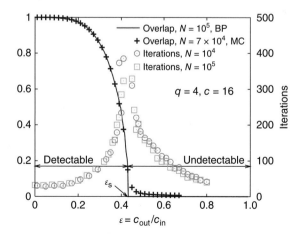

**Fig. 2.2** Second-order phase transition for $k \leq 4$. From Decelle *et al.* (2011). ©2011 American Physical Society.

Let us study the linear stability of this fixed point. To do so, we introduce a small perturbation in the following way:

$$\mu_s^{i \to j} = \frac{1}{k} + \epsilon_s^{k \to i}.$$

The linearization of the BP equations then takes the form

$$\epsilon^{i \to j} = \sum_{\substack{k \sim i \\ k \neq j}} T \epsilon^{k \to i},$$

where $T$ is a $k \times k$ matrix with components

$$T_{rs} = \frac{1}{k} \left( \frac{c_{rs}}{c} - 1 \right).$$

The eigenvalues of this matrix are 0 and

$$\lambda = \frac{c_{\text{in}} - c_{\text{out}}}{kc}.$$

If $\lambda$ is too small, the uniform distribution is a stable fixed point of BP, and we will not learn anything by running it: it will simply conclude that every node is equally likely to be in every group.

More precisely, if we assume that distant nodes are independent, then the perturbation $\epsilon$ gets multiplied by a factor $(\lambda \sqrt{c})^d$ at level $d$ of the tree. The stability condition of the uniform distribution is therefore $\lambda \sqrt{c} < 1$, and the transition occurs at

$$c_{\text{in}} - c_{\text{out}} = k \sqrt{c}.$$

Thus, BP fails if $\lambda \sqrt{c} < 1$, labeling the nodes no better than chance. Our claim is that any other algorithm will also fail in this region: the randomness in the block model "washes away" the information of the underlying types.

Indeed, after this argument was presented by Decelle *et al.* (2011), it was shown rigorously and independently by Mossel *et al.* (2013) and Massoulié (2013) in the case of $k = 2$ groups of equal size that if $\lambda \sqrt{c} < 1$, the marginal distributions of the nodes approach the uniform distribution. In fact, the ensemble of graphs generated by the stochastic block model is contiguous to the Erdős–Renyí ensemble, meaning that one graph is not enough to distinguish the two ensembles: there is no statistical test that determines, with probability approaching 1, whether communities exist or not. Conversely, if $\lambda \sqrt{c} > 1$, then a BP-like algorithm can indeed label the nodes better than chance.

One may then ask what happens if we have more than two groups. It turns out that for $k > 4$, the situation is different, and the transition is of first order (see Fig. 2.3). The curve $\times$ - - - $\times$ indicates the probability of detection when BP starts from a random initial condition. This is called robust reconstruction, but it fails at an "easy/hard transition" well above the detectability threshold $c_d$. On the other hand, if we give BP a hint, starting from a configuration not too far from the true

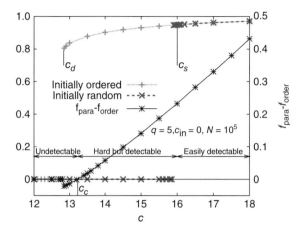

**Fig. 2.3** First-order phase transition for $k > 4$. From Decelle *et al.* (2011). ©2011 American Physical Society.

assignment, then it achieves the detectability threshold (the curve $+ \cdots +$). In between the detectability transition and the easy/hard transition, there are two fixed points of BP: the uniform one and an accurate one. The Bethe free energy of the accurate one is lower, so we would choose it if we knew it; however, its basin of attraction is exponentially small. Thus, we have a "hard but detectable" regime where detection is information-theoretically possible, but (we believe) exponentially hard.

### 2.5.2 Spectral methods

We now turn to another family of algorithms for community detection. These are called spectral methods, and aim at inferring types of nodes by diagonalizing a matrix that encodes the graph we are given. Popular choices for this matrix include the adjacency matrix $A$, the random walk matrix $D^{-1}A$ (where $D$ is the diagonal matrix where $D_{ii}$ is the degree $d_i$ of node $i$), the Laplacian $D - A$, and the symmetrically normalized Laplacian $D^{-1/2}(D - A)D^{-1/2}$.

Generically, the largest eigenvalues of these matrices will be correlated with the community structure—except for the Laplacian, where one has to look at the smallest positive eigenvalues. More precisely, the largest eigenvector depends on the node degree or other notions of "centrality" (see, e.g., Newman, 2010), while (for $k = 2$ groups) the second eigenvector will be positive in one group and negative in the other. However, for the matrices listed above, this program will work up to a value of $c_{in} - c_{out}$ slightly above the BP threshold. Can we understand why?

The problem is that the spectrum of sparse random matrices is more difficult to study than that of dense matrices. If a random matrix is dense enough, with the average degree growing at least like $\log n$, then the spectrum can be modeled as a discrete part related to the community structure, and a continuous "bulk" that follows Wigner's semicircle law, lying in the interval $[-2\sqrt{c}, 2\sqrt{c}]$

(see Nadakuditi and Newman, 2012). However, if the average degree is $O(1)$, the distribution of eigenvalues differs from Wigner's semicircle law. In particular, tails appear at the edges of the semicircle due to the high-degree vertices, which might drown the informative eigenvalue.

To make this argument more precise, it is easy to see that the adjacency matrix $A$ (for example) has an eigenvalue $\lambda$ such that

$$|\lambda| \geq \sqrt{d_{\max}},$$

where $d_{\max}$ is the largest degree of a node in the graph. To see this, consider a node $i$ with degree $d_{\max}$, and let $e_i$ be the vector with components equal to 1 at $i$ and zero elsewhere. Note that $(A^k)_{ij}$ is the number of paths from $i$ to $j$ in the graph of exactly $k$ steps. Since there are $d_i = d_{\max}$ ways to leave $i$ and return with two steps, we have $A^2_{i,i} = d_{\max}$. Taking inner products, we have

$$\langle e_i, A^2 e_i \rangle = d_{\max}.$$

Since $A^2$ is symmetric and $e_i$ can be orthogonally decomposed as a linear combination of eigenvectors, it follows that $A^2$ has an eigenvector at least $d_{\max}$, and $A$ has an eigenvector that is at least $\sqrt{d_{\max}}$ in absolute value.

In the Erdős–Renyí ensemble, we have $d_{\max} = O(\log n / \log \log n)$. Thus, for sufficiently large $n$, $|\lambda|$ becomes bigger than the edge of the Wigner semicircle. As a consequence, spectral methods based on $A$ or $L$ will tend to find localized eigenvectors, clustered around high-degree vertices, as opposed to finding those correlated with the community structure. The normalized versions of $A$ and $L$ have analogous problems, where they get stuck in the trees of size $O(\log n)$ that dangle from the giant component. Thus, in the sparse case where the average degree is $O(1)$, all these methods fail significantly above the detectability threshold.

We therefore see that if we want to achieve optimal community detection by spectral methods, we need to design a matrix that forbids going back and forth on the same edge. The simplest way to do this is to define a matrix $B$ that describes a non-backtracking walk on the edges of the network:

$$B^{\text{non-backtracking}}_{i \to j, k \to \ell} = \delta_{jk}(1 - \delta_{i\ell}).$$

Krzakala *et al.* (2013) conjectured that even in the sparse case, all of $B$'s eigenvalues are inside a circle of radius $\sqrt{c}$ in the complex plane, except for the leading eigenvalue and those correlated with community structure. This conjecture is supported by extensive numerical experiments (see Fig. 2.4). Moreover, $B$ describes to first order how messages propagate in BP, and in particular the linear stability of the uniform fixed point discussed above. Thus, if this conjecture is true, spectral clustering with $B$ succeeds all the way down to the detectability transition for $k = 2$, and to the easy/hard transition for larger $k$.

As a concluding remark, we should keep in mind that this discussion has assumed that the network is generated by the stochastic block model. In fact, for many networks this model is inaccurate; for instance, it does not produce heavy-tailed degree

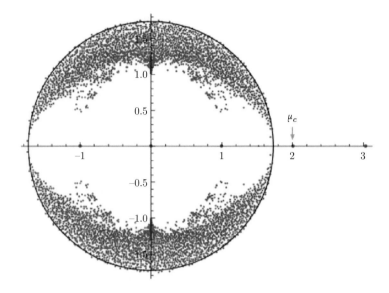

**Fig. 2.4** The eigenvalues of the non-backtracking matrix. From Krzakala *et al.* (2013).

distributions, which are common in real networks. The good news is that many of these techniques, including belief propagation, the Bethe free energy, and so on, are easy to extend to more elaborate models, such as the degree-corrected block model of Karrer and Newman (2011).

## 2.6 Appendix

### 2.6.1 Definition of perfect matching

A matching on a graph $G$ is a subset of the edges of $G$ such that no two edges share a vertex. A perfect matching is a matching such that all vertices are covered.

### 2.6.2 Definition of max-flow

Consider a directed graph $G = (V, E)$, with a source node $s$ and a sink node $t$. Each edge $e$ has a given capacity $c(e)$, which is a non-negative integer. A flow assigns a number $f(e)$ to each edge so that the total flow in and out of each node is conserved, except for the source and sink. We require that $0 \leq f(e) \leq c(e)$, and we seek to maximize the total flow, i.e., the sum of $f(e)$ over all the outgoing edges of $s$ (or the incoming edges of $t$).

### 2.6.3 Definition of $k$-SAT

An instance of $k$-SAT is a formula involving $n$ Boolean variables $x_1, \ldots, x_n$. There are many variations such as NAESAT, 1-in-$k$ SAT, and so on, but in the standard version of $k$-SAT the formula is in "conjunctive normal form." That is, it is the AND of a set of clauses; each clause is the OR of a set of literals; and each literal is either a variable $x_i$ or its negation $\overline{x}_i$. Thus, a truth assignment, i.e., an assignment of true/false values to the variables, satisfies a clause if it makes at least one of its literals true, and it satisfies a formula if and only if it satisfies all its clauses. A formula is satisfiable if a satisfying assignment exists.

## References

Decelle, A., Krzakala, F., Moore, C., and Zdeborová, L. (2011). Inference and phase transitions in the detection of modules in sparse networks. *Physical Review Letters*, **107**(6), 065701.

Karrer, B. and Newman, M. E. J. (2011). Stochastic blockmodels and community structure in networks. *Physical Review E*, **83**(1), 016107.

Krzakala, F., Moore, C., Mossel, E., Neeman, J., Sly, A., Zdeborová, L., and Zhang, P. (2013). Spectral redemption in clustering sparse networks. *Proceedings of the National Academy of Sciences*, **110**(52), 20935–20940.

Massoulié, L. (2013). Community detection thresholds and the weak Ramanujan property. arXiv:1311.3085.

Moore, C. and Mertens, S. (2011). *The Nature of Computation*. Oxford University Press.

Mossel, E., Neeman, J., and Sly, A. (2013). A proof of the block model threshold conjecture. arXiv:1311.4115.

Nadakuditi, R. R. and Newman, M. E. J. (2012). Graph spectra and the detectability of community structure in networks. *Physical Review Letters*, **108**(18), 188701.

Newman, M. (2010). *Networks: An Introduction*. Oxford University Press.

Pittel, B., Spencer, J., and Wormald, N. (1996). Sudden emergence of a giant $k$-core in a random graph. *Journal of Combinatorial Theory, Series B*, **67**(1), 111–151.

Yedidia, J. S., Freeman, W. T., Weiss, Y. et al. (2000). Generalized belief propagation. In *NIPS*, Volume 13, pp. 689–695.

Zdeborová, L. and Boettcher, S. (2010). A conjecture on the maximum cut and bisection width in random regular graphs. *Journal of Statistical Mechanics: Theory and Experiment*, **2010**(02), P02020.

# 3
# Replica theory and spin glasses

## Giorgio PARISI

Dipartimento di Fisica
"La Sapienza" Università di Roma,
P.le A. Moro 2,
00185 Roma, Italy

*Lecture notes taken by*
Flaviano Morone, City College of New York, New York
Francesco Caltagirone, CEA Saclay, France
Elizabeth Harrison, Aston University, UK

*Statistical Physics, Optimization, Inference, and Message-Passing Algorithms.* First Edition.
F. Krzakala et al. © Oxford University Press 2016. Published in 2016 by Oxford University Press.

# Chapter Contents

## 3.1   Introduction to the replica method: the Wigner law

We introduce the basic idea of the replica method by computing the spectrum of random matrices. To be concrete, we explore the case of real symmetric matrices of large order $N$ having random elements $J_{ij}$ that are independently distributed with zero mean and variance $\overline{J^2} = N^{-1}$. We will always denote the average over the $J$s with an overbar. This problem has been widely studied: many rigorous results are known and many different techniques can be used to compute the spectrum of these matrices.

We also assume that $\overline{J^4}/\left(\overline{J^2}\right)^2 < \infty$ (i.e., the probability distribution of $J$ has no fat tails): less stringent conditions may be imposed, but we are not interested here in finding the optimal condition.

Before describing the replica method, we will perform the calculations using a probabilistic approach, which is at the heart of the cavity method.

### 3.1.1   Computing the spectrum with the cavity method

The resolvent $\hat{R}$ of the $N \times N$ matrix $\hat{J}$ is defined as follows:[1]

$$\hat{R}(\mathcal{E}) = (\mathcal{E}\mathbb{1} - \hat{J})^{-1} \,, \tag{3.1}$$

where $\mathbb{1}$ is the $N \times N$ identity matrix and $\mathcal{E}$ is a complex number (only at the end of the computation shall we take the limit where $\mathcal{E}$ is real).

We are interested in computing the trace of $\hat{R}(\mathcal{E})$,

$$\mathrm{Tr}\big[\hat{R}(\mathcal{E})\big] = \sum_k \frac{1}{\mathcal{E} - \mathcal{E}_k} \,, \tag{3.2}$$

in the large-$N$ limit. We denote with $\mathcal{E}_k$ the eigenvalues of $\hat{J}$. Many physically interesting quantities may be extracted from knowledge of the trace of the resolvent, the simplest (and the only one we shall consider here) being the spectral density.

As $\hat{J}$ is a random matrix, the trace of the resolvent is also a random quantity. We will proceed by first writing exact relations at fixed $\hat{J}$, and at a later stage we shall perform the average for the random matrices.

The idea behind cavity computation consists in finding a recursion relation between $\mathrm{Tr}\big[\hat{R}^{(N)}\big]$ and $\mathrm{Tr}\big[\hat{R}^{(N+1)}\big]$, where $\hat{R}^{(N)}$ is the $N \times N$ resolvent matrix. To do this, it is sufficient to write the matrix element $R_{N+1,N+1}^{(N+1)}$ as a function of the matrix elements $R_{ij}^{(N)}$.

Let us define for simplicity the $N \times N$ matrix $\hat{M}^{(N)} \equiv \mathcal{E}\mathbb{1} - \hat{J}$, where $J$ is a matrix with dimension larger than $N+1$ and the equality holds only in $N$-dimensional space. It is easy to verify that

$$R_{N+1,N+1}^{(N+1)} = \frac{\det[\hat{M}^{(N)}]}{\det[\hat{M}^{(N+1)}]} \,. \tag{3.3}$$

---

[1] Here we adopt the convention of denoting the matrices with a hat: "ˆ", while, when we refer to particular matrix elements, we omit the "hat" symbol.

On the other hand, the determinant $\det[\hat{M}^{(N+1)}]$ can be Laplace-expanded first along the last row, and then on the last column, thereby giving

$$\det[\hat{M}^{(N+1)}] = M_{N+1,N+1}^{(N+1)} \det[\hat{M}^{(N)}] - \sum_{k,\ell=1}^{N} M_{N+1,k}^{(N+1)} M_{\ell,N+1}^{(N+1)} C^{(N)}(\ell,k), \quad (3.4)$$

where $C^{(N)}(\ell,k)$ is the $(\ell,k)$ cofactor of the matrix $\hat{M}^{(N)}$. Dividing the previous expression by $\det[\hat{M}^{(N)}]$ and recalling the definition of $\hat{M}^{(N)}$, we get

$$\frac{1}{R_{N+1,N+1}^{(N+1)}} = \mathcal{E} - J_{N+1,N+1}^{(N+1)} - \sum_{k,\ell=1}^{N} J_{N+1,k}^{(N+1)} J_{\ell,N+1}^{(N+1)} R_{k\ell}^{(N)}. \quad (3.5)$$

Now we must make the crucial assumption that the off-diagonal elements of the resolvent are $O(N^{-1/2})$ (as we shall see later, this is equivalent to assuming the validity of replica symmetry). The motivations for this assumption (which can be rigorously proved) are as follows:

- For large values of $\mathcal{E}$, the resolvent can be expanded in inverse powers of $\mathcal{E}$ and it possible to check at each order in this expansion that the off-diagonal elements of the resolvent are $O(N^{-1/2})$.

- If we assume that $|R_{k\ell}^{(N)}|^2 = A/N$, then a computation similar to the previous one (going from $N$ to $N+2$) predicts consistently a finite value for $A$, at least for $\mathcal{E}$ not on the real axis. (In replica language, this statement is equivalent to the condition of stability of the replica-symmetric fixed point.)

If we assume that the off-diagonal elements of the resolvent are $O(N^{-1/2})$, then in the large-$N$ limit the recursion equation (3.5) becomes

$$R_{N+1,N+1}^{(N+1)} = \frac{1}{\mathcal{E} - N^{-1} \sum_{k=1}^{N} R_{kk}^{(N)}} + O(N^{-1/2}). \quad (3.6)$$

In the last step, we have used the fact that $J^{(N+1)}$ and $R^{(N)}$ are uncorrelated quantities.

At this point, a little thought should be enough to convince us that the matrix elements $R_{N+1,N+1}^{(N+1)}$ and $R_{kk}^{(N)}$ are all identically distributed as a consequence of the fact that the $J$s are identically distributed and uncorrelated. Taking the average on both sides of (3.6), together with the limit $N \to \infty$, we obtain the following fixed-point equation:

$$R(\mathcal{E}) = \frac{1}{\mathcal{E} - R(\mathcal{E})}, \quad (3.7)$$

where $R(\mathcal{E}) \equiv \overline{R_{ii}(\mathcal{E})}$. Moreover, the dependence on $k$ of $R_{kk}^{(N)}$ should disappear in the limit where $N$ goes to infinity and the diagonal elements do not fluctuate.

The solution of (3.7) is given by

$$R(\mathcal{E}) = \frac{\mathcal{E} \pm \sqrt{\mathcal{E}^2 - 4}}{2}. \quad (3.8)$$

Which sign should we take for the square root? Let us consider what happens on the real line: for large $|\mathcal{E}|$, we must have $R(\mathcal{E})$ go to zero like $1/R(\mathcal{E})$. Therefore, in the region $\mathcal{E} > 2$, we must take the negative sign, while for $\mathcal{E} < -2$, we must take the positive sign. This may look strange; however, the function $R(\mathcal{E})$ has a cut on the real axis from $-2$ to $2$: the choice of the sign of the square root for negative values of $\mathcal{E}$ is exactly what we get if we start from $(\mathcal{E} \pm \sqrt{\mathcal{E}^2 - 4})/2$ and perform an analytic continuation from positive to negative $\mathcal{E}$, avoiding the cut on the real axis.

Let us now introduce the density of states $\rho(\lambda)$, which is defined as follows:

$$\rho(\lambda) \equiv \frac{1}{N} \sum_{k=1}^{N} \delta(\lambda - \mathcal{E}_k). \tag{3.9}$$

Using the density of states $\rho(\lambda)$, we can rewrite the trace of $\hat{R}(\mathcal{E})$ as

$$\mathrm{Tr}[\hat{R}(\mathcal{E})] = \sum_{k=1}^{N} \frac{1}{\mathcal{E} - \mathcal{E}_k} = N \lim_{\epsilon \to 0} \int d\lambda \, \frac{\rho(\lambda)}{\mathcal{E} - \lambda - i\epsilon}$$

$$= N \left[ \int d\lambda \, \rho(\lambda) \, \mathrm{P}\left(\frac{1}{\mathcal{E} - \lambda}\right) + i\pi\rho(\mathcal{E}) \right],$$

where the distribution $\mathrm{P}(1/x)$ denotes the principal value of $1/x$. Now, since $\overline{\mathrm{Tr}[\hat{R}(\mathcal{E})]} = NR(\mathcal{E})$, we can relate the density of states $\rho(\lambda)$ to the imaginary part of $R(\mathcal{E})$ via the following equation:

$$\rho(\mathcal{E}) = \frac{1}{\pi} \mathrm{Im}[R(\mathcal{E})], \tag{3.10}$$

from which we get the Wigner semicircle law:

$$\rho(\mathcal{E}) = \left(\frac{1}{2\pi} \sqrt{4 - \mathcal{E}^2}\right) \Theta(2 - |\mathcal{E}|). \tag{3.11}$$

In a nutshell, the function $R(\mathcal{E})$ is a real analytic function with a cut on the real line and the discontinuity on the cut is the density of states. It should be stressed that these results are valid for the resolvent when we are a little away from the cut. Near the cut, new effects arise and the computation becomes more involved.

### 3.1.2  Computing the spectrum with replicas

We now perform the same calculation using the replica method. The object in which we are interested is the average of the trace of the resolvent for an $N \times N$ random matrix, which can be written as

$$\overline{\mathrm{Tr}[\hat{R}(\mathcal{E})]} = \overline{\mathrm{Tr}\left(\frac{1}{\mathcal{E}\mathbb{1} - \hat{J}}\right)} = \frac{d}{d\mathcal{E}} \overline{\mathrm{Tr}\, \log(\mathcal{E}\mathbb{1} - \hat{J})} = \frac{d}{d\mathcal{E}} \overline{\log \det(\mathcal{E}\mathbb{1} - \hat{J})}. \tag{3.12}$$

Now we can write the average of the logarithm of the determinant as follows:

$$\overline{\log \det(\mathcal{E}\mathbb{1} - \hat{J})} = -2 \lim_{n \to 0} \frac{d}{dn} \overline{\left(\frac{1}{\det(\mathcal{E}\mathbb{1} - \hat{J})}\right)^{n/2}}. \tag{3.13}$$

This is the famous replica trick. Operatively, we compute the quantity on the right-hand side for integer (positive) values of the *replica* number $n$, then we analytically continue the result to real values of $n$ and eventually take the limit $n \to 0$. The main assumption concerning the analytic continuation relies on the possibility of uniquely determining an analytic function knowing it only on a countable (infinite) set of points (the positive integers in our case). This is not a real point of concern: the problems will arise in exchanging the limits $N \to \infty$ and $n \to 0$.

To go further in the calculation, we represent the determinant on the right-hand side of (3.13) using a Gaussian integral:

$$\overline{\left(\frac{1}{\det(\mathcal{E}\mathbb{1} - \hat{J})}\right)^{n/2}} = \int \left(\prod_{i=1}^{N} \prod_{a=1}^{n} \frac{d\phi_i^a}{\sqrt{2\pi}}\right) \exp\left[-\frac{1}{2} \sum_{i,k=1}^{N} \sum_{a=1}^{n} \phi_i^a \left(\mathcal{E}\delta_{ik} - J_{ik}\right) \phi_k^a\right]$$

$$= \int \left(\prod_{i=1}^{N} \prod_{a=1}^{n} \frac{d\phi_i^a}{\sqrt{2\pi}}\right) \exp\left[-\frac{1}{2} \sum_{i=1}^{N} \sum_{a=1}^{n} \mathcal{E} \, (\phi_i^a)^2 - \frac{1}{4N} \sum_{i,k=1}^{N} \sum_{a,b=1}^{n} \phi_i^a \phi_k^a \phi_i^b \phi_k^b\right], \tag{3.14}$$

where terms $O(1)$ in the argument of the exponential have been neglected. Note that if the distribution of the matrix elements $J_{ij}$ were Gaussian, this representation would be exact.

To decouple the sites $i$, we introduce the following *order parameter*:

$$q_{ab} = \frac{1}{N} \sum_{i=1}^{N} \phi_i^a \phi_i^b. \tag{3.15}$$

Inserting a $\delta$-function into (3.14) to enforce the constraint (3.15), we obtain

$$\int \left(\prod_{i=1}^{N} \prod_{a=1}^{n} \frac{d\phi_i^a}{\sqrt{2\pi}}\right) \exp\left[-\frac{1}{2} \sum_{i=1}^{N} \sum_{a=1}^{n} \mathcal{E} \, (\phi_i^a)^2 + \frac{1}{4N} \sum_{i,k=1}^{N} \sum_{a,b=1}^{n} \phi_i^a \phi_k^a \phi_i^b \phi_k^b\right]$$

$$= \int \left(\prod_{a,b=1}^{n} dq_{ab}\right) \exp\left[-\frac{1}{2}\mathcal{E}N \sum_{a=1}^{n} q_{aa} + \frac{1}{4}N \sum_{a,b=1}^{n} q_{ab}^2\right] \tag{3.16}$$

$$\times \int \left(\prod_{i=1}^{N} \prod_{a=1}^{n} \frac{d\phi_i^a}{\sqrt{2\pi}}\right) \prod_{a,b} \delta\left(Nq_{ab} - \sum_{i=1}^{N} \phi_i^a \phi_i^b\right).$$

In the second line of (3.16), we have omitted a prefactor independent of $q_{ab}$. In the following, we will also neglect any multiplicative constants. Let us consider the last integral in (3.16). Using the integral representation of the $\delta$-function, we find

$$
\int \left( \prod_{i=1}^{N} \prod_{a=1}^{n} \frac{d\phi_i^a}{\sqrt{2\pi}} \right) \prod_{a,b} \delta \left( N q_{ab} - \sum_{i=1}^{N} \phi_i^a \phi_i^b \right)
$$

$$
= \int \left( \prod_{a,b=1}^{n} d\psi_{ab} \right) \int \left( \prod_{i=1}^{N} \prod_{a=1}^{n} \frac{d\phi_i^a}{\sqrt{2\pi}} \right) \exp \left( \frac{N}{2} \sum_{a,b=1}^{n} q_{ab}\psi_{ab} - \frac{1}{2} \sum_{i=1}^{N} \sum_{a,b=1}^{n} \phi_i^a \psi_{ab} \phi_i^b \right)
$$

$$
= \int \left( \prod_{a,b=1}^{n} d\psi_{ab}\, e^{(N/2)q_{ab}\psi_{ab}} \right) \left[ \int \left( \prod_{a=1}^{n} \frac{d\phi^a}{\sqrt{2\pi}} \right) \exp \left( -\frac{1}{2} \sum_{a,b=1}^{n} \phi^a \psi_{ab} \phi^b \right) \right]^N
$$

$$
= \int \left( \prod_{a,b=1}^{n} d\psi_{ab} \right) \exp \left( \frac{N}{2} \sum_{a,b=1}^{n} q_{ab}\psi_{ab} - \frac{N}{2} \operatorname{Tr} \log(\hat{\psi}) \right), \tag{3.17}
$$

where $\hat{\psi}$ is the matrix with entries $\psi_{ab}$, and the integral over $\psi_{ab}$ is performed on the imaginary axis.

Since we are interested in the limit $N \to \infty$, we can evaluate the integral using the method of steepest descent, and we obtain the following saddle point equations:

$$
q_{ab} = (\hat{\psi}^{-1})_{ab}. \tag{3.18}
$$

Evaluating the integral at the saddle point, we finally obtain (always discarding terms that do not depend on $\hat{q}$)

$$
\int \left( \prod_{i=1}^{N} \prod_{a=1}^{n} \frac{d\phi_i^a}{\sqrt{2\pi}} \right) \prod_{a,b} \delta \left( N q_{ab} - \sum_{i=1}^{N} \phi_i^a \phi_i^b \right) = \exp \left( \frac{N}{2} \operatorname{Tr} \log(\hat{q}) \right). \tag{3.19}
$$

Coming back to (3.16), and using the result (3.19), we find

$$
\left( \frac{1}{\det(\mathcal{E}\mathbb{1} - \hat{J})} \right)^{n/2} = \int \left( \prod_{a,b=1}^{n} dq_{ab} \right) \exp \left( -\frac{1}{2} N f_n[\hat{q}] \right), \tag{3.20}
$$

where the *free energy* function $f_n[\hat{q}]$ is given by

$$
f_n[\hat{q}] = \mathcal{E} \sum_{a=1}^{n} q_{aa} - \frac{1}{2} \sum_{a,b=1}^{n} q_{ab}^2 - \operatorname{Tr} \log(\hat{q}). \tag{3.21}
$$

The integral on the right-hand side of (3.20) can still be evaluated via the saddle point method in the limit $N \to \infty$. The key issue is to identify the correct saddle point. Usually, if the free energy is invariant under the action of a group, the first choice is a saddle point that is invariant under that group.

The most obvious symmetry group of the function $f_n[\hat{q}]$ is the permutation group of $n$ elements;[2] i.e., we can permute simultaneously the rows and the columns of the matrix $q$ (this point will be discussed at length later in this chapter).

We now make the *replica-symmetric* ansatz on the form of the saddle point solution; i.e., we assume that the *free energy* $f_n[\hat{q}]$ has a minimum on the subspace of matrices $\hat{q}$ that are invariant under the permutation group. These matrices are of the form[3]

$$q_{ab} = q\delta_{ab} + p(1 - \delta_{ab}).\tag{3.22}$$

The saddle point equations then become

$$\frac{\partial f_n(q,p)}{\partial q} = n\left[\mathcal{E} - q - \frac{1}{q-p} + \frac{p}{(q-p)^2}\right] = 0,$$
$$\frac{\partial f_n(q,p)}{\partial p} = n\left[p - \frac{p}{(q-p)^2}\right] = 0.\tag{3.23}$$

The meaningful solution is given by

$$q^* = \frac{\mathcal{E} - \sqrt{\mathcal{E}^2 - 4}}{2},\tag{3.24}$$
$$p^* = 0.$$

Evaluating the integrand of (3.20) at the saddle point, we obtain the following result:

$$\overline{\left(\frac{1}{\det(\mathcal{E}\mathbb{1} - \hat{J})}\right)}^{n/2} = \exp\left(-\frac{1}{2}Nf_n(q^*,p^*)\right).\tag{3.25}$$

The function $f_n(q^*,p^*)$, in the small-$n$ limit, can be written as $f_n(q^*,p^*) = nf(q^*,p^*) + O(n^2)$, where the function $f(q,p)$ is given by

$$f(q,p) = \mathcal{E}q - \frac{1}{2}(q^2 - p^2) - \log(q-p) - \frac{p}{q-p}.\tag{3.26}$$

From the expression (3.25), using (3.12) and (3.13), we can easily compute the trace of the resolvent in the limit $N \to \infty$:

$$\lim_{N\to\infty}\frac{1}{N}\overline{\mathrm{Tr}[\hat{R}(\mathcal{E})]} = -2\frac{d}{d\mathcal{E}}\lim_{n\to 0}\frac{d}{dn}\left[\lim_{N\to\infty}\frac{1}{N}\overline{\left(\frac{1}{\det(\mathcal{E}\mathbb{1} - \hat{J})}\right)}^{n/2}\right]$$
$$= \frac{d}{d\mathcal{E}}f(q^*,p^*).\tag{3.27}$$

---

[2]  In this case, but not in the spin glass case, the symmetry group is larger, namely, the group $O(n)$.
[3]  The $O(n)$ symmetry would imply $p = 0$.

The right-hand side of (3.27) is given by

$$\frac{\mathrm{d}}{\mathrm{d}\mathcal{E}}\, f(q^*, p^*) = \frac{\partial}{\partial\mathcal{E}}\, f(q^*, p^*) = q^*\,, \tag{3.28}$$

because of the stationarity of $f(q, p)$ with respect to variations in $q$ and $p$. So, we finally find

$$\lim_{N\to\infty} \frac{1}{N}\, \overline{\mathrm{Tr}[\hat{R}(\mathcal{E})]} = \frac{\mathcal{E} - \sqrt{\mathcal{E}^2 - 4}}{2}\,. \tag{3.29}$$

We can now follow exactly the same lines as after (3.8) to recover the Wigner distribution $\rho(\mathcal{E})$, given by (3.11).

## 3.2   The Sherrington–Kirkpatrick model

Mean-field spin glass models are spin systems with random interactions $(J)$ in which there is no notion of space or distance and each degree of freedom interacts with all the others. Here we consider the paradigmatic example of the Sherrington–Kirkpatrick model, defined by the Hamiltonian

$$\mathcal{H}[\sigma] = -\sum_{\langle i,j\rangle}^{N} J_{ij}\sigma_i\sigma_j - h\sum_{i}^{N} \sigma_i\,, \tag{3.30}$$

where $\sigma_i \in \{-1, 1\}$, $h$ is the uniform magnetic field and the couplings are random variables extracted from the distribution

$$P(J) = \sqrt{\frac{N}{2\pi}}\, \exp\left(-\frac{NJ^2}{2}\right). \tag{3.31}$$

The randomicity of the interaction is called *quenched disorder*, where the word *quenched* underlines the fact that the thermodynamic properties have to be computed at a fixed instance of the disorder. When the size of the system goes to infinity, thanks to the *self-averaging* property, the free energy of the single sample (specific realization of the disorder) is given by the average over the disorder of the $J$-dependent free energy; i.e., we have

$$\lim_{N\to\infty} f_J^{(N)} = f = \lim_{N\to\infty} \overline{f_J^{(N)}}\,, \tag{3.32}$$

where the overbar denotes the average over the disorder according to its distribution (Gaussian or bimodal for example). This relation is satisfied with probability one: there are sequences of the $J$s such that the result does not hold (e.g., all $J_{i,k}^N = 1$), but these sequences have zero measure.

Therefore, the solution of the statics of such a system in the presence of quenched disorder requires the computation of the average of the logarithm of the (sample-dependent) partition function

$$Z_J = \sum_{\{\sigma\}} \exp\left(-\beta \mathcal{H}[\sigma]\right) = \sum_{\{\sigma\}} \exp\left(+\beta \sum_{\langle i,j \rangle} J_{ij}\sigma_i\sigma_j + \beta h \sum_i \sigma_i\right), \qquad (3.33)$$

which unfortunately cannot be done analytically.

The so-called *replica trick* allows us to overcome this difficulty; in fact, the following identity holds:

$$\lim_{n \to 0} \frac{\overline{Z^n} - 1}{n} = \overline{\log Z}, \qquad (3.34)$$

or, equivalently,

$$\lim_{n \to 0} \frac{\log(\overline{Z^n})}{n} = \overline{\log Z}. \qquad (3.35)$$

Computing the average of the $n$th power of the partition function is a much easier task than computing the average of its logarithm. One actually computes the quantity for an integer $n$ and then takes a (a posteriori) harmless analytic continuation to real $n$ in order to reach the limit zero:

$$\overline{Z_J^n} = \int \left(\prod_{i<j} dJ_{ij}\, P(J_{ij})\right) \sum_{\{\underline{\sigma}\}} \exp\left(\beta \sum_a \sum_{\langle i,j \rangle} J_{ij}\sigma_i^a\sigma_j^a + \beta h \sum_a \sum_i \sigma_i^a\right). \qquad (3.36)$$

When $n$ is an integer, we actually deal with the partition function of $n$ non-interacting copies (replicas) of the system of interacting spins living in the same realization of the disorder. At the end of the computation, after the disorder average has been taken, we will end up with an effective action that must be *minimized* with respect to an order parameter that is an $n \times n$ symmetric matrix. Let us now proceed with the computation of the replicated partition function.

In the following, we will largely make use of the identities

$$\int dx\, \exp(-Ax^2 + Bx) = \sqrt{\frac{\pi}{A}} \exp\left(\frac{B^2}{4A}\right) \qquad (3.37)$$

and

$$\int d^M x\, \exp\left(-\sum_{i,k=1,M} A_{i,k}x_ix_k + \sum_{i,M} B_i x_i\right)$$

$$= \sqrt{\frac{\pi}{\det(A)}} \exp\left(\frac{1}{4} \sum_{i,k=1,M} (\hat{A}^{-1})_{i,k}B_iB_k\right), \qquad (3.38)$$

the latter of which is a Gaussian integration if considered from left to right or a so-called *Hubbard–Stratonovich* (H–S) transformation if considered from right to left.

If we consider (3.38), we can easily get rid of the disorder average, obtaining

$$\overline{Z_J^n} = \sum_{\{\sigma\}} \exp\left(\beta h \sum_i \sum_a \sigma_i^a + \frac{\beta^2}{2N} \sum_{i<j} \sum_{a,b} \sigma_i^a \sigma_i^b \sigma_j^a \sigma_j^b\right). \tag{3.39}$$

Note that by performing the Gaussian integration, we have introduced an "interaction" between replicas, which are therefore no longer independent. The interaction term can be rewritten as follows:

$$\sum_{i<j} \sum_{a,b} \sigma_i^a \sigma_i^b \sigma_j^a \sigma_j^b = N^2 \sum_{a<b} \left(\frac{1}{N} \sum_i \sigma_i^a \sigma_i^b\right)^2 + \frac{N^2 n - N n^2}{2}. \tag{3.40}$$

The replicated partition function reads

$$\overline{Z_J^n} = \sum_{\{\sigma\}} \exp\left(\beta h \sum_i \sum_a \sigma_i^a\right) \exp\left(\frac{\beta^2}{4}(Nn - n^2)\right) \prod_{a<b} \exp\left(\frac{\beta^2}{2N}\left(\sum_i \sigma_i^a \sigma_i^b\right)^2\right). \tag{3.41}$$

At this point, we perform an H–S transformation (3.38) with

$$4A = 2N\beta^2, \quad B^2 = \left(\beta^2 \sum_i \sigma_i^a \sigma_i^b\right)^2, \quad C = 0, \tag{3.42}$$

so that the partition function becomes

$$\overline{Z_J^n} = \sum_{\{\sigma\}} \exp\left(\beta h \sum_i \sum_a \sigma_i^a + \frac{\beta^2}{4}(Nn - n^2)\right) \left(\frac{2\pi\beta^2}{N}\right)^{n(n-1)/2}$$
$$\times \prod_{a<b} \int dQ_{ab} \exp\left(-\frac{N}{2}\beta^2 Q_{ab}^2 + \beta^2 \sum_i \sigma_i^a \sigma_i^b Q_{ab}\right), \tag{3.43}$$

where finally replicas are coupled and spins inside one replica have been decoupled at the price of introducing an integration over the matrix $Q_{ab}$. This formula can be also written more explicitly as

$$\overline{Z_J^n} = \exp\left(\frac{\beta^2}{4}(Nn - n^2)\right) \left(\frac{2\pi\beta^2}{N}\right)^{n(n-1)/2}$$
$$\times \int dQ \exp\left(-\frac{N}{2}\beta^2 \sum_{a<b} Q_{ab}^2\right) \left(\sum_{\{\sigma\}} \exp\left(\beta h \sum_a \sigma^a + \beta^2 \sum_{a<b} \sigma^a \sigma^b Q_{ab}\right)\right)^N. \tag{3.44}$$

We are now interested in computing the free energy

$$f(\beta, h) = \lim_{n \to 0} \lim_{N \to \infty} \left( -\frac{1}{\beta N n} \ln \overline{Z_J^n} \right). \tag{3.45}$$

Note that, in principle, the two limits should have been taken in the opposite order; however, the calculation would be impossible with the right order of the limits. For this reason we "blindly" exchange the order assuming that this operation is harmless. Since $N \to \infty$, we will perform a saddle point calculation; therefore, in the expression for the replicated partition function (3.44), we can discard all the terms that are not exponential in $N$. In addition, we retain only terms exponential in $n$, dropping terms that are exponential in $n^2$ since the latter would vanish in the final limit $n \to 0$. Given these comments, we can express the replicated partition function as follows:

$$\overline{Z_J^n} \propto \int dQ \, \exp\left(-N\mathcal{S}[Q, h]\right), \tag{3.46}$$

with

$$\mathcal{S}[Q, h] = -\frac{\beta^2 n}{4} + \frac{\beta^2}{2} \sum_{a<b} Q_{ab}^2 - \mathcal{W}[Q] \tag{3.47}$$

and

$$\mathcal{W}[Q] = \ln \sum_{\{\sigma\}} \exp\left(\beta h \sum_a \sigma^a + \beta^2 \sum_{a<b} \sigma^a \sigma^b Q_{ab}\right). \tag{3.48}$$

Taking the saddle point with respect to the order parameter $Q$, we find that the free energy can be written as

$$f(\beta, h) = \lim_{n \to 0} \frac{1}{\beta n} \, \text{extr}_Q \, \mathcal{S}[Q, h], \tag{3.49}$$

where we have used *extremal point* instead of *minimum* because, in the $n \to 0$ limit, minima become maxima, but the discussion of this topic is beyond the scope of this chapter (see Mézard *et al.* (1987) and Guerra (2003) and references therein). At the end of the day, one finds that a good saddle point must satisfy the condition that the Hessian matrix

$$M_{(ab),(cd)} = \frac{\partial^2 W}{\partial Q_{a,b} \, \partial Q_{c,d}} \tag{3.50}$$

has non-negative eigenvalues. In a *bona fide* Hilbert space, this condition is a prerequisite for having a minimum.

The minimization of the "effective action" $\mathcal{S}$ leads to a self-consistency equation for the order parameter, namely,

$$Q_{ab} = \langle\langle s^a s^b \rangle\rangle, \tag{3.51}$$

where $\langle\langle \cdots \rangle\rangle$ stands for the expectation value taken with respect to the probability measure

$$\mu[\underline{\sigma}] = \frac{\exp\left(\beta h \sum_a \sigma^a + \beta^2 \sum_{a<b} \sigma^a \sigma^b Q_{ab}\right)}{\sum_{\{\tau\}} \exp\left(\beta h \sum_a \tau^a + \beta^2 \sum_{a<b} \tau^a \tau^b Q_{ab}\right)}. \tag{3.52}$$

It should be evident at this point that performing the minimization of the effective action with a generic $n \times n$ symmetric matrix is an impossible task to accomplish. We need to parametrize the matrix in such a way that makes explicit the $n$ dependence and makes it possible to take the $n \to 0$ limit. Note also that one important requirement on the parametrization we choose is that the effective action contains terms that goes to zero *at least* linearly in $n$, otherwise the free energy would diverge in the zero-replicas limit. In the next section, we show the simplest parametrization we can choose and the results to which it leads.

### 3.2.1 The replica-symmetric solution

Let us start with an elementary example that introduces us to the more difficult task we are trying to accomplish. Consider a function $g$ of two variables, such that

$$g(x, y) = g(y, x); \tag{3.53}$$

i.e., the function is symmetric under the exchange of the two arguments. In this simple case, it is easy to understand the consequences of such a symmetry in the process of minimizing (or in general extremizing) the function $g$. When looking for a point of minimum $(x^*, y^*)$ of this function, we clearly have two possibilities:

- The first possibility is that $x^* = y^*$, which means that the extremal point is *invariant* under the action of the symmetry of the problem, namely, the exchange of the two variables. In order to find such an extremal point, we could easily restrict to the line $x = y$ and minimize $g$ in this one-dimensional subspace.
- The second possibility is that $x^* \neq y^*$. If this is the case, then both $(x^*, y^*)$ and $(y^*, x^*)$ are necessarily minimum points. If this happens, the symmetry is said to be *broken* and the action of the symmetry group of the problem transforms one extremal point into the other and vice versa. The way in which the symmetry must be broken, as we will see, is a completely nontrivial issue in many situations.

The problem we are facing is, despite its more intricate nature, absolutely analogous. In fact, the effective action (3.47) is symmetric under the permutation group over replicas, which means that

$$\mathcal{S}[\{Q_{ab}\}] = \mathcal{S}[\{Q_{\pi(a)\pi(b)}\}], \tag{3.54}$$

where $\pi$ is an arbitrary permutation of replica indices. As in the example above, we want to start with the simplest (and historically the first) ansatz possible: we assume that the solution is invariant under the symmetry group of the action; namely, if $Q^*$ is a minimum point, we have

$$\{Q^*_{ab}\} = \{Q^*_{\pi(a)\pi(b)}\}, \tag{3.55}$$

i.e., the matrix does not change under an arbitrary renumbering of replicas. It is easily shown that the only invariant matrix is the so-called replica-symmetric (RS) one

$$Q_{aa} = 0 \,, \tag{3.56}$$
$$Q_{ab} = q \,, \quad a \neq b \,,$$

where all the diagonal elements are zero and all the off-diagonal elements take the same value $q$.

We can plug this ansatz into the effective action (3.47), obtaining the following result:

$$
\begin{aligned}
S[q, h] &= -\frac{n\beta^2}{4} - \frac{n\beta^2 q^2}{4} - \ln \sum_{\{\sigma\}} \exp \left( \beta h \sum_a \sigma^a + \frac{\beta^2}{2} q \sum_{a \neq b} \sigma^a \sigma^b \right) \\
&= -\frac{n\beta^2}{4} - \frac{n\beta^2 q^2}{4} - \ln \sum_{\{\sigma\}} \exp \left( \beta h \sum_a \sigma^a \right) \exp \left( -\frac{n\beta^2 q}{2} \right) \left( \frac{\beta^2}{2} q \left( \sum_a \sigma^a \right)^2 \right) \\
&= -\frac{n\beta^2}{4} - \frac{n\beta^2 q^2}{4} - \ln \sum_{\{\sigma\}} e^{\left( \beta h \sum_a \sigma^a - n\beta^2 q/2 \right)} \int \frac{dz}{\sqrt{2\pi}} \exp \left( -\frac{z^2}{2} + \beta \sqrt{q} z \sum_a \sigma^a \right) \\
&= -\frac{n\beta^2}{4} - \frac{n\beta^2 q^2}{4} + \frac{n\beta^2 q}{2} - \ln \int \frac{dz}{\sqrt{2\pi}} \exp \left( -\frac{z^2}{2} \right) (2 \cosh(\beta h + \beta \sqrt{q} z))^n \\
&= -\frac{n\beta^2}{4} - \frac{n\beta^2 q^2}{4} + \frac{n\beta^2 q}{2} - \ln \int \frac{dz}{\sqrt{2\pi}} \exp \left( -\frac{z^2}{2} \right) \exp \left( n \ln \left( 2 \cosh(\beta h + \beta \sqrt{q} z) \right) \right) \\
&= -\frac{n\beta^2}{4} - \frac{n\beta^2 q^2}{4} + \frac{n\beta^2 q}{2} - \ln \left( 1 + n \int \frac{dz}{\sqrt{2\pi}} e^{-z^2/2} \ln \left( 2 \cosh(\beta h + \beta \sqrt{q} z) \right) \right) \\
&= -\frac{n\beta^2}{4} - \frac{n\beta^2 q^2}{4} + \frac{n\beta^2 q}{2} - n \int \frac{dz}{\sqrt{2\pi}} \exp \left( -\frac{z^2}{2} \right) \ln \left( 2 \cosh(\beta h + \beta \sqrt{q} z) \right),
\end{aligned}
\tag{3.57}
$$

where again, in the third line, we have introduced an H–S transformation and all the steps are valid only in the $n \to 0$ regime. Dividing by $\beta$ and $n$ and taking the zero-replicas limit, we obtain the Gibbs free energy to be minimized with respect to $q$. The final Gibbs free energy reads

$$\tilde{f}(q, h) = -\frac{\beta}{4}(1 - q)^2 - \frac{1}{\beta} \int \frac{dz}{\sqrt{2\pi}} \exp \left( -\frac{z^2}{2} \right) \ln \left( 2 \cosh(\beta h + \beta \sqrt{q} z) \right). \tag{3.58}$$

Imposing the stationarity condition, we obtain

$$\frac{\partial \tilde{f}(q, h)}{\partial q} = \frac{\beta}{2}(1 - q) - \int \frac{dz}{\sqrt{2\pi}} \exp \left( -\frac{z^2}{2} \right) \tanh(\beta h + \beta \sqrt{q} z) \frac{z}{2\sqrt{q}} = 0 \,. \tag{3.59}$$

With one integration by parts and some simple algebraic manipulation, we obtain the following canonical form for the self-consistency equation:

$$q = \int \frac{dz}{\sqrt{2\pi}} \exp\left(-\frac{z^2}{2}\right) \tanh^2(\beta h + \beta\sqrt{q}z).$$ (3.60)

At zero magnetic field, (3.60) has a unique solution $q = 0$ for $\beta < \beta_c = 1$, while for $\beta > \beta_c$ another (physical) solution appears with $q \neq 0$. Therefore, there is a phase transition at $h = 0$ and $\beta = 1$, while there is no phase transition for finite $h$. Although everything seems to be consistent, an accurate computation shows that the entropy becomes negative at low temperature (in a discrete system, this is forbidden by definition); in particular, the zero-temperature entropy is $S(0) = -1/(2\pi) \approx -0.17$. This result is evidence that our assumption of replica symmetry is wrong at low temperatures; therefore, in the next section, we perform the stability analysis of the RS solution in some detail.

### 3.2.2 Stability

The stability of the replica-symmetric solution was first studied in de Almeida and Thouless (1978).

In order to analyze the stability of the solution in the full replica space, we consider the Hessian, which is an $\frac{1}{2}n(n-1) \times \frac{1}{2}n(n-1)$ symmetric matrix of the form

$$M_{(ab),(cd)} = \frac{\partial^2 W}{\partial Q_{ab} \partial Q_{cd}}$$

$$= \frac{\partial}{\partial Q_{cd}} \left[ \beta^2 Q_{ab} - \beta^2 \frac{\sum_{\{\sigma\}} \sigma^a \sigma^b \exp\left(\beta h \sum_f \sigma^f + \beta^2 \sum_{f<g} \sigma^f \sigma^g Q_{fg}\right)}{\sum_{\{\sigma\}} \exp\left(\beta h \sum_f \sigma^f + \beta^2 \sum_{f<g} \sigma^f \sigma^g Q_{fg}\right)} \right]$$

$$= \frac{\partial}{\partial Q_{cd}} \left[ \beta^2 Q_{ab} - \beta^2 \langle\langle \sigma^a \sigma^b \rangle\rangle \right]$$

$$= \beta^2 \delta_{(ab),(cd)} - \beta^4 \left[ \langle\langle \sigma^a \sigma^b \sigma^c \sigma^d \rangle\rangle - \langle\langle \sigma^a \sigma^b \rangle\rangle \langle\langle \sigma^c \sigma^d \rangle\rangle \right],$$ (3.61)

where again the double angular brackets stand for the average with respect to the measure (3.52).

Since we are computing the Hessian at an RS point, it is easy to see that it contains three kinds of elements, namely,

$$M_{(ab),(cd)} = \begin{cases} M_1, & (ab) = (cd), \\ M_2, & a = c, b \neq d \text{ or } a \neq c, b = d, \\ M_3, & a \neq c, b \neq d, \end{cases}$$ (3.62)

which we can compute explicitly through steps similar to those in (3.57), obtaining

$$\begin{aligned} M_1 &= \beta^2 - \beta^4(1 - q^2), \\ M_2 &= -\beta^4(q - q^2), \\ M_3 &= -\beta^4(r - q^2), \end{aligned}$$ (3.63)

where

$$r = \int \frac{dz}{\sqrt{2\pi}} \exp\left(-\frac{z^2}{2}\right) \tanh^4(\beta h + \beta\sqrt{q}z) . \tag{3.64}$$

Now we have to compute the eigenvalues of the stability matrix. The eigenvector equation reads

$$\sum_{(cd)} M_{(ab),(cd)} v_{(cd)} = m v_{(ab)} . \tag{3.65}$$

The first straightforward thing to notice is that all the rows of the Hessian matrix have the same sum, which means that the subspace spanned by the vector with all components equal is a good eigenspace (we indicate this subspace of dimension 1 as the "longitudinal" (scalar) subspace). In fact,

$$\left(M_1 + 2(n-2)M_2 + \frac{(n-2)(n-3)}{2}M_3\right) v = m_L v \tag{3.66}$$

gives the equation for the first eigenvalue $m_L$, where $L$ stands for longitudinal. The reader can easily check that the subspace orthogonal to the longitudinal one can be divided into another two orthogonal eigenspaces called the "anomalous" (vectorial) and "replicon" (tensorial) subspaces, respectively. The subspaces are defined by

$$\begin{aligned} \text{longitudinal} \quad & v_{(ab)} = v , \\ \text{anomalous} \quad & v_{(ab)} = \frac{1}{2}(v_a + v_b) , \quad \sum_a v_a = 0 , \\ \text{replicon} \quad & v_{(ab)} , \quad \sum_b v_{ab} = 0 . \end{aligned} \tag{3.67}$$

Note that the dimensionality of the anomalous space is $n-1$, while the replicon has dimension $n(n-3)/2$ and, as it should be, the sum of the dimensionality of the three subspaces is $n(n-1)/2$. The eigenvalues are

$$\begin{aligned} m_L &= M_1 + 2(n-2)M_2 + \frac{(n-2)(n-3)}{2}M_3 , \\ m_A &= M_1 + (n-4)M_2 - (n-3)M_3 , \\ m_R &= M_1 - 2M_2 + M_3 . \end{aligned} \tag{3.68}$$

In the limit $n \to 0$,

$$\begin{aligned} m_L &= M_1 - 4M_2 + 3M_3 , \\ m_A &= M_1 - 4M_2 + 3M_3 , \\ m_R &= M_1 - 2M_2 + M_3 , \end{aligned} \tag{3.69}$$

where we notice that the longitudinal and anomalous spaces are degenerate. The onset of the instability is given by the first eigenvalue that becomes zero, i.e. (the reader can verify this) the replicon. An explicit computation gives

$$m_R = \beta^2 - \beta^4 \left[ -(1-q^2) + 2(q-q^2) - (r-q^2) \right].$$  (3.70)

Combining the equation $m_R = 0$ with the self-consistency (3.60), we obtain the definition of the so-called de Almeida–Thouless (dAT) line, the line of instability of the RS solution in the $\beta$–$h$, plane, namely,

$$1 = \beta^2 \int \mathcal{D}z \, \mathrm{sech}^4(\beta h + \beta\sqrt{q}z),$$

$$q = \int \mathcal{D}z \, \tanh^2(\beta h + \beta\sqrt{q}z),$$  (3.71)

where

$$\mathcal{D}z = \frac{dz}{\sqrt{2\pi}} \exp\left(-\frac{z^2}{2}\right).$$  (3.72)

On the left of the dAT line, the replica-symmetric ansatz is not valid and replica symmetry must be broken. We will see in the next section the Parisi breaking scheme that has recently been proved to be the one that provides the correct solution to the SK model.

### 3.2.3 Breaking replica symmetry

In this section, we introduce in a formal way the iterative replica symmetry breaking (RSB) scheme proposed by Parisi (1980*a,b*), while we will give a more physical interpretation of the RSB phenomenon in the next section. The Parisi solution is a sequence of ansatzes that approximate progressively better the true solution. At the end of the procedure, the solution can be formulated in terms of a continuous function, defined in the limit of the infinite sequence. We consider now, step by step, the way in which the solution is obtained. The first stage (1-RSB) goes as follows: the $n$ replicas are divided into $n/m$ groups of $m$ replicas, and if two replicas $a$ and $b$ belong to the same group, then the overlap matrix takes a value $q_1$, while if they belong to different groups, then it takes a value $q_0$. A compact way to express this is as follows:

$$Q_{ab} = \begin{cases} q_1, & I(a/m) = I(b/m), \\ q_0, & I(a/m) \neq I(b/m), \end{cases}$$  (3.73)

where $I(z)$ is the integer part of $z$. A pictorial representation of the 1-RSB scheme is given in Fig. 3.1. With computations similar to (3.57) (technically slightly more

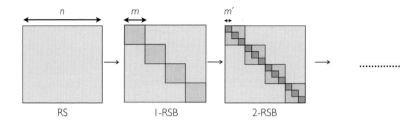

**Fig. 3.1**   A schematic representation of the Parisi replica symmetry breaking scheme.

complicated), we can obtain the 1-RSB free energy, which is now a function of the three parameters $q_1$, $q_0$, and $m$. The Gibbs free energy reads

$$\tilde{f}(q_1, q_0, m) = -\frac{1}{4}\beta\left[1 + mq_0^2 + (1-m)q_1^2 - 2q_1\right] - \frac{1}{m\beta}\int\frac{dz}{\sqrt{2\pi q_0}}\exp\left(-\frac{z^2}{2q_0}\right)$$

$$\times \ln\left[\int\frac{dy}{\sqrt{2\pi(q_1 - q_0)}}\exp\left(-\frac{y^2}{2(q_1 - q_0)}\right)(2\cosh(\beta(z + y)))^m\right].$$

(3.74)

In the process of taking the $n \to 0$ limit, we also have to promote the group size $m$, which was originally an integer between 1 and $n$, to be a real number $m \in [0, 1]$. Again, in the next section, we will provide a more physical explanation for this fact. The resulting self-consistency equations for $q_1$, $q_0$, and $m$ can be obtained by extremizing the free energy (3.74). Notice that in the two extreme limits when $m = 0$ and $m = 1$, we actually recover a replica-symmetric solution with respectively $q = q_0$ or $q = q_1$. Three comments are in order at this point:

- When minimizing the free energy in the region of parameters where the replica symmetry is supposed to be broken (below the dAT line), $m$ is neither zero nor one, as we expect.
- The zero-temperature entropy is increased with respect to the replica-symmetric entropy; in particular, one can obtain $S(0) \approx -0.01$.
- Computation of the Hessian eigenvalues shows that close to the zero-field critical point, the replicon eigenvalue is less negative with respect to the RS case; more precisely, $m_R = -C\tau^2/9$, compared with $m_R = -C\tau^2$, where $\tau = (1 - T)$.

All these facts suggest that even though we have not found the correct solution, in some sense we have taken the right path. The procedure we have just described can be iterated and the single group of $m$ replicas can be divided into $m/m'$ groups of $m'$ replicas. At this point, the matrix element between replicas belonging to the same subgroup will take the value $q_2$, that between replicas belonging to the same group but not the same subgroup will take the value $q_1$, and, finally, that between replicas belonging to different groups will take the value $q_0$. A pictorial representation of a 2-RSB solution can also be found in Fig. 3.1. We can now imagine repeating the procedure $k$ times, obtaining the following definition for the matrix $Q$:

$$Q_{ab} = q_i, \quad I(a/m_i) = (b/m_i) \text{ and } I(a/m_{i+1} = b/m_{i+1}): \ i = 1, 2, \ldots, k+1. \tag{3.75}$$

In principle, for each of these schemes, the computation of the free energy and the subsequent minimization can be repeated with respect to the parameters $q_1, \ldots, q_{k+1}, m_1, \ldots, m_k$. The solution improves progressively with increasing number of steps, meaning that the zero-temperature entropy becomes less and less negative and the instability at the transition weaker and weaker. This suggests a "natural" step of sending the number of breakings to infinity.

In order to do this, we define the following multistep function in the interval $x \in [0, 1]$:

$$q(x) = q_i, \quad 0 \le m_i < x < m_{i+1} \le 1: \ i = 1, 2, \ldots, k. \tag{3.76}$$

Taking the $k \to \infty$ limit, the free energy becomes a functional of $q(x)$ and the problem reduces to extremizing this functional with respect to all the possible functions $q$ with support in the unit interval, i.e., equating the functional derivative to zero:

$$\frac{\mathrm{d}f[q]}{\mathrm{d}q(x)} = 0. \tag{3.77}$$

Generically, the solution can be found numerically with a high degree of approximation (e.g., 8 decimal figures), but very close to the critical temperature an analytical treatment is possible by expanding the free energy around $q(x) = 0$.

With some effort, one can derive the algebra (Parisi, 1980$a$) that is associated with the matrices parametrized with the Parisi scheme (for any $n < 1$ and $n \le x \le 1$); here we just present it briefly without derivation. A Parisi matrix is fully defined by its diagonal element and its off-diagonal function, namely,

$$A \to (\tilde{a}, a(x)). \tag{3.78}$$

Now, consider two Parisi matrices and their Hadamard (dot) product $(A \cdot B)_{ab} = A_{ab}B_{ab}$. It is easy to prove that

$$A \cdot B \to (\tilde{a}\tilde{b}, a(x)b(x)). \tag{3.79}$$

Considering instead the usual matrix product $C = AB$, with some algebra, it can be proved that

$$\tilde{a}\tilde{b} - \langle ab \rangle \tag{3.80}$$

and

$$c(x) = -na(x)b(x) + [\tilde{a} - \langle a \rangle]b(x) + [\tilde{b} - \langle b \rangle]a(x) - \int_n^x \mathrm{d}y\, [a(x) - a(y)][b(x) - b(y)], \tag{3.81}$$

where the compact notation

$$\langle a \rangle = \int_n^1 \mathrm{d}x \, a(x) \tag{3.82}$$

has been used.

Here we simply introduce the free energy in its full-RSB form (the derivation can be found in Parisi (1980a)). The full-RSB free energy reads

$$f = -\frac{\beta}{4}\left[1 + \int_0^1 \mathrm{d}x \, q^2(x) - 2q(1)\right] - \frac{1}{\beta}\int \mathrm{D}u \, f_0(0, \sqrt{q(0)}u)\,, \tag{3.83}$$

where $f_0$ is the solution to the nonlinear antiparabolic equation

$$\frac{\partial f_0(x, h)}{\partial x} = -\frac{1}{2}\frac{\mathrm{d}q}{\mathrm{d}x}\left[\frac{\partial^2 f_0}{\partial h^2} + x\left(\frac{\partial f_0}{\partial h}\right)^2\right], \tag{3.84}$$

with the initial condition

$$f_0(1, h) = \ln(2\cosh h). \tag{3.85}$$

A nice collection of examples of solutions $q(x)$ in different situations can be found in Mézard *et al.* (1987, pp. 40–43).

We remark that in spite of numerical evidence that the solution of the stationary equation is unique, only recently has it been proved that the free energy in (3.83) is convex and therefore its solution is unique (Auffinger and Chen, 2014).

## 3.3   Pure states, ultrametricity, and stochastic stability in the RSB phase of the SK model

The spin glass solution of the SK model is characterized by a spontaneous breaking of the replica symmetry, which corresponds to a nontrivial, i.e., non-replica-symmetric, form of the order parameter matrix $Q_{ab}$. As a consequence, there are many solutions to the saddle point equations. This is due to the symmetry of the replica free energy function with respect to replica permutations. In other words, if there exists a particular solution for the matrix $Q_{ab}$ with an RSB structure, then any other matrix obtained via a renumbering of the replica indices in $Q_{ab}$ will be also a solution. It is also worth noting that the free energy barriers separating the corresponding RSB states must be infinite in the thermodynamic limit, since the mean-field free energy is proportional to the volume of the system. Since all these states are stabilized in the thermodynamic limit, we could call them the *pure states* of the spin glass phase. On the other hand, the Gibbs state is obtained by summing over all the pure states of the system, each pure state being taken with its statistical weight, which is defined by the value of the corresponding free energy. If we label the pure states with an index $\alpha$, we can define the corresponding thermodynamic weights $w_\alpha$ as (Mézard *et al.*, 1984; Dotsenko, 2005)

$$w_\alpha = \frac{e^{-\beta F_\alpha}}{\sum_\alpha e^{-\beta F_\alpha}} \,. \tag{3.86}$$

By definition, a pure state is a state in which the clustering property is satisfied, i.e., all correlation functions must factorize in the long-distance limit. For example, the two-point correlation function must satisfy the aforementioned clustering property, which, in the case of the fully connected SK model, reads $\langle \sigma_i \sigma_j \rangle_\alpha = \langle \sigma_i \rangle_\alpha \langle \sigma_j \rangle_\alpha$. The symbol $\langle \cdot \rangle_\alpha$ means that the Gibbs measure has to be restricted to the pure state $\alpha$ only.

Let us now consider two different pure states of our system, labeled by two indices $\alpha$ and $\beta$, and let us define their mutual overlap $q_{\alpha\beta}$ as

$$q_{\alpha\beta} \equiv \frac{1}{N} \sum_{i=1}^{N} \langle \sigma_i \rangle_\alpha \langle \sigma_i \rangle_\beta \,. \tag{3.87}$$

To study the statistics of the whole set of possible overlaps $\{q_{\alpha\beta}\}$, it is useful to introduce the following probability distribution function:

$$P_J(q) = \sum_{\alpha\beta} w_\alpha w_\beta \, \delta(q_{\alpha\beta} - q) \,. \tag{3.88}$$

This distribution is sample-dependent, i.e., it depends on the specific realization of the random couplings $J_{ij}$. So, if we take the average of $P_J(q)$ over the disorder, we get the *physical* distribution function

$$P(q) = \overline{P_J(q)} \,, \tag{3.89}$$

which gives the probability of finding a pair of pure states having overlap equal to $q$. The distribution $P(q)$ can be considered as the physical order parameter. The fact that it is a function is a manifestation of the phenomenon that for the description of the spin glass phase one needs an infinite number of order parameters. Now we want to establish contact between the physical order parameter $P(q)$ and the replica world. To this end, let us first consider the following correlation function:

$$q_J^{(k)} = \frac{1}{N^k} \sum_{i_1 \ldots i_k} \langle \sigma_{i_1} \cdots \sigma_{i_k} \rangle^2 \,. \tag{3.90}$$

Using the representation of the Gibbs average in terms of the pure states, it is not difficult to show that

$$q_J^{(k)} = \int \mathrm{d}q \, P_J(q) \, q^k \,. \tag{3.91}$$

Taking the average over the disorder, we get

$$q^{(k)} = \overline{q_J^{(k)}} = \int \mathrm{d}q \, P(q) q^k \,. \tag{3.92}$$

We see that the function $P(q)$, originally defined to describe the statistics of *pure states*, can be practically calculated from the multipoint correlation functions in the *Gibbs state*. Now it should be clear how to bridge replicas and physics. If we calculate the multipoint correlation functions using the replica approach, the connection of the physical order parameter with the Parisi RSB pattern will be established. The moment $q^{(k)}$ can be represented as follows:

$$q^{(k)} = \lim_{n \to 0} \left[ \overline{\langle \sigma_i^a \sigma_i^b \rangle} \right]^k = \lim_{n \to 0} (Q_{ab})^k , \qquad (3.93)$$

where $Q_{ab} = \overline{\langle \sigma_i^a \sigma_i^b \rangle}$ is the replica order parameter matrix, which is obtained from the saddle point equation for the replica free energy. Notice that in the RSB phase, the entries of $Q_{ab}$ are not equivalent, and one has to sum over all the saddle point solutions for the matrix $Q_{ab}$ to perform the Gibbs average. Such solutions can be obtained from one of the RSB solutions by applying all possible permutations of rows and columns in $Q_{ab}$. The summation over all these permutations corresponds to the summation over the replica indices $a$ and $b$ of the matrix $Q_{ab}$. As a consequence, the moment $q^{(k)}$ should be computed as follows:

$$q^{(k)} = \lim_{n \to 0} \frac{1}{n(n-1)} \sum_{a \neq b} (Q_{ab})^k , \qquad (3.94)$$

where the factor $n(n-1)$ is precisely the number of different replica permutations. We then get the following explicit expression for the distribution function $P(q)$:

$$P(q) = \lim_{n \to 0} \frac{1}{n(n-1)} \sum_{a \neq b} \delta(Q_{ab} - q) . \qquad (3.95)$$

In the continuum $n \to 0$ limit, this result can be rewritten as

$$P(q) = \int_0^1 dx \, \delta(q(x) - q) . \qquad (3.96)$$

Assuming that the function $q(x)$ is monotonic (which is the case for the RSB solution we are considering here), we can introduce the inverse function $x(q)$, and from (3.96) we find

$$P(q) = \frac{dx(q)}{dq} . \qquad (3.97)$$

This result is a key point: it defines the physical order parameter, i.e., the distribution function $P(q)$, in terms of the formal saddle point function $q(x)$ (Parisi, 2013). By representing $x(q)$ in the integral form

$$x(q) = \int_0^q dq' \, P(q'), \qquad (3.98)$$

we can assign to this function a clear physical meaning: the function $x(q)$, inverse to $q(x)$, gives the probability of finding a pair of pure states with an overlap less than $q$.

### 3.3.1   Fluctuations of $P_J(q)$

Let us consider the function $P_J(q)$:

$$P_J(q) = \sum_{\alpha\beta} w_\alpha w_\beta \delta(q_{\alpha\beta} - q) . \qquad (3.99)$$

One reasonable problem to address is whether the probability $P_J(q)$ approaches a definite limit or fluctuates when the number $N$ of spins becomes infinite. To have an idea of why the latter case could be possible, let us consider, for the moment, a homogeneous ferromagnetic system (i.e., in the absence of disorder). If we start from the high-temperature phase and cool down below the Curie temperature, the probabilities $w_+$ and $w_-$ of arriving at a state of magnetization $+m$ or $-m$ are well known to depend on the boundary conditions we have imposed on the system.

So, even in this simple case, $w_\alpha$, and therefore $P_J(q)$, are not good extensive quantities. The logical suspicion is that the same is true even in the more complicated spin glass case. Indeed, what happens is that $P_J(q)$ depends on the realization of the couplings, even in the thermodynamic limit. In other words, it is not a self-averaging quantity, in the sense that, as far as $P_J(q)$ is concerned, an increasing size of the sample does not imply an average over all disorder configurations.

To quantify the fluctuations of $P_J(q)$, we now compute $\overline{P_J(q_1)P_J(q_2)} - P(q_1)P(q_2)$. To this end, we make use of the Laplace transform $g(y)$:

$$g(y) = \int dq \, \overline{P_J(q)} e^{yq} = \int dq \, P(q) e^{yq} \,. \tag{3.100}$$

Using (3.95) and (3.96), we can write $g(y)$ as follows:

$$g(y) = \frac{1}{n(n-1)} \sum_{a \neq b} e^{yQ_{ab}} \xrightarrow[n \to 0]{} \int_0^1 dx \, e^{yq(x)} \,. \tag{3.101}$$

The computation of $\overline{P_J(q_1)P_J(q_2)}$ requires a slightly generalized Laplace transform $g(y_1, y_2)$:

$$g(y_1, y_2) = \int dq_1 \, dq_2 \, \overline{P_J(q_1)P_J(q_2)} e^{y_1 q_1 + y_2 q_2} = \int dq_1 \, dq_2 \, P(q_1, q_2) e^{y_1 q_1 + y_2 q_2} \,. \tag{3.102}$$

$P(q_1, q_2)$ is the probability (averaged over $J$) to have overlaps $q_{\alpha_1 \alpha_2} = q_1$ and $q_{\alpha_3 \alpha_4} = q_2$ between four pure states $\alpha_1, \alpha_2, \alpha_3, \alpha_4$. The function $g(y_1, y_2)$ can be computed by means of replicas using the following formula:

$$g(y_1, y_2) = \frac{1}{n(n-1)(n-2)(n-3)} \sum_{a \neq b \neq c \neq d=1}^{n} e^{y_1 Q_{ab} + y_2 Q_{cd}} \,, \tag{3.103}$$

where the sum is restricted to quadruplets of replicas $a, b, c, d$ that are all different. Using the RSB parametrization for $Q_{ab}$, and finally taking the limit $n \to 0$, we get (Mézard *et al.*, 1984; Ghirlanda and Guerra, 1998)

$$\overline{P_J(q_1)P_J(q_2)} = \frac{1}{3} P(q_1)\delta(q_1 - q_2) + \frac{2}{3} P(q_1)P(q_2) \,. \tag{3.104}$$

This formula is evidence that $P_J(q)$ fluctuates with $J$ even after the thermodynamic limit is taken. There are infinite many relations of this kind, as we will see in the next section (Ghirlanda and Guerra, 1998; Aizenman and Contucci, 1998; Parisi, 1998; Contucci and Giardinà, 2005).

### 3.3.2 Stochastic stability

The principle of stochastic stability assumes that the distribution of the free energies of the various pure states is stable under independent random increments. The reason why we require such a principle is that the perturbations we use are random and they are not correlated with the original Hamiltonian. As a consequence, they should change the free energies of the pure states by random amounts, leaving their distribution unaffected.

We present here the simplest nontrivial case of a stochastically stable system (Aizenman and Contucci, 1998), where the overlaps have only two possible values: $q_0$ among different states and $q_1$ among the same state. In this case (which is usually called 1-step replica symmetry breaking), we have only to specify the weight of each state, which is given by

$$w_\alpha \propto e^{-\beta F_\alpha} \,. \tag{3.105}$$

Let us consider the case where the $F$s are independent random variables. The number of values of $F$ in the interval $[F, F + dF]$ is given by

$$\rho(F)dF \,. \tag{3.106}$$

Notice that because the number of states is infinite, the function $\rho(F)$ has a divergent integral.

Let us now consider the effect of a perturbation of strength $\epsilon$ on the free energy of a state, say $\alpha$. The unperturbed value of the free energy is denoted by $F_\alpha$. The new value of the free energy is given by $F'_\alpha = F_\alpha + \epsilon r_\alpha$, where $r_\alpha$ are identically distributed uncorrelated random variables. Stochastic stability implies that the distribution $\rho(F')$ is the same as $\rho(F)$. Expanding to second order in $\epsilon$, this condition leads to

$$\frac{d\rho}{dF} \propto \frac{d^2\rho}{dF^2} \,. \tag{3.107}$$

The only physical solution (apart the trivial one $\rho(F) = 0$, corresponding to non spin glass systems) is given by

$$\rho(F) \propto \exp(\beta m F) \,, \tag{3.108}$$

with an appropriate value of $m$. The parameter $m$ must satisfy the condition $m < 1$, in order for the sum $\sum_\alpha \exp(-\beta F_\alpha)$ to be convergent. We see that stochastic stability fixes the form of the distribution $\rho$, and hence connects the low and high free energy parts of the function $\rho$. In other words, stochastic stability relates the property of the low-lying configurations (which dominate the Gibbs measure) to those of the configurations much higher in free energy (which usually dominate the dynamics). Accordingly, the requirement of stochastic stability gives also, nearly for free, some information on the dynamics in the aging regime. In the dynamical evolution from a higher-temperature initial state, the difference between the total free energy at time $t$ and the equilibrium value will always be of order $N$, with a prefactor going to zero when the time goes to infinity. So, one could argue that dynamics probes the behavior of the function $\rho(F)$ at very large argument, and should not be related to the static properties, which, on the contrary, depend on the distribution $\rho$ for small values of the argument. However, stochastic stability forces the function $\rho(F)$ to be of the form (3.108), also in the range where $F$ is extensive but small (let us say of order $\epsilon N$), and the previous objection can be discarded.

The function $P(q)$, in the 1-step RSB case, is given by

$$P(q) = m\delta(q - q_0) + (1 - m)\delta(q - q_1). \tag{3.109}$$

So we see that the same parameter $m$ enters both in the form of the function $P(q)$, which is dominated by the lowest values of $F$ (i.e., those producing the largest $w$s), and in the form of the function $\rho(F)$ at large values of $F$. This result is related in a deep manner to the existence of only one family of stochastic stable systems with uncorrelated variables $F$.

In the case of full-RSB, the construction is more complex and is described in detail by Mézard *et al.* (1987) and Ruelle (1987).

### 3.3.3   Ultrametricity

The order parameter $P(q)$, far from being a parameter, indicates that the structure of the space of the spin glass pure states must be highly nontrivial. However, the distribution function $P(q)$ of the pure state overlaps does not give enough information about this structure. To gain insight into the topology of the space of pure states, one needs to know the properties of the higher-order correlations of the overlaps. So, let us consider any three pure states $\alpha_1$, $\alpha_2$, and $\alpha_3$, with $P_J(q_1, q_2, q_3)$ being the probability for them to have overlaps $q_1 = q_{\alpha_2\alpha_3}$, $q_2 = q_{\alpha_3\alpha_1}$, and $q_3 = q_{\alpha_1\alpha_2}$. To compute $P_J(q_1, q_2, q_3)$, we first consider the Laplace transform $g_J(y_1, y_2, y_3)$, and then use replicas to calculate the average over the $J$s (Mézard *et al.*, 1984, 1987):

$$g(y_1, y_2, y_3) = \overline{g_J(y_1, y_2, y_3)} = \frac{1}{n(n-1)(n-2)} \sum_{a\neq b\neq c=1}^{n} e^{y_1 Q_{ab} + y_2 Q_{bc} + y_3 Q_{ca}}, \tag{3.110}$$

where the sum must run over triplets of replicas $a, b, c$ that are all different. Taking the limit $n \to 0$, after some algebra, we obtain the following result for $P(q_1, q_2, q_3) = \overline{P_J(q_1, q_2, q_3)}$:

$$
\begin{aligned}
P(q_1, q_2, q_3) = {}&\frac{1}{2}P(q_1)x(q_1)\delta(q_1 - q_2)\delta(q_1 - q_3) \\
&+ \frac{1}{2}P(q_1)P(q_2)\theta(q_1 - q_2)\delta(q_2 - q_3) \\
&+ \frac{1}{2}P(q_2)P(q_3)\theta(q_2 - q_3)\delta(q_3 - q_1) \\
&+ \frac{1}{2}P(q_3)P(q_1)\theta(q_3 - q_1)\delta(q_1 - q_2).
\end{aligned}
\tag{3.111}
$$

From this equation, one can extract the following crucial property of the function $P(q_1, q_2, q_3)$: it is nonzero only in the three cases

$$q_1 = q_2 \leq q_3 \,,$$
$$q_1 = q_3 \leq q_2 \,, \qquad (3.112)$$
$$q_3 = q_2 \leq q_1 \,.$$

In all other cases, $P(q_1, q_2, q_3)$ is identically zero. In other words, this function is non-zero only if at least two of the three overlaps are equal, and their common value is not larger than the third. This means that in the space of spin glass pure states, there exist no scalene triangles. More precisely, if we sample three configurations independently with respect to their common Gibbs distribution, and average over the disorder, the distribution of the (Hamming) distances among them is supported, in the limit of very large system sizes, only on equilateral and isosceles triangles, with no contribution from scalene triangles. Spaces having this metric property are called *ultrametric*.

It was once believed that stochastic stability could be an independent property from ultrametricity; however, there have been many recent papers suggesting the contrary, and this line of research has culminated in the general proof of Panchenko (2013) that stochastic stability *does* imply ultrametricity.

We conclude this section with the following observation. In the case where the function $P_J(q)$ fluctuates when we change the parameters of the system, we can define its functional probability distribution $\mathcal{P}[P]$. This functional order parameter is a description of the probability of $P_J(q)$ when we change the disorder.[4] The functional probability distribution $\mathcal{P}[P]$ is an object that should have an infinite-volume limit, namely, $\mathcal{P}^\infty[P]$.

## 3.4 Thermodynamic limit in the SK model and exactness of the hierarchical RSB solution: rigorous results

### 3.4.1 Existence of the thermodynamic limit

The rigorous control of the infinite-volume limit in the SK model can be very difficult, owing to the effects of very large fluctuations produced by external noise (the random couplings $J_{ij}$). Nonetheless, Guerra and Toninelli (2002) have introduced a very simple strategy for control of the infinite-volume limit. The main idea is to split a large system, made of $N$ spins, into two subsystems, made of $N_1$ and $N_2$ sites, respectively, where each subsystem is subject to some external noise, similar to but independent of the noise acting on the large system. By a smooth interpolation between the system and the subsystems, one can show the *subadditivity* of the quenched average of the free energy, with respect to the size of the system, and therefore obtain complete control of the infinite-volume limit.

---

[4] It may be possible that also for nonrandom systems we have a similar description where the average over the number of degrees of freedom plays the same role as the average over the disorder.

The Hamiltonian of the SK model, in a uniform external field of strength $h$, is given by

$$H_N(\sigma, J, h) = -\frac{1}{\sqrt{N}} \sum_{j>i=1}^{N} J_{ij} \sigma_i \sigma_j - h \sum_{i=1}^{N} \sigma_i \,, \tag{3.113}$$

where the quenched disorder is given by the $N(N-1)/2$ independent and identically distributed random variables $J_{ij}$. Let us assume each $J_{ij}$ to be a Gaussian random variable with zero mean and unit variance.

For a given realization of the sample $J_{ij}$, the disorder-dependent partition function $Z_N(\beta, J, h)$, at the inverse temperature $\beta$, is given by

$$Z_N(\beta, J, h) = \sum_{\{\sigma\}} \exp\left[-\beta H_N(\sigma, J, h)\right] \,. \tag{3.114}$$

The quenched average of the free energy per spin $f_N(\beta, h)$ is

$$f_N(\beta, h) = -\frac{1}{\beta N} \mathbb{E}_J \log Z_N(\beta, J, h) \,. \tag{3.115}$$

Another important concept to recall is that of replicas. These are independent copies of the system, say $n$ in number, characterized by the spin variables $\sigma_i^1, \sigma_i^2, \dots, \sigma_i^n$. All the replicas are subject to the same sample $J$ of the quenched disorder. The Boltzmann factor for the replicated system is given by

$$\exp\left[-\beta H_N(\sigma^1, J, h) - \beta H_N(\sigma^2, J, h) - \dots - \beta H_N(\sigma^n, J, h)\right]. \tag{3.116}$$

The overlaps between any two replicas $a$ and $b$ are defined as

$$q_{ab} = \frac{1}{N} \sum_i \sigma_i^a \sigma_i^b \,, \tag{3.117}$$

and they satisfy the bounds $-1 \le q_{ab} \le 1$. For a generic smooth function $F$ of the overlaps, we define the $\overline{\;\cdot\;}$ averages

$$\overline{F(q_{12}, q_{13}, \dots)} = \mathbb{E}\langle F(q_{12}, q_{13}, \dots)\rangle_J \,, \tag{3.118}$$

where the thermal average $\langle \cdot \rangle_J$ is performed using the Boltzmann factor for the replicated system, (3.116), and $\mathbb{E}$ is the average with respect to the quenched disorder $J$.

Now we explain the main idea behind the Guerra–Toninelli method to control the infinite-volume limit. Let us divide the $N$ spins into two blocks of $N_1$ and $N_2$ spins, with $N_1 + N_2 = N$, and define

$$Z_N(t) = \sum_{\{s\}} \exp\left(\beta\sqrt{\frac{t}{N}} \sum_{j>i=1}^{N} J_{ij}\sigma_i\sigma_j + \beta\sqrt{\frac{1-t}{N_1}} \sum_{j>i=1}^{N_1} J'_{ij}\sigma_i\sigma_j \right.$$
$$\left. + \beta\sqrt{\frac{1-t}{N_2}} \sum_{j>i=N_1+1}^{N_1} J''_{ij}\sigma_i\sigma_j + \beta h \sum_{i=1}^{N} \sigma_i \right), \tag{3.119}$$

with $0 \leq t \leq 1$. The quenched disorder is represented by the independent families of random variables $J$, $J'$ and $J''$. The two subsystems are subjected to a different external noise, with respect to the original system, but, of course, the probability distributions are the same. The parameter $t$ allows us to interpolate between the original $N$ spin system at $t = 1$ and a system composed of two non-interacting parts at $t = 0$. So we have

$$
\begin{aligned}
Z_N(1) &= Z_N(\beta, h, J), \\
Z_N(0) &= Z_{N_1}(\beta, h, J')Z_{N_2}(\beta, h, J'').
\end{aligned}
\tag{3.120}
$$

As a consequence, remembering (3.115), we have

$$
\begin{aligned}
-\beta^{-1} \, \mathbb{E}Z_N(1) &= N f_N(\beta, h), \\
-\beta^{-1} \, \mathbb{E}Z_N(0) &= N_1 f_{N_1}(\beta, h) + N_2 f_{N_2}(\beta, h).
\end{aligned}
\tag{3.121}
$$

By taking the derivative of $-(N\beta)^{-1}\mathbb{E}Z_N(t)$ and performing a standard integration by parts on the Gaussian disorder, we obtain

$$
-\frac{1}{N\beta}\frac{\mathrm{d}}{\mathrm{d}t}\mathbb{E}Z_N(t) = \frac{\beta}{4}\left\langle q_{12}^2 - \frac{N_1}{N}\left(q_{12}^{(1)}\right)^2 - \frac{N_2}{N}\left(q_{12}^{(2)}\right)^2 \right\rangle,
\tag{3.122}
$$

where

$$
\begin{aligned}
q_{12}^{(1)} &= \frac{1}{N_1}\sum_{i=1}^{N_1}\sigma_i^1\sigma_i^2, \\
q_{12}^{(2)} &= \frac{1}{N_2}\sum_{i=N_1+1}^{N}\sigma_i^1\sigma_i^2.
\end{aligned}
\tag{3.123}
$$

On the other hand, $q_{12}$ is a linear combination of $q_{12}^{(1)}$ and $q_{12}^{(2)}$ with positive coefficients in the form

$$
q_{12} = \frac{N_1}{N}q_{12}^{(1)} + \frac{N_2}{N}q_{12}^{(2)}.
\tag{3.124}
$$

Owing to the convexity of the function $f : x \mapsto x^2$, we have the inequality

$$
\left\langle q_{12}^2 - \frac{N_1}{N}\left(q_{12}^{(1)}\right)^2 - \frac{N_2}{N}\left(q_{12}^{(2)}\right)^2 \right\rangle \leq 0,
\tag{3.125}
$$

From this result we have a simple cascade of consequences. First of all, the quenched average of the logarithm of the interpolating partition function, defined in (3.119), is increasing in $t$:

$$
\frac{1}{N}\frac{\mathrm{d}}{\mathrm{d}t}\mathbb{E}\log Z_N(t) \geq 0.
\tag{3.126}
$$

By integrating over $t$ and recalling the boundary conditions (3.120), we have

$$N f_N(\beta, h) \leq N_1 f_{N_1}(\beta, h) + N_2 f_{N_2}(\beta, h) \,. \tag{3.127}$$

This inequality express the *subadditivity* of the quenched average of the free energy of the SK model. Since for every subadditive sequence $\{F_N\}_{N=1}^{\infty}$, the limit $\lim_{N \to \infty} F_N/N$ exists and is equal to $\inf_N F_N/N$ (Fekete's subadditive lemma), we have the following theorem:

**Theorem 3.1 (Guerra–Toninelli 1)** *The infinite-volume limit for $f_N(\beta, h)$ exists and is equals to its infimum:*

$$\lim_{N \to \infty} f_N(\beta, h) = \inf_N f_N(\beta, h) \equiv f(\beta, h) \,. \tag{3.128}$$

After proving the existence of the thermodynamic limit for the quenched average, the result can be extended to prove that convergence holds for almost every disorder realization $J$. The following theorem is also due to Guerra and Toninelli:

**Theorem 3.2 (Guerra–Toninelli 2)** *The infinite-volume limit*

$$\lim_{N \to \infty} \frac{1}{N} \log Z_N(\beta, h) = f(\beta, h) \tag{3.129}$$

*exists $J$-almost surely.*

The proof is based on the fact that the fluctuations of the free energy per spin vanish exponentially rapidly as $N$ grows:

$$P \left( \left| \frac{1}{\beta N} \log Z_N(\beta, h, J) - \frac{1}{\beta N} \mathbb{E} \log Z_N(\beta, h, J) \right| \geq C \right) \leq e^{-NC^2/2} \,. \tag{3.130}$$

Then, noticing that the right-hand side of (3.130) is summable in $N$ for every fixed $C$, the Borel–Cantelli lemma and the convergence given by (3.128) imply (3.129).

### 3.4.2 Exactness of the broken replica symmetry solution

After having established the existence of the thermodynamic limit in the SK model, we turn to the problem of proving the correctness of the Parisi mechanism for the phenomenon of spontaneous replica symmetry breaking.

The proof is based on two main ingredients. The first is an interpolation method, invented and described in the marvelous paper by Guerra (2003), which allows us to prove that the RSB ansatz is a rigorous lower bound for the quenched average of the free energy per spin, uniformly in the size of the system. The second idea is due to Talagrand, who in 1998 made the observation that in order to prove an upper bound for the quenched average free energy, it is sufficient to prove a lower bound on a similar quantity that involves two copies of the system (two real replicas). This observation was not very useful at the time, since there was no method to prove the lower bound. Soon after the discovery of the interpolating method, Talagrand combined Guerra's method of proving lower bounds with his method to turn lower bounds into upper bounds, thus obtaining the proof.

We give here a brief sketch of Guerra's interpolating method and we refer to the original work by Guerra (2003) for the details and to the work by Talagrand (2006) for a complete proof.

Following Guerra, we first formulate the RSB ansatz without using replicas. Therefore, let us consider the (convex) space $\aleph$ of the functional order parameter $x(q)$, as a nondecreasing function of $q$, with both $x$ and $q$ taking values on the interval $[0, 1]$; i.e.,

$$x : [0, 1] \ni q \mapsto x(q) \in [0, 1], \quad x \in \aleph . \tag{3.131}$$

It is useful to consider the case of piecewise-constant functional order parameters, characterized by an integer $K$, and two sequences $q_0, q_1, \ldots, q_K$ and $m_1, m_2, \ldots, m_K$ of numbers satisfying

$$0 = q_0 \leq q_1 \leq \cdots \leq q_{K-1} \leq q_K = 1,$$
$$0 \leq m_1 \leq m_2 \leq \cdots \leq m_K \leq 1, \tag{3.132}$$

such that

$$x(q) = \begin{cases} m_1 & \text{for} \quad q_0 \leq q \leq q_1 , \\ m_2 & \text{for} \quad q_1 \leq q \leq q_2 , \\ \vdots \\ m_K & \text{for} \quad q_{K-1} \leq q \leq q_K. \end{cases} \tag{3.133}$$

The replica-symmetric case corresponds to

$$K = 2, \ q_1 = \bar{q}, \ m_1 = 0, \ m_2 = 1 . \tag{3.134}$$

The case $K = 3$ is the first level of replica symmetry breaking, and so on.

Let us now introduce the function $f_x(q, y)$ of the variables $q \in [0, 1]$, $y \in \mathbb{R}$, depending also on the functional order parameter $x(q)$, defined as the solution of the nonlinear antiparabolic equation

$$2\frac{\partial f}{\partial q} + \frac{\partial^2 f}{\partial y^2} + x(q) \left(\frac{\partial f}{\partial y}\right)^2 = 0 , \tag{3.135}$$

with final condition $f_x(1, y) = \log \cosh(\beta y)$. In the following, we will omit, for simplicity, the dependence of $f$ on $\beta$.

It turns out that the function $f_x(q, y)$ is monotonic in $x$, in the sense that $x(q) \leq \bar{x}(q)$ for all $0 \leq q \leq 1$ implies $f_x(q, y) \leq f_{\bar{x}}(q, y)$ for any $0 \leq q \leq 1$, $y \in \mathbb{R}$. Moreover, $f_x(q, y)$ is pointwise-continuous, so that, for generic $x$ and $\bar{x}$, we have

$$|f_x(q, y) - f_{\bar{x}}(q, y)| \leq \frac{\beta^2}{2} \int_q^1 |x(q') - \bar{x}(q')| \, dq' . \tag{3.136}$$

This result is very important. Indeed, any functional order parameter can be approximated by a piecewise-constant one. The pointwise continuity allows us to deal mostly with piecewise-constant order parameters.

Let us now define the trial *free entropy* function, depending on the functional order parameter $x(q)$, as follows:

$$\bar{\alpha}_x(\beta, h) \equiv \log 2 + f_x(0, h) - \frac{\beta^2}{2} \int_0^1 dq \, q x(q) \,. \tag{3.137}$$

When multiplied by $-1/\beta$, this becomes the trial *free energy* function. Notice that in this expression the function $f$ appears evaluated at $q = 0$ and $y = h$, where $h$ is the value of the external magnetic field.

The Parisi RSB solution is defined by

$$\bar{\alpha}_{\mathrm{RSB}}(\beta, h) \equiv \inf_x \bar{\alpha}_x(\beta, h) \,, \tag{3.138}$$

where the infimum is taken with respect to all functional order parameters $x(q)$. The main result of Guerra's work can be summarized in the following theorem:

**Theorem 3.3 (Guerra)** *For all values of the inverse temperature $\beta$ and the external magnetic field $h$, and for any functional order parameter $x(q)$, the following bound holds:*

$$\frac{1}{N} \, \mathbb{E} \log Z_N(\beta, h, J) \leq \bar{\alpha}_x(\beta, h) \,, \tag{3.139}$$

*uniformly in $N$. Consequently, we have also*

$$\frac{1}{N} \, \mathbb{E} \log Z_N(\beta, h, J) \leq \bar{\alpha}_{\mathrm{RSB}}(\beta, h) \,, \tag{3.140}$$

*uniformly in $N$. Moreover, for the thermodynamic limit, we have*

$$\lim_{N \to \infty} \frac{1}{N} \mathbb{E} \log Z_N(\beta, h, J) \equiv \alpha(\beta, h) \leq \bar{\alpha}_{\mathrm{RSB}}(\beta, h) \tag{3.141}$$

*and*

$$\lim_{N \to \infty} \frac{1}{N} \log Z_N(\beta, h, J) \equiv \alpha(\beta, h) \leq \bar{\alpha}_{\mathrm{RSB}}(\beta, h) \,, \tag{3.142}$$

*J-almost surely.*

The proof of the theorem is long, and we refer to the original paper by Guerra for an exhaustive presentation. Here we only sketch the main ideas of the proof.

Consider a generic piecewise-constant functional order parameter $x(q)$, as in (3.133), and define the following interpolating partition function $\tilde{Z}(t; x)$:

$$\tilde{Z}(t; x) \equiv \sum_{\{\sigma\}} \exp[\beta H_t(\sigma)],$$

$$H_t(\sigma) = \sqrt{\frac{t}{N}} \sum_{j>i=1}^{N} J_{ij}\sigma_i\sigma_j + h \sum_{i=1}^{N} \sigma_i + \sqrt{1-t} \sum_{a=1}^{K} \sqrt{q_a - q_{a-1}} \sum_{i=1}^{N} J_i^a \sigma_i, \tag{3.143}$$

where we have omitted the dependence of $\tilde{Z}(t; x)$ on $\beta$, $h$, $J$, and $N$. The numbers $J_i^a$ are additional independent centered unit Gaussian random variables, and the parameter $t$ runs in the interval $[0, 1]$.

For $a = 1, \dots, K$, let us denote by $\mathbb{E}_a$ the average with respect to all random variables $J_i^a$, $i = 1, \dots, N$. Moreover, we denote by $\mathbb{E}_0$ the average with respect to all $J_{ij}$, and by $\mathbb{E}$ the average with respect to all $J$ random variables.

Now we define recursively the random variables

$$\begin{aligned} Z_K &= \tilde{Z}(t; x), \\ Z_{K-1} &= \mathbb{E}_K Z_K, \\ &\;\vdots \\ Z_0 &= \mathbb{E}_1 Z_1 \end{aligned} \tag{3.144}$$

and the auxiliary function

$$\tilde{\alpha}_N(t) = \frac{1}{N} \mathbb{E}_0 \log Z_0. \tag{3.145}$$

Because of the partial integrations, any $Z_a$ depends only on the $J_{ij}$ and on the $J_i^b$ with $b \leq a$, while in $\tilde{\alpha}_N(t)$ all $J$ have been completely averaged out.

At the extreme values of the interpolating parameter $t$, the function $\tilde{\alpha}_N(t)$ evaluates

$$\begin{aligned} \tilde{\alpha}_N(1) &= \frac{1}{N} \mathbb{E} \log Z_N(\beta, h, J), \\ \tilde{\alpha}_N(0) &= \log 2 + f_q(0, h). \end{aligned} \tag{3.146}$$

What we want to calculate is the $t$ derivative of $\tilde{\alpha}_N(t)$. Before doing this, we need a few additional definitions. So let us introduce the random variables

$$f_a = \frac{Z_a^{m_a}}{\mathbb{E}_a(Z_a^{m_a})}, \quad a = 1, \dots, K, \tag{3.147}$$

and notice that they depend only on the $J_i^b$ with $b \leq a$, and are normalized: $\mathbb{E}(f_a) = 1$. Following Guerra, we consider the $t$-dependent state $\langle \cdot \rangle$ associated with the Boltzmann factor in (3.143),

$$\langle \cdot \rangle = \frac{\sum_{\{\sigma\}} (\cdot) \exp[\beta H_t(\sigma)]}{\tilde{Z}(t; x)}, \tag{3.148}$$

and the replicated state

$$\langle \cdot \rangle^{(s)} = \frac{\sum_{\{\sigma^1 \dots \sigma^s\}} (\cdot) \exp[\beta H_t(\sigma^1) + \dots + \beta H_t(\sigma^s)]}{[\tilde{Z}(t; x)]^s}. \tag{3.149}$$

Then we define the states $\langle \cdot \rangle_a$, $a = 0, \dots, K$, as

$$\langle \cdot \rangle_K = \langle \cdot \rangle, \\ \langle \cdot \rangle_a = \mathbb{E}_{a+1} \cdots \mathbb{E}_K [f_{a+1} \cdots f_K \langle \cdot \rangle], \tag{3.150}$$

and the replicated states as

$$\langle \cdot \rangle_K^{(s)} = \langle \cdot \rangle^{(s)}, \\ \langle \cdot \rangle_a^{(s)} = \mathbb{E}_{a+1} \cdots \mathbb{E}_K [f_{a+1} \cdots f_K \langle \cdot \rangle^{(s)}]. \tag{3.151}$$

Finally, we define the $\overline{\cdot}^a$ averages as

$$\overline{\cdot}^a = \mathbb{E} \left[ f_1 \cdots f_a \langle \cdot \rangle_a^{(s)} \right]. \tag{3.152}$$

The idea behind the definition of the $\overline{\cdot}^a$ averages is that they are able, in a sense, to concentrate the overlap fluctuations around the value $q_a$.

With these definitions, one can show that the derivative of $\tilde{\alpha}_N(t)$ is given by

$$\frac{\mathrm{d}}{\mathrm{d}t} \tilde{\alpha}_N(t) = -\frac{\beta^2}{4} \left[ 1 - \sum_{a=0}^K (m_{a+1} - m_a) \left( q_a^2 - \overline{(q_{12} - q_a)^2}^a \right) \right], \tag{3.153}$$

where $q_{12}$ is the overlap defined in (3.117), i.e., $q_{12} = N^{-1} \sum_i \sigma_i^1 \sigma_i^2$.

By integrating with respect to $t$, and taking into account the boundary values (3.146), we find *Guerra's sum rule*

$$\tilde{\alpha}_x(\beta, h) = \frac{1}{N} \mathbb{E} \log Z_N(\beta, h, J) + \frac{\beta^2}{4} \sum_{a=0}^K (m_{a+1} - m_a) \int_0^1 \overline{(q_{12} - q_a)^2}^a(t) \, \mathrm{d}t. \tag{3.154}$$

All terms in the sum are non-negative, since $m_{a+1} \geq m_a$, and the validity of the theorem is established.

# References

Aizenman, M. and Contucci, P. (1998). On the stability of the quenched state in mean-field spin-glass models. *J. Stat. Phys.*, **92**(5–6), 765.

Auffinger, A. and Chen, W. K. (2015). The Parisi formula has a unique minimizer. *Comm. Math. Phys.*, 335, 1429–1444.

Contucci, P. and Giardinà, P. (2005). Spin-glass stochastic stability: a rigorous proof. *Ann. Henri Poincaré*, **6**(5), 915–923.

de Almeida, J. R. L. and Thouless, D. J. (1978). Stability of the Sherrington–Kirkpatrick solution of a spin glass model. *J. Phys. A*, **11**(5), 983–990.

Dotsenko, V. (2005). *Introduction to the Replica Theory of Disordered Statistical Systems.* Cambridge University Press.

Ghirlanda, S. and Guerra, F. (1998). General properties of overlap probability distributions in disordered spin systems. towards Parisi ultrametricity. *J. Phys. A: Math. Gen.*, **31**(46), 9149–9155.

Guerra, F. (2003). Broken replica symmetry bounds in the mean field spin glass model. *Commun. Math. Phys.*, **233**(1), 1–12.

Guerra, F. and Toninelli, F. L. (2002). The thermodynamic limit in mean field spin glass models. *Commun. Math. Phys.*, **230**(1), 71–79.

Mézard, M., Parisi, G., Sourlas, N., Toulouse, G., and Virasoro, M. A. (1984). Replica symmetry breaking and the nature of the spin glass phase. *J. Phys. Paris*, **45**(5), 843–854.

Mézard, M., Parisi, G., and Virasoro, M. A. (1987). *Spin Glass Theory and Beyond.* World Scientific.

Panchenko, D. (2013). Spin glass models from the point of view of spin distributions. *Ann. Probab.*, **41**(3A), 1315–1361.

Parisi, G. (1980*a*). Magnetic properties of spin glasses in a new mean field theory. *J. Phys. A*, **13**(5), I101–I112.

Parisi, G. (1980*b*). A sequence of approximated solutions to the S–K model for spin glasses. *J. Phys. A*, **13**(4), L115–L121.

Parisi, G. (1998). On the probabilistic formulation of the replica approach to spin glasses. arXiv cond-mat/9801081.

Parisi, G. (2013). The overlap in glassy systems. arXiv:1310.5354.

Ruelle, D. (1987). A mathematical reformulation of Derrida's REM and GREM. *Commun. Math. Phys.*, **108**(2), 225–239.

Talagrand, M. (2006). The Parisi formula. *Ann. Math.*, **163**(1), 221–263.

# 4
# Cavity method: message-passing from a physics perspective

## Marc MÉZARD

Université Paris-Sud & CNRS, LPTMS,
UMR8626, Bât. 100,
91405 Orsay, France

*Lecture notes taken by*
Gino Del Ferraro, KTH Royal Institute of Technology, Sweden
Chuang Wang, Institute of Theoretical Physics, Chinese Academy of Sciences, China
Dani Martí, École Normale Supérieure & Inserm, France

*Statistical Physics, Optimization, Inference, and Message-Passing Algorithms.* First Edition.
F. Krzakala et al. © Oxford University Press 2016. Published in 2016 by Oxford University Press.

# Chapter Contents

## 4.1  Replica solution without replicas

### 4.1.1  The Sherrington–Kirkpatrick model

The Sherrington–Kirkpatrick (SK) model (Sherrington and Kirkpatrick, 1975) is a mean-field version of the Edwards–Anderson model (Edwards and Anderson, 1975) and is defined by a system of $N$ Ising spins $\sigma = (\sigma_1, \sigma_2, \dots, \sigma_N)$ taking values $\pm 1$ placed on the vertices of a lattice. In the SK mean-field description, the model is fully connected: every spin interacts with every other one, and the couplings $J_{ij}$ are chosen to be independent and identically distributed according to a Gaussian probability distribution such that the probability distribution of the whole set of couplings reads

$$P(J) = \prod_{i<j} P(J_{ij}) \propto \exp\left(-\frac{N}{2}\sum_{i<j} J_{ij}^2\right).$$

The $J_{ij}$ variables are assumed to be symmetric and without self-interaction terms, i.e., $J_{ij} = J_{ji}$ and $J_{ii} = 0$; we stress here that physically they play the role of quenched disorder among each pair of spins in the system. By quenched disorder, we mean that the couplings $J$ exert a stochastic external influence on the system, but they do not participate in the thermal equilibrium. The Hamiltonian of the system, given a particular configuration $\sigma$, is

$$H_J(\sigma) = -\sum_{i<j} J_{ij}\sigma_i\sigma_j - h\sum_i \sigma_i,$$

where $h$ is the homogeneous external magnetic field on each site $i$, and the couplings $J_{ij}$ are of the order of $1/\sqrt{N}$ to ensure correct thermodynamic behavior of the free energy. In this chapter, we will be interested in equilibrium properties of the system; the probability distribution at equilibrium is then given by the Boltzmann–Gibbs distribution

$$P_J(\sigma) = \frac{1}{Z}\exp\left(-\beta H_J(\sigma)\right),$$

where we have introduced the *partition function*

$$Z = \sum_{\{\sigma\}}\exp\left(-\beta H_J(\sigma)\right),$$

which includes a sum over all possible spin configurations, which we denote by $\{\sigma\}$.

The phase diagram of $h$ versus $T$ for this problem, relative to the stability of the replica-symmetric (RS) solution, was found by de Almeida and Thouless (1978) and is shown in Fig. 4.1. We observe that there are two phases: in the high-temperature regime, there is a paramagnetic phase and in the low-temperature regime, there is a spin glass phase where the RS solution is unstable. The transition line between these two phases is called the de Almeida–Thouless line. We can then define an order parameter that allows us to distinguish between these two phases. Let us consider two copies of the same system, which are two different spin configurations $\sigma$ and $\tau$ with associated probability $P_J(\sigma)$ and $P_J(\tau)$. Then, defining the overlap between these two

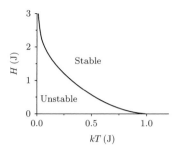

**Fig. 4.1** Phase diagram showing the limit of stability of the Sherrington–Kirkpatrick solution for the paramagnetic phase in the presence of a magnetic field $h$.

configurations as $q_{\sigma\tau} = (1/N)\sum_i \sigma_i\tau_i$, it is possible to compute the probability that this overlap is equal to $q$ as follows:

$$P_J(q) = \lim_{N\to\infty} \sum_{\sigma\tau} P_J(\sigma)P_J(\tau)\delta(q_{\sigma\tau} - q).$$

In principle, the probability of having a given overlap configuration depends on the sample, i.e., on the disorder, which means that we need to take the average over the disorder to remove this dependence, namely $P(q) = \mathbb{E}_J P_J(q)$, where $\mathbb{E}_J$ is the average over the disorder. The probability distribution $P(q)$ in the case of the replica symmetry breaking (RSB) ansatz is shown in Fig. 4.2.

### Pure states

The RSB solution of the SK model is characterized by the order parameter matrix $Q_{ab}$ (as shown by Parisi in Chapter 3). Since this system presents spontaneous symmetry breaking, if there is a particular solution for the matrix $\hat{Q}$ with the RSB, then any other matrix obtained via any permutation of the replica indices in $\hat{Q}$ will also be a solution. On the other hand, within the mean-field approximation, because the total free energy is proportional to the volume of the system, the energy barriers separating the corresponding ground states must be infinite in the thermodynamic limit. As a consequence, once the system is found to be in one of these states, it will never be able to jump into another one in a finite time. In this sense, the observable state is

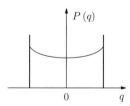

**Fig. 4.2** Distribution function $P(q)$ of the SK model with a full-RSB ansatz, i.e., a system with multi-valley structure.

not the Gibbs one, but one of these states. To distinguish them from the Gibbs states, they can be called *pure states* and the probability measure can be decomposed as the sum of the measures over the pure states. According to this definition, the average of any observable $\mathcal{O}$ can be taken as the sum of the averages in each of the pure states, as follows:

$$\langle \mathcal{O} \rangle = \sum_\alpha w_\alpha \mathcal{O}, \quad \text{with} \quad w_\alpha = \frac{e^{-\beta F_\alpha}}{\sum_\alpha e^{-\beta F_\alpha}},$$

where $F_\alpha$ is the free energy associated with the pure state $\alpha$. More formally, the pure states can be defined as those in which the correlation function of two spin variables belonging to the same pure states tends to zero in the thermodynamic limit, i.e., $\langle \sigma_i \sigma_j \rangle_\alpha - \langle \sigma_i \rangle_\alpha \langle \sigma_j \rangle_\alpha \to 0$ as $N \to \infty$.

### *The cavity method in the RS case*

We now investigate an alternative method with respect to the replica trick used so far to investigate the SK model from which it is possible to recover all the results at the RS level (Mézard *et al.*, 1986). This method can be also viewed as an analytic ansatz to derive and analyze the Thouless–Anderson–Palmer (TAP) equations (Thouless *et al.*, 1977). The basic idea is to go from an SK system $\Sigma_N$ composed of $N$ spins to a $\Sigma_{N+1}$ system that has $N + 1$ spins, assuming that the thermodynamic limit exists, or, in other words, assuming that in the thermodynamic limit there is no difference between observables computed in both systems (e.g., the free energy). We shall make some physical assumption about the organization of the configuration of $\Sigma_N$ inspired from the results obtained in the SK model with the RSB ansatz by using replicas (Parisi, 1979, 1980): the ultrametric organization of the states and the independent exponential distribution of their free energies. Once these properties are assumed to be valid in $\Sigma_N$, we will show that they are valid also for $\Sigma_{N+1}$, and so, for instance, $\overline{\langle \sigma_i \rangle_N^2} = \overline{\langle \sigma_i \rangle_{N+1}^2}$ as $N \to \infty$, where the bar denotes average over $J$. Let us assume that $\sigma_0$ is the spin added to the system of $N$ spins to create the $(N + 1)$-spin system. The probability distributions of disorder in each of them are respectively

$$P_N(J) = \prod_i P_N(J_{ij}) \propto \exp\left( -\frac{N}{2} \sum_{i<j} J_{ij}^2 \right),$$

$$P_N(J, J_0) = \prod_{j, i<j} P_{N+1}(J_{ij}, J_{0j}) = \prod_j P_{N+1}(J_{0j}) \prod_{i<j} P_{N+1}(J_{ij})$$

$$\propto \exp\left( -\frac{N+1}{2} \left[ \sum_{i<j} J_{ij}^2 + \sum_j J_{0j} \right] \right),$$

where $J_{0j}$ is the coupling between the added spin $\sigma_0$ and all the other spins in the $\Sigma_{N+1}$ system and we also note that there is a small change in scale of $J$ (from $N \to N + 1$ in the exponent). Then the probability distribution of a certain configuration of spins in the $N + 1$ system is given by

$$P_N(\sigma, \sigma_0) = \exp\left( -\beta H_N(\sigma) + \beta \sum_j J_{0j} \sigma_0 \sigma_j \right),$$

where $\sigma = (\sigma_1, \ldots, \sigma_N)$, and $h^c \equiv \sum_j J_{0j}\sigma_j$ is the local field felt by all the other spins in the $\Sigma_{N+1}$ system because of the presence of $\sigma_0$. The index $c$ indicates the "cavity," since $h^c$ is usually called the "cavity field." In the following, we want to compute the probability distribution of $h^c$. To do this, we will compute all the moments of the distribution. Let us start by defining the *nonlinear susceptibility* as

$$\chi = \frac{1}{N}\sum_{i<j}\big(\langle\sigma_i\sigma_j\rangle - \langle\sigma_i\rangle\langle\sigma_j\rangle\big)^2 \tag{4.1}$$

and computing the expectation and variance of the cavity field:

$$\langle h^c\rangle_N = \sum_i J_{0i}\langle\sigma_i\rangle_N \xrightarrow{N\to\infty} h, \tag{4.2}$$

$$\langle (h^c)^2\rangle_N - \langle h^c\rangle_N^2 = \sum_{i,j} J_{0i}J_{0j}\big(\langle\sigma_i\sigma_j\rangle_N - \langle\sigma_i\rangle_N\langle\sigma_j\rangle_N\big)^2 \tag{4.3}$$

The assumption of the cavity method at the RS level is that the susceptibility (4.1) has to be finite. Because $J_{ij}$ is of the order of $1/\sqrt{N}$ and because the sum over $i,j$ involves $N^2$ terms, $\chi$ will be finite as long as the connected correlation of $\sigma_i$ and $\sigma_j$, namely, $\langle\sigma_i\sigma_j\rangle_N - \langle\sigma_i\rangle_N\langle\sigma_j\rangle_N$, is of order $1/\sqrt{N}$. Then, if we take the sum in (4.3), it will be dominated by the term $i=j$:

$$\text{for } i=j: \quad \langle (h^c)^2\rangle_N - \langle h^c\rangle_N^2 = \sum_i J_{0i}^2(1 - \langle\sigma_i\rangle_N^2)$$

$$= 1 - \frac{1}{N}\sum_i\langle\sigma_i\rangle^2 = 1 - \overline{\langle\sigma_i\rangle^2} = 1 - q,$$

where in the second equality we have used $J_{ij} \sim 1/\sqrt{N}$, while in the third we have substituted the sum over all sites with the average over the disorder at a single site, because they are equivalent. Finally, we have used the definition of the Edwards–Anderson order parameter $\overline{\langle\sigma_i\rangle^2} = q$. Using similar reasoning, one can compute the fourth moment

$$\Big\langle (h^c - \langle h^c\rangle)^4\Big\rangle = \sum_{i,j,k,l} J_{0i}J_{0j}J_{0k}J_{0l}\langle(\sigma_i - \langle\sigma_i\rangle)(\sigma_j - \langle\sigma_j\rangle)(\sigma_k - \langle\sigma_k\rangle)(\sigma_l - \langle\sigma_l\rangle)\rangle$$

$$= 3(1-q)^2.$$

Iterating this computation and applying similar considerations, we claim that all odd moments bigger than the first are zero, while all even moments are given by the expression

$$\Big\langle (h^c)^{2p}\Big\rangle = (2p-1)!!(1-q)^p. \tag{4.4}$$

These are the moments of a Gaussian distribution with variance $1-q$, and therefore the probability distribution of the cavity field in the $\Sigma_N$ system is given by

$$P_N(h^c) \cong \exp\left(-\frac{(h^c - h)^2}{2(1-q)}\right), \tag{4.5}$$

where $\cong$ means "equal up to a normalization constant," and $h = \sum_i J_{0i}\langle\sigma_i\rangle_N = \langle h^c\rangle$ is the average value of the cavity field. We stress that the only assumption made so far in computing these moments has been that the connected correlation function is of order $1/\sqrt{N}$. Now we can consider the probability distribution of $h^c$ in the $\Sigma_{N+1}$ system, which is built by adding the spin $\sigma_0$ to the system $\Sigma_N$:

$$P_{N+1}(h^c, \sigma_0) \cong \exp\left(-\frac{(h^c - h)^2}{2(1-q)} + \beta\sigma_0 h^c\right). \tag{4.6}$$

With this joint distribution, it is finally possible to compute many things, such as the expectation value of the spin $\sigma_0$ in the $\Sigma_{N+1}$ system:

$$\langle\sigma_0\rangle_{N+1} = \tanh(\beta h) = \tanh\left(\beta\sum_{i=1}^{N} J_{0i}\langle\sigma_i\rangle_N\right), \tag{4.7}$$

where the average is taken with respect to the probability density (4.6), integrating over the cavity field $h^c$. This is one of the first results where there is an evident connection, a mathematical relation, between the system $\Sigma_{N+1}$ and the system $\Sigma_N$. Let us then compute the order parameter $q$ from its definition, by using the probability density in (4.6):

$$q = \overline{\langle\sigma_0\rangle_{N+1}^2} = \overline{\tanh(\beta h)^2} \tag{4.8}$$

where the second equality comes from (4.7). To compute this average, we need to derive the probability distribution of the cavity field, $P(h)$. Let us compute its moments. The averaged field reads

$$\overline{h} = \sum_i \overline{J_{0i}\langle\sigma_i\rangle_N} = 0, \tag{4.9}$$

which is equal to zero because the average of the couplings $J$ is zero. The average squared field reads

$$\overline{h^2} = \sum_{i,j} \overline{J_{0i}J_{0j}\langle\sigma_i\rangle_N\langle\sigma_j\rangle_N} = \frac{1}{N}\sum_i \langle\sigma_i\rangle_N^2 = q. \tag{4.10}$$

By computing all the higher-order moments, it is possible to show that all the odd moments are zero, while all the even ones obey a similar relation to that seen in (4.4). We can thus conclude that $h$ is Gaussian-distributed.

Therefore, we get

$$q = \int \frac{dh}{\sqrt{2\pi q}} \exp\left(-\frac{h^2}{2q}\right) \tanh(\beta h)^2. \tag{4.11}$$

This is the self-consistent equation for the $q$ order parameter, originally found by Sherrington and Kirkpatrick (1975). It tells us that there is a phase transition at temperature $T = 1$, but this solution is unfortunately wrong. This can be shown by looking at the thermodynamics; in particular, it is possible to show that the entropy of the system, computed with this method and under its assumptions, is negative, which is unphysical. This inconsistency arises because the approach followed is equivalent to the RS assumption when one uses replicas, which is not a correct ansatz to solve the model.

We now go back to the initial assumption that the susceptibility is finite in the thermodynamic limit. To check the validity of this assumption, we will compute $\chi$ in a system $\Sigma_{N+2}$ composed of $N+2$ spins, and we will check the region where the assumption is valid, or, more precisely, the region where $\chi$ remains finite. Since we are dealing with a $\Sigma_{N+2}$ system, we will have to deal with two cavity fields. The probability measure in this system reads

$$P_{N+2}(\sigma_0, \sigma_{0'}, \sigma) \cong \exp\left(-\beta H_N(\sigma) + \beta h^c \sigma_0 + \beta h^{c'} \sigma_{0'} + \beta J_{00'} \sigma_0 \sigma_{0'}\right),$$

where $h^c = \sum_i J_{0i} \sigma_i$ and $h^{c'} = \sum_i J_{0'i} \sigma_i$ are the cavity fields acting on $\sigma_0$ and $\sigma_{0'}$, respectively. The term $J_{00'} \sigma_0 \sigma_{0'}$ corresponds to the interaction between the two spins where the cavity has been made. First of all, we start by computing the part of the susceptibility containing the correlation between the spins $\sigma_0$ and $\sigma_{0'}$: $\chi_{\text{nl}} = N(\langle \sigma_0 \sigma_{0'} \rangle - \langle \sigma_0 \rangle \langle \sigma_{0'} \rangle)^2$, where the label "nl" means nonlinear. To compute this correlation, we need to keep in mind that the terms inside the bracket are of order $1/\sqrt{N}$, and then keep all the terms of this order. Before computing the averages using the cavity method, we need to derive the probability density $P(h^c, h^{c'})$. To this end, we need to compute the second-order moment, i.e., the 2-point correlator $\langle (h^c - \langle h^c \rangle)(h^{c'} - \langle h^{c'} \rangle) \rangle = \sum_{i,j} J_{0i} J_{0j} (\langle \sigma_i \sigma_j \rangle - \langle \sigma_i \rangle \langle \sigma_j \rangle)^2$, which is of order $1/\sqrt{N}$, but this time we keep the terms of this order because we are interested in correlations that are exactly of order $1/\sqrt{N}$:

$$P_N(h^c, h^{c'}) \cong \exp\left(-\frac{(h^c - h)^2}{2(1-q)} - \frac{(h^c - h')^2}{2(1-q)} + \epsilon(h^c - h)(h^{c'} - h')\right), \qquad (4.12)$$

where $\epsilon(h^c - h)(h^{c'} - h')$ represents the correlation term between the two fields and $\epsilon$ is a small parameter, of order $1/\sqrt{N}$. By using (4.12), it is possible to derive the following marginal joint probability distribution, which depends on the cavity fields and explicitly on the two cavity spins:

$$P_N(h^c, h^{c'}, \sigma_0, \sigma_{0'}) \propto P_N(h^c, h^{c'}) \exp\left(\beta h^c \sigma_0 + \beta h^{c'} \sigma_{0'} + \beta J_{00'} \sigma_0 \sigma_{0'}\right).$$

With this marginal, it is finally possible to compute the susceptibility introduced above, namely, $\chi_{\text{nl}}$. The computation follows the same lines as above, and we give here only the final result, which is

$$\chi_{\text{nl}} = \frac{\beta^2 A^2}{1 - \beta^2 A}, \qquad \text{with} \qquad A = \int \frac{dh}{\sqrt{2\pi q}} \exp\left(-\frac{h^2}{2q}\right)(1 - \tanh(\beta h)^2)^2,$$

and shows how the nonlinear susceptibility is related to the $q$ order parameter. We observe that $\chi_{\text{nl}}$ diverges as soon as $\beta^2 A = 1$, and therefore we can make the system eventually reach this point by increasing $\beta$ and, because of this divergence, our initial assumption for the susceptibility is wrong around this point. The assumption of a finite $\chi$ is then valid only for high temperatures or, rather, as long as $\beta^2 A < 1$. This is precisely the location of the AT line. This result is thus consistent with what we mentioned above: the cavity method shown so far is equivalent to the RS approach, because the RS solution also is only valid for high temperatures. In addition, we can also give a physical meaning to the RS ansatz: it corresponds to assuming that the 2-point correlation function is small (leading to a finite $\chi$).

## *Derivation of the TAP equation*

Now, let us go back to the probability measure for the cavity field in the system $\Sigma_N$:

$$P_{N+1}(h^c, \sigma_0) \cong \exp\left(-\frac{(h^c - h)^2}{2(1-q)} + \beta\sigma_0 h^c\right).$$

With the previous measure, we can compute the expectation for the cavity field in the $\Sigma_{N+1}$ system,

$$\langle h^c \rangle_{N+1} = \sum_i J_{0i}\langle\sigma_i\rangle_{N+1} = h + \beta(1-q)\langle\sigma_0\rangle_{N+1}, \tag{4.13}$$

and also the expectation value of $\sigma_0$ in the same system,

$$\langle\sigma_0\rangle_{N+1} = \tanh\left(\beta\sum_i J_{0i}\langle\sigma_i\rangle_N\right). \tag{4.14}$$

Multiplying (4.13) by $\beta$, we get $\beta h = \beta\sum_i J_{0i}\langle\sigma_i\rangle_{N+1} - \beta^2(1-q)\langle\sigma_0\rangle_{N+1}$, which, after applying tanh($\cdot$) to both sides of the equation and making use of (4.14), gives rise to the TAP equation (Thouless *et al.*, 1977),

$$\langle\sigma_0\rangle = \tanh\left(\beta\sum_i J_{0i}\langle\sigma_i\rangle - \beta^2(1-q)\langle\sigma_0\rangle\right),$$

where we have generalized the result by omitting the label $N+1$ on the averaged terms. The first term in the argument of tanh($\cdot$) is the effect of all the spins except $\sigma_0$ on $\sigma_0$, while the second term is a correction called Onsager's reaction term. Physically speaking, the reaction term arises because the presence of $\sigma_0$, when we consider the whole system without any preformed cavity, it takes into account the reaction of the neighbors of $\sigma_0$ due to the correlation created by the presence $\sigma_0$. The TAP equation as derived above is correct as long as the connected correlation between spins is small, i.e., of the order of $1/\sqrt{N}$, which is the only assumption made to derive the equation. From the replica point of view, the assumption of small connected correlations is equivalent to an RS ansatz, and we then can conclude that the TAP equation is correct only in the high-temperature regime.

## 4.2   Cavity method for diluted graph models

### 4.2.1   Replica symmetry breaking and pure states

The cavity method applied to the SK model, within the RS assumptions, assumes that the two-point correlation function between spins is small, i.e., $c_{ij} = \langle\sigma_i\sigma_j\rangle - \langle\sigma_i\rangle\langle\sigma_j\rangle$ is of the order of $1/\sqrt{N}$. When the system falls into the spin glass phase, the configuration space decomposes into many *pure states*. The probability of a given configuration $\sigma = (\sigma_1, \ldots, \sigma_N)$ can then be decomposed as a sum over pure states,

$$P(\sigma) = \sum_\alpha w_\alpha \mu_\alpha(\sigma),$$

where $\mu_\alpha(\cdot)$ is the measure within the pure state, which determines how configurations are weighted in one particular pure state, and $w_\alpha$ is the weight of the pure state $\alpha$, given by

$$w_\alpha = \frac{e^{-\beta N f_\alpha}}{\sum_{\alpha'} e^{-\beta N f_{\alpha'}}}.$$

where $f_\alpha$ called *the free energy density* of the pure state $\alpha$. (Some authors prefer to use the free entropy density, defined as $\phi = \log(w)/N = -\beta f$.) Physical quantities depend on the pure state $\alpha$ that the system is in. For instance, the single-spin magnetization at the pure state $\alpha$ is

$$\langle \sigma_i \rangle_\alpha = \sum_{\sigma \in \alpha} \sigma_i \mu_\alpha(\sigma).$$

In greater generality, the average value of any observable $\mathcal{O}$ within the pure state $\alpha$ is given by $\langle \mathcal{O} \rangle_\alpha = \sum_{\sigma \in \alpha} \mathcal{O}(\sigma) \mu_\alpha(\sigma)$. The average magnetization over all the pure states is simply the weighted sum

$$\langle \sigma_i \rangle = \sum_\alpha w_\alpha \langle \sigma_i \rangle_\alpha.$$

The decomposition in pure states is justified because the escape time from a pure state grows exponentially long with the system size $N$.

In the replica method described in Chapter 3, we saw that pure states are grouped hierarchically. At the 1-step replica symmetry breaking (1RSB) level, all the states are equally separated from each other, i.e., the overlap between two replica systems in any two different pure states is the same. At the 2RSB level, some pure states are closer than others, forming a larger cluster structure, but the distance between any two larger clusters of pure states is the same. This hierarchical structure is also present in the cavity method. Instead, one assumes that within a pure state $\alpha$ the correlation $c_{ij}$ is weak at the 1RSB level, while the overall correlation may be strong.

If we know one pure state, we can use a set of external auxiliary fields $\{B_i^\alpha\}$ to quench the system into a particular pure state $\alpha$. In that case, the measure within the pure state $\alpha$ is obtained as the limit, when $B_i^{(\alpha)}$ goes to 0, of

$$P_{B_\alpha}(\sigma) \cong \exp\left[\beta \sum J_{ij}\sigma_i\sigma_j + \sum_i B_i^{(\alpha)}\sigma_i\right].$$

The cavity method at RS level, as showed in Section 4.1, can be applied within a given pure state. The self-consistency equation for the magnetization is

$$\langle \sigma_i \rangle_\alpha = \tanh\left[\beta \sum_j J_{ij} \langle \sigma_j \rangle_\alpha - \beta(1-q)\langle \sigma_i \rangle_\alpha\right].$$

One can write all the above equations for each pure state and the problem will be solved at 1RSB level. However, we know nothing about the details of pure states except that they exist. Fortunately, this fact, together with the weak correlation assumption within a pure state, is enough to write a self-consistency equation for the 1RSB cavity method.

Solving the SK model at 1RSB and 2RSB levels can be done, although it is rather involved (Mézard *et al.*, 1986). The intricate part is that one needs to deal with the reshuffling of the pure states' weights after adding one node to the $N$-system:

$$\left\{ w_\alpha^{(N)} \right\} \longrightarrow \left\{ w_\alpha^{(N+1)} \right\}$$

Solving the self-consistency equation of the cavity method, finally, gives the same equation as obtained from the saddle point equation in the replica method.

In this section, another type of system is used to illustrate the cavity method, namely, the dilute graph model, which has a wide application in random constraint satisfaction problems. The 1RSB of such a system is stable, so there is no need for a higher-level symmetry breaking.

### 4.2.2   Counting the pure states at 1RSB level

Let us denote by $\Omega(f)$ the number of pure states with weight $w = \mathrm{e}^{-\beta N f}$. In the large-$N$ limit, we are interested in its leading exponential order, which we assume to be of the form

$$\Omega(f) = \mathrm{e}^{N\Sigma(f)}, \tag{4.15}$$

where $\Sigma(f)$ is the *complexity*, or *configurational entropy*.

Define the *grand partition function* with a re-weighting parameter $m$ of pure states,

$$\mathcal{Z}(m, \beta) = \sum_\alpha \exp\left(-\beta m N f_\alpha\right) = \int \exp\left(N[\Sigma(f) - \beta m f]\right) \mathrm{d}f = \mathrm{e}^{N\Phi(m,\beta)},$$

where $\Phi(m, \beta)$ is called the *grand free entropy*. As $N \to \infty$, this integral is dominated by the largest exponential term:

$$f^* = \arg\max_f \left[\Sigma(f) - \beta m f\right],$$

$$\Phi(m, \beta) = \Sigma(f^*) - \beta m f^*. \tag{4.16}$$

$\Phi(m, \beta)$ is the Legendre transform of $\Sigma(f)$. For given $m$ and $\beta$, $\Phi(m, \beta)$ can be derived with the 1RSB cavity method. It is assumed that $\Sigma(f)$ is a concave function. The complexity $\Sigma(f)$ can then be computed with an inverse Legendre transform. We can also compute the average free energy density over all the pure states, which is equal to the dominating value $f^*$. The complexity $\Sigma(f)$ can then be obtained from (4.16).

From a physical standpoint, we should require the complexity to be non-negative, because otherwise there would be an exponentially small number of pure states with free energy density $f$. In the large-$N$ limit, that would mean no such pure states at all. In any case, the grand partition function is dominated by the existing pure state with largest weight $w$, i.e., with the smallest free energy density. The phenomenon by which the measure is dominated by sub-exponentially many states is called *condensation*.

The original system $Z(\beta)$ is related to $\mathcal{Z}(m,\beta)$ at $m = 1$ if $\Sigma(f^*) \geq 0$, where $f^*$ satisfies

$$\left.\frac{\mathrm{d}\Sigma}{\mathrm{d}f}\right|_{f^*} = \beta m \,.$$

If $\Sigma(f^*) < 0$, the original system should correspond to the largest $m$ such that $\Sigma(f^*) = 0$. We are left with two 1RSB phases. When $\Sigma(f^*) > 0$, we are in the so-called dynamic 1RSB (*cluster phase*), and the system is dominated by exponentially many pure states. When $\Sigma(f^*) = 0$, we are in the static 1RSB (*condensed phase*), and the system is dominated by sub-exponentially many pure states.

Computing the complexity is analogous to computing the entropy of a new system in which each microstate (each configuration) is a pure state $\alpha$, and where the free energy of the microstate is $f_\alpha$. The computation of the complexity versus the free energy density of pure states by a Legendre transform is the topic of large-deviation theory. A general review of this subject has been given by Touchette (2009).

### 4.2.3 Randomly diluted graphical models

The factor graph $\mathcal{F}(V, F, E)$ (Fig. 4.3) is a bipartite graph with two types of nodes: variable nodes and factor nodes. Each variable node is associated with a random variable $x_i$, $i \in V$, and each factor node is associated with a factor, a non-negative function $\psi_a(x_{\partial a})$, where $a = 1, \ldots, F$ and $\partial a$ represents the set of neighbor variable nodes of the factor node $a$.

The joint probability of $x = (x_1, \ldots, x_N)$ is expressed as

$$p(x) = \frac{1}{Z} \prod_{a \in F} \psi_a(x_{\partial a}), \tag{4.17}$$

where $Z$ is the partition function. In such a context, we may want to answer different questions. For example, we may want to compute the marginal probability $p_i(x_i) = \sum_{x_{V \setminus \{i\}}} p(x)$. Another example would be determining the partition function $Z$ or, rather, its first leading exponential order, $\phi = (1/N) \log Z$. We might also want to find a particular configuration of the variables such that $p(x) \neq 0$, which is the situation encountered in constraint satisfaction problems.

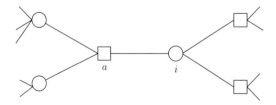

**Fig. 4.3** Factor graph: a circle node represents a variable node; a square represents a factor node.

## *Examples*

1. Ising spin glass: $x_i \in \{+1, -1\}$, $a = (i, j)$, where $(i, j)$ is an edge of the lattice. Here $\psi_a = e^{\beta J_{ij} x_i x_j}$.
2. Coloring problem: Given a set of $q$ colors and a graph $\mathcal{G}(V, F)$, label each node with a color $x_i \in \{1, 2, \dots, q\}$, such that no neighboring nodes have the same color. Each constraint is defined on the edges and has the form $\psi_{(ij)} = 1 - \delta_{x_i, x_j}$, or the soft constraint version $\psi_{(ij)} = e^{-\beta \delta_{x_i, x_j}}$. The inverse temperature $\beta$ alters the tolerance to the presence of neighbor nodes sharing the same color.
3. $K$-SAT problem: Given $N$ Boolean variables $x_i \in \{0, 1\}$, with $i = 1, 2, \dots, N$, and $M$ $K$-clauses in conjunctive normal form (a $K$-clause is a logical expression involving $K$ variables, or their negation, which are connected with logical ORs), find an assignment of Boolean variables $\{x_i\}$ that satisfies all the $M$ clauses. In the corresponding graphical model, the factor is an indicator function, which is 1 when the clause is satisfied and 0 otherwise. In other words, $\psi_a(x_{\partial a}) = \mathbb{I}[\text{clause } a \text{ is satisfied}]$. We will study $K$-SAT problems in more detail in Section 4.3.
4. $K$-XORSAT problem: A variant of the former problem, where the $K$ variables are now connected with logical XOR. These are fundamental elements in error correcting codes (see the low-density parity check codes in Chapter 6).

## *The structure of a factor graph*

1. Line or cylinder: This case can be solved exactly by the transfer matrix method.
2. Tree: BP or cavity method is exact on tree.
3. Random hypergraph: An extension of the random Erdős–Rényi graph to a factor graph. There are $N$ variable nodes and $M$ factor nodes. The factor node has a fixed degree $K$, which is randomly chosen from $\binom{N}{K}$ $K$-tuples. The degree of variable node follows the Poisson distribution $P_c(d) = c^d e^{-c}/d!$. The length of a typical loop is of the order of $\log N$.

### 4.2.4   Cavity method at the RS level, for general graphical models

#### *Calculating the marginal distribution*

We consider a random hypergraph with $N$ variables and $\alpha N$ factors, where $\alpha$ is the *constraint density* in $K$-SAT. The cavity method is illustrated in Fig. 4.4. The system with $N + 1$ variable nodes is generated by adding a new variable $x_0$ and $d$ factors, where $d$ is a random integer drawn from a Poisson distribution with mean $c = \alpha K$, the mean degree of a variable node. Each new factor is connected to $x_0$, and $K - 1$ variables randomly chosen from the $N$-variable system. Note that the constraint density $\alpha$ of the $(N+1)$-system is slightly changed. While it does not affect the marginal distribution, it should be taken into account when computing the free energy density.

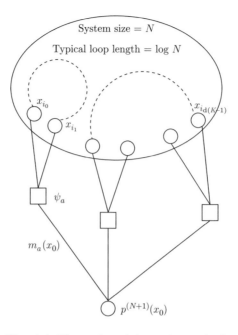

**Fig. 4.4** Illustration of the cavity method.

The assumption of the cavity method states that the joint probability of a constant number of variables chosen randomly is factorized, because the typical distance between any two variable nodes is of the order of $\log N$:

$$P(x_{i_1}, x_{i_2}, \ldots, x_{i_{d(K-1)}}) \approx \prod_{j=1}^{d(K-1)} P(x_{i_j}). \tag{4.18}$$

The joint marginal probability of $x_0$ and the $d(K-1)$ variables connected to the new $d$ factors is

$$P^{(N+1)}(x_0, x_{i_1}, x_{i_2}, \ldots, x_{i_{d(K-1)}})$$

$$\cong \prod_{a=1}^{d} \psi_a(x_0, x_{i_{a(K-1)+1}}, x_{i_{a(K-1)+2}}, \ldots) P^{(N)}(x_{i_1}, x_{i_2}, \ldots, x_{i_{d(K-1)}})$$

$$\approx \prod_{a=1}^{d} \left[ \psi_a(x_0, x_{i_{a(K-1)+1}}, x_{i_{a(K-1)+2}}, \ldots) \prod_{k=a(K-1)+1}^{(a+1)(K-1)} P_{i_k}^{(N)}(x_{i_k}) \right].$$

The marginal probability $P^{(N+1)}(x_0)$ of the newly added variable is

$$P^{(N+1)}(x_0) \cong \prod_{a=1}^{d} \hat{m}_a(x_0),$$

where

$$\hat{m}_a(x_0)$$

$$\cong \sum_{x_{i_{a(K-1)+1}},\ldots,x_{i_{(a+1)(K-1)}}} \psi_a(x_0, x_{i_{a(K-1)+1}}, \ldots, x_{i_{(a+1)(K-1)}}) \prod_{k=a(K-1)+1}^{(a+1)(K-1)} P_{i_k}^{(N)}(x_{i_k}).$$

The system with $N$ variable nodes can be considered as a system with $N + 1$ variable nodes in which one node $x_i$ is absent. The *cavity probability* $m_{i \to a}(x_i)$ denotes the marginal probability of $x_i$, when the factor node $a$ is absent. $P_{i_k}^{(N)}(x_{i_k})$ can be considered as the cavity probability in the system with $N+1$ variables when the node $x_0$ and its neighboring factor nodes are absent. The self-consistent equations for the cavity probabilities are obtained by considering that the $x_0$ node is also a cavity node when one of its neighbor variables and its neighbor factors is absent:

$$\hat{m}_{a \to i}(x_i) \cong \sum_{x_{\partial a \setminus \{i\}}} \psi_a(x_{\partial a}) \prod_{j \in \partial a \setminus \{i\}} m_{j \to a}(x_j), \tag{4.19}$$

$$m_{i \to b}(x_i) \cong \prod_{a \in \partial i \setminus \{b\}} \hat{m}_{a \to i}(x_i). \tag{4.20}$$

These equations are the same as the BP equations, but here messages are cavity probabilities. The marginal probability of a node $x_i$ is then expressed as the cavity probability

$$m_i(x_i) \cong \prod_{a \in \partial i} \hat{m}_{a \to i}(x_i).$$

### The Bethe free energy

The Bethe free energy can be derived in the cavity method by considering the free energy shift $f_{i+\partial i}$ on adding a variable $i$ and its neighbor factors $a \in \partial i$. One has to be careful, though, because the constraint density $\alpha$ will change slightly. This effect is eliminated by subtracting $K - 1$ times the free energy shift $f_a$ when adding a single factor $a$. For a given instance, the Bethe free energy is

$$Nf = \sum_i f_{i+\partial i} - (K - 1) \sum_a f_a. \tag{4.21}$$

The RS cavity independence assumption postulates that on removing a node $i$ and its neighbor factor $a \in \partial i$, the partition function of the cavity system with fixed cavity variable $x_j$ $j \in \partial a \setminus i$, $a \in \partial i$ can be factorized as

$$Z_{\setminus i, \partial i}(x_{j:j \in \partial a \setminus i, a \in \partial i}) \approx \prod_{a \in \partial i} \prod_{j \in \partial a \setminus i} Z_{j \to a}(x_j).$$

Here $Z_{j \to a}(x_j)$ is the partition function of the subsystem connected to $x_j$ with fixed value $x_j$ when the factor $a$ is absent.

The free energy shift $f_{i+\partial i}$ due to adding a node $i$ and its neighbor factors is

$$
\begin{aligned}
f_{i+\partial i} &= -\frac{1}{\beta} \log \frac{Z}{Z_{\setminus i, \partial i}} \\
&= -\frac{1}{\beta} \log \frac{\sum_{x_i, x_{j:j \in \partial a \setminus i, a \in \partial i}} \prod_{a \in \partial i} \left[ \psi_a(x_{\partial a}) \prod_{j \in \partial a \setminus i} Z_{j \to a}(x_j) \right]}{\sum_{x_{j:j \in \partial a \setminus i, a \in \partial i}} \prod_{a \in \partial i} \prod_{j \in \partial a \setminus i} Z_{j \to a}(x_j)} \\
&= -\frac{1}{\beta} \log \sum_{x_i} \left\{ \prod_{a \in \partial i} \left[ \sum_{x_{\partial a \setminus i}} \psi_a(x_{\partial a}) \prod_{j \in \partial a \setminus i} \frac{Z_{j \to a}(x_j)}{\sum_{x_j'} Z_{j \to a}(x_j')} \right] \right\} \\
&= -\frac{1}{\beta} \log \sum_{x_i} \left\{ \prod_{a \in \partial i} \left[ \sum_{x_{\partial a \setminus i}} \psi_a(x_{\partial a}) \prod_{j \in \partial a \setminus i} m_{j \to a}(x_j) \right] \right\}.
\end{aligned}
\tag{4.22}
$$

Similarly, the free energy shift $f_a$ caused by the addition of a factor node $a$ is

$$
\begin{aligned}
f_a &= -\frac{1}{\beta} \log \frac{Z}{Z_{\setminus a}} \\
&= -\frac{1}{\beta} \log \frac{\sum_{x_{\partial a}} \psi_a(x_{\partial a}) \prod_{j \in \partial a} Z_{j \to a}(x_j)}{\sum_{x_{\partial a}} \prod_{j \in \partial a} Z_{j \to a}(x_j)} \\
&= -\frac{1}{\beta} \log \sum_{x_{\partial a}} \psi_a(x_{\partial a}) \prod_{j \in \partial a} \frac{Z_{j \to a}(x_j)}{\sum_{x_j'} Z_{j \to a}(x_j')} \\
&= -\frac{1}{\beta} \log \sum_{x_{\partial a}} \psi_a(x_{\partial a}) \prod_{j \in \partial a} m_{j \to a}(x_j).
\end{aligned}
\tag{4.23}
$$

Now, the Bethe free energy can be computed with (4.21). The expression for the Bethe free energy has several variants, for example

$$
Nf = \sum_i f_i + \sum_a f_a - \sum_{(ia)} f_{ia},
\tag{4.24}
$$

where

$$
f_i = -\frac{1}{\beta} \log \sum_{x_i} \prod_{a \in \partial i} \hat{m}_{a \to i}(x_i),
\tag{4.25}
$$

$$
f_{ia} = -\frac{1}{\beta} \log \sum_{x_i} m_{i \to b}(x_i) \hat{m}_{i \to b}(x_i).
\tag{4.26}
$$

One can prove that the two Bethe free energy expressions (4.21) and (4.24) are equivalent when the cavity probability satisfies (4.19)–(4.20).

A comprehensive derivation of Bethe free energy by the cavity method can be found in Mézard and Parisi (2003), which also shows the 1RSB cavity method in a special simple case (in which the temperature $T = 1/\beta$ and the Parisi parameter $m$ are both 0). A review has been given by Mézard and Montanari (2009).

### Average over the disorder and the graph ensemble

To calculate the free energy average over the disorder and the graph ensemble, one should solve self-consistent integral equations for the distribution of the cavity probabilities $P[m]$ and $\hat{P}[\hat{m}]$:

$$P[m] = \sum_{d=1}^{\infty} P_c(d) \int \prod_{a=1}^{d} \left[ \mathrm{d}\hat{m}_a \hat{P}(\hat{m}_a) \right] \delta \left[ m - \mathbf{m}_{i \to b}[\{m_a\}] \right],$$

$$\hat{P}[\hat{m}] = \int \mathrm{d}\psi_a P_J(\psi_a) \int \prod_{i=1}^{K-1} \left[ \mathrm{d}m_i P(m_i) \right] \delta \left[ \hat{m} - \hat{\mathbf{m}}_{a \to i}[\psi_a, \{\hat{m}_i\}] \right],$$

where $\mathbf{m}_{i \to b}$ and $\hat{\mathbf{m}}_{a \to i}$ are the functionals of the BP equations (4.19) and (4.20), respectively, and $P_c(d)$ is the degree distribution of a cavity variable node, which is still a Poisson distribution with $c = \alpha K$ for a random hypergraph. The function $P_J(\psi_a)$ is the distribution of the disorder, which depends on the concrete model. For instance, in the random $K$-SAT problem, $\psi_a$ is parametrized as $J_a^i$ randomly chosen from $\{+1, -1\}$ with equal probability. The average free energy shift on adding a factor is given by

$$\bar{f}_a = \int \mathrm{d}\psi_a \, P_J(\psi_a) \int \prod_{i=1}^{K} \left[ \mathrm{d}m_i P(m_i) \right] f_i(\psi_a, \{\hat{m}_i\}),$$

where $f_i(\psi_a, \{\hat{m}_i\})$ is defined by (4.25). Other average free energy shifts could be written down in a similar way. The average free energy density over the disorder and the graph ensemble is

$$\bar{f} = \bar{f}_{i+\partial i} - \alpha(K-1)\bar{f}_a. \tag{4.27}$$

In general, it is hard or impossible to get an analytical solution of the above equation, but one can use numerical simulations to solve it. The appropriate algorithm is called *population dynamics*, or density evolution.

> *Initialization:* Set an array $P$ to store the messages $\{m_{i \to a}\}$. (Note that if $x_i$ is an Ising variable, then $m_{i \to a}(x_i)$ can be parametrized by a single real number.)
>
> 1. An integer $d$ is randomly assigned following the Poisson distribution $d \sim P_c(d)$.
> 2. Pick $(K-1)d$ messages randomly from the array $P$.
> 3. Generate $d$ $\psi_a$'s following $P_J(\psi_a)$.
> 4. Compute a new message $\hat{m}'$ with (4.19)–(4.20), and compute $f_{i+\partial i}$ with (4.22).
> 5. Choose a message randomly in $P$ and replace it by the new one $\hat{m}'$.
> 6. Pick $K$ messages randomly from the array $P$, and generate a factor $\psi_a$ following $P_J(\psi_a)$. Compute $f_a$ with (4.23).
> 7. Repeat 1–5 until a stable distribution $P(\hat{m})$ is obtained. Then, keep repeating 1–6 to obtain the mean $\bar{f}_{i+\partial a}, \bar{f}_a$, and calculate $\bar{f}$ with (4.27).

For more discussion on BP free energy in average cases, see Mézard and Montanari (2009, pp. 322–325).

### 4.2.5    Cavity method at 1RSB level

Something may go wrong for the Bethe independence hypothesis (4.18), and there are two possible reasons for this. The first possibility is that (4.18) holds only when the system is infinitely large, $\log N \to \infty$. For a finite system, (4.18) is only an approximation. The other possible reason is that when the constraint density $\alpha$ is high or the temperature is low, the Bethe hypothesis may fail even for an infinitely large system. In this latter case, the whole probability distribution no longer factorizes, $P(x_1, x_2, \ldots, x_n) \neq p_1(x_1)p_2(x_2) \cdots p_n(x_n)$, and so we need to make a more accurate assumption. As proposed by Mézard and Parisi (2001), we invoke the 1RSB approximation, by which the probability distribution factorizes within each pure state $\alpha$, but not globally. More specifically, because of the presence of pure states, the whole Gibbs measure splits into many states $\alpha$, and within the measure $\mu_\alpha(\cdot)$ of a pure state, the independence hypothesis still holds:

$$\mu_\alpha(x_1, x_2, \ldots, x_n) \approx \mu_\alpha(x_1)\mu_\alpha(x_2) \ldots \mu_\alpha(x_n). \tag{4.28}$$

Furthermore, it is assumed that the numbers of pure states and fixed points of BP solutions are the same up to the first exponential leading order. The leading exponential order of the number of pure states with free energy density $f$ is $\Sigma(f)$, as defined in (4.15). The grand partition function is expressed as

$$\mathcal{Z}(m, \beta) = \sum_\alpha e^{-\beta m N f_\alpha}$$

$$= \sum_{\{m_{i \to a}\text{'s are fixed point}\}} e^{-\beta m N f_\alpha[\{m_{i \to a}\}]}$$

$$= \int_{\{m_{i \to a}, \hat{m}_{a \to i}\}} d\{m_{i \to a}\} d\{\hat{m}_{a \to i}\} \prod_{(i,a)} \delta\left[m_{i \to a} - \mathbf{m}_{i \to a}[\{\hat{m}_{\text{input msgs}}\}]\right]$$

$$\times \prod_{(i,a)} \delta\left[\hat{\mathbf{m}}_{a \to i} - \hat{p}_{a \to i}[m_{\text{input msgs}}]\right]$$

$$\times \prod_i e^{-\beta m f_i[\cdot]} \prod_a e^{-\beta m f_a[\cdot]} \prod_{(i,a)} e^{\beta m f_{(ia)}[\cdot]},$$

where $\hat{\mathbf{m}}_{i \to a}[\cdot]$ , $\mathbf{m}_{a \to i}[\cdot]$ are the functionals defined by (4.19)–(4.20), and $f_i[\cdot]$, $f_a[\cdot]$, and $f_{(ia)}[\cdot]$ are defined in (4.25)–(4.26). The delta function ensures that the messages satisfy the BP iteration (4.19)–(4.20), so the integral means that it sums over all the BP fixed point with weight $w = e^{-\beta m f_{\text{BP}}}$.

The above expression is precisely another graphical model defined on a new factor graph, shown in Fig. 4.5. The joint probability is still factorized and defined on the factor graph with the same topological structure. So the sparsity condition of the graph still holds. The Bethe approximation on the new graphical model is the assumption of the 1RSB cavity method. Computation of the graph partition function, the complexity, or any other physical quantity follows the same lines as the computations at RS level. The only difference is that now the variables we operate with are functions (a cavity probability at RS level), and factors are functionals. More details of the 1RSB cavity method can be found in Mézard and Montanari (2009, Chapter 19).

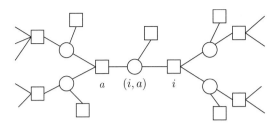

**Fig. 4.5** Computing the grand partition function using a new graphical model.

## 4.3   An example: random $K$-SAT problem

### 4.3.1   Cavity method and random $K$-satisfiability

In the previous section, we saw that the RS cavity method leads to BP equations, and that we can average the BP equations to obtain the density evolution description of the BP equation. We also saw that, at an abstract level, 1RSB is associated with proliferation of states, and that there is a whole hierarchy of such transitions.

In this section, we will show how the cavity method works in practice. Although the cavity method has been used in the SK model up to two-step replica symmetric breaking (2RSB) (Mézard *et al.*, 1986), the derivation becomes too technical and is not particularly enlightening. The random $K$-SAT problem provides another, more workable, example in which to use message-passing techniques. We shall start with a short summary of the problem, to set the notation.

*Definitions and notation*

We consider $N$ Boolean variables $x_i \in \{0, 1\}$, with $i = 1, \ldots, N$. In our representation, the value 0 corresponds to "false," while the value 1 corresponds to "true." A satisfiability problem is defined as a set of logical constraints that these random variables have to satisfy. Each logical constraint is called a *clause* and is expressed as a logical OR of a subset of the Boolean variables that may or not be negated. The negation of variable $x_i$ is denoted by $\bar{x}_i \equiv 1 - x_i$. An example of a 2-clause is "either $x_1$ is true or $x_2$ is false," expressed more succinctly as $x_1 \vee \bar{x}_2$, where $\vee$ denotes the logical OR. Another example is the 3-clause $x_1 \vee x_2 \vee \bar{x}_3$, which is satisfied by all configurations of $x_1, x_2, x_3$ except for $\{x_1 = 0, x_2 = 0, x_3 = 1\}$. In general, a satisfiability problem consists of a set of $M$ clauses $C_1, C_2, \ldots, C_M$ that have to be satisfied simultaneously. The problem is satisfiable if there is at least one choice of the Boolean variables $x = (x_1, \ldots, x_N)$, also called an *assignment*, that satisfies the logical formula

$$F = C_1 \wedge C_2 \wedge \cdots \wedge C_M, \tag{4.29}$$

where $\wedge$ is the logical AND.

In a $K$-SAT problem, each clause consists of exactly $K$ variables. We consider *random* $K$-SAT problems, where each clause $C_a$, $a = 1, \ldots, M$, contains exactly three variables chosen randomly in $\{x_1, \ldots, x_N\}$, and each variable is negated randomly with probability $1/2$. In other words, each clause is drawn with uniform distribution from the set of all the $\binom{N}{K}2^K$ clauses of length $K$.

An instance of a $K$-SAT problem can be represented by a factor graph, where variable nodes correspond to the Boolean variables and factor nodes correspond to clauses. When the variable $x_i$ (or its negation) appears in clause $a = 1, \ldots, M$, the node $i$ is connected to the clause factor $a$. It is useful to use a slightly modified version of the standard factor graph, in which the edge between $i$ and $a$ is is plotted with either a solid or a dashed line depending on whether the variable $i$ appears unnegated or negated in clause $a$ (see Fig. 4.6 for an example). With this modification, there is a one-to-one correspondence between a $K$-SAT problem and a factor graph. For consistency, we carry over the notation and use the indices $i, j, \ldots$ for variable nodes and indices $a, b, \ldots$ for factor nodes.

Note that each factor node has a fixed degree $K$, but the degree of a variable node is random. More specifically, because a randomly chosen $K$-tuple contains the variable $i$ with probability $K/N$, the degree of the variable node $i$ is a binomial random variable with parameters $M$ and $p = K/N$. In the limit of large $N$, the binomial distribution can be safely approximated by a Poisson distribution with parameter $\alpha K$, i.e., $\Pr(\text{degree}_i = n) = e^{-K\alpha}(K\alpha)^n/n!$.

The crucial parameter that characterizes random $K$-SAT problems is the clause density $\alpha \equiv M/N$, which sets the ratio of constraints per variable. Intuitively, one expects that for small $\alpha$ most of the instances will be satisfiable, while for large enough $\alpha$ most of the instances will be unsatisfiable. Numerical experiments confirm this intuition (see Fig. 4.7a). The probability that a random instance is SAT drops from values close to 1 to values close to 0 as crosses the value $\alpha_c \approx 4.3$, and this transition becomes sharper the larger the number of variables $N$ is. This is the characteristic behavior of a phase transition, and as such it has been analyzed using the methods of statistical physics.

The clause density also determines how hard the problem is. The difficulty of the problem can be quantified by the time taken by an algorithm to decide whether a typical instance is satisfiable or not. It turns out that a problem is easy when $\alpha$ is well below the critical value $\alpha_c$, it becomes harder as $\alpha$ approaches $\alpha_c$ (see Fig. 4.7a), and less hard when $\alpha$ is much larger than $\alpha_c$. In other words, the region around the phase transition is the hardest from a computational point of view. In the following,

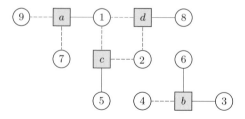

**Fig. 4.6**  Example of factor graph with nine variable nodes, $i = 1, \ldots, 9$ and four factor nodes $a, b, c, d$, The factor graph encodes the formula $F = (x_1 \vee \bar{x}_7 \vee \bar{x}_9) \wedge (x_3 \vee \bar{x}_4 \vee x_6) \wedge (\bar{x}_1 \vee \bar{x}_2 \vee x_5) \wedge (\bar{x}_1 \vee \bar{x}_2 \vee x_8)$.

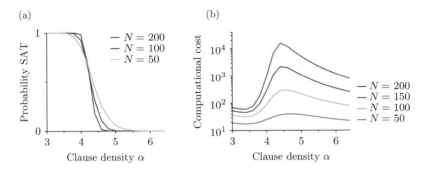

**Fig. 4.7** (a) Probability that a formula drawn from the random 3-SAT ensemble is satisfiable, as a function of the clause density $\alpha = M/N$. (b) Computational time (in arbitrary units) required to either find a solution or prove that there is none, as a function of the clause density. Figures adapted from Mézard and Mora (2009). Reproduced with permission from Elsevier.

we will define the thermodynamic limit as $M \to \infty$ and $N \to \infty$ while keeping the clause density $\alpha$ constant.

### Belief propagation

Each variable $i$ appears in a random set of clauses. We denote by $\partial i$ the set of indices of the clauses where $i$ appears. In the factor graph, $\partial i$ is the set of factor nodes adjacent to the variable node $i$. Similarly, we denote by $\partial a$ the indices of the $K$ variables appearing in clause $a$, and by $x_{\partial a}$ the corresponding variables, i.e., $x_{\partial a} \equiv \{x_i \mid i \in \partial a\}$. For later convenience, we define the number

$$J_{ai} = \begin{cases} 0 & \text{if } x_i \in C_a, \\ 1 & \text{if } \bar{x}_i \in C_a. \end{cases}$$

We will also distinguish the neighbors of $i$, $a \in \partial i$, according to the values of $J_{ai}$, and define $\partial_0 i = \{a \in \partial i \mid J_{ai} = 0\}$ and $\partial_1 i = \{a \in \partial i \mid J_{ai} = 1\}$.

Given the edge between the factor node $a$ and the variable node $i$, it is useful to distinguish the set of all remaining edges of $i$ according to whether or not their associated $J$s coincide with $J_{ai}$:

$$\mathcal{S}_{ia} \equiv \{b \in \partial i \backslash a \mid J_{bi} = J_{ai}\},$$
$$\mathcal{U}_{ia} \equiv \{b \in \partial i \backslash a \mid J_{bi} = 1 - J_{ai}\},$$

where $\partial i \backslash a$ means the set of all factors connected to $i$, excluding $a$. It follows from these definitions that the neighborhood of $i$ is partitioned as $\partial i = \{a\} \cup \mathcal{S}_{ia} \cup \mathcal{U}_{ai}$. Figure 4.8 summarizes our notation and conventions.

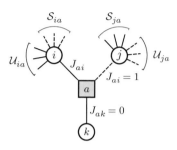

**Fig. 4.8** Factor graph associated with the single 3-clause $x_i \vee \bar{x}_j \vee x_k$. For clarity, we show only the subsets $\mathcal{U}$ and $\mathcal{S}$ associated with nodes $i$ and $j$.

Given the satisfiability formula $F$ in (4.29), we consider the uniform probability distribution $P_{\text{SAT}}(x)$ over the truth assignments $x = (x_1, \ldots, x_N) \in \{0, 1\}^N$ that satisfy $F$, assuming they exist. This probability can be written as

$$P_{\text{SAT}}(x) \cong \prod_{a=1}^{M} \psi_a(x_{\partial a}). \tag{4.30}$$

Each factor $\psi_a(x_{\partial a})$ is 1 if clause $a$ is satisfied by the assignment $x$ and 0 otherwise. Put differently,

$$\psi_a(x_{\partial a}) = \mathbb{I}(x_{\partial a} \text{ satisfies } C_a), \tag{4.31}$$

with $\mathbb{I}$ being the indicator function.

### The belief propagation equations

BP is an iterative algorithm that operates on "messages" associated with the directed edges of a factor graph. For each edge $(i, a)$, there exist two messages $\hat{m}_{a \to i}(x_i)$, $m_{i \to a}(x_i)$, defined in the space of probability distributions on the set $\{0, 1\}$: their values lie the interval $[0, 1]$ and satisfy $\sum_{x_i} m_{i \to a}(x_i) = 1$. Messages are updated according to

$$\hat{m}_{a \to i}^{(t)}(x_i) \cong \sum_{x_{\partial a \setminus i}} \psi_a(x_{\partial a}) \prod_{k \in \partial a \setminus i} m_{k \to a}^{(t)}(x_k), \tag{4.32}$$

$$m_{i \to a}^{(t+1)}(x_i) \cong \prod_{b \in \partial i \setminus a} \hat{m}_{b \to i}^{(t)}(x_i). \tag{4.33}$$

These are the BP, or *sum–product*, update rules. In tree-like graphical models the messages converge to fixed-point values. The resulting message $m_{i \to a}^{(\infty)}(x_i)$ is the marginal distribution of variable $x_i$ in a modified graphical model that does not include the factor $a$. Analogously, $\hat{m}_{a \to i}^{(\infty)}(x_i)$ is the marginal distribution of $x_i$ in a graphical model where all factors $\partial i$ but $a$ have been removed.

We can simplify the formulation of the BP equations for $K$-SAT, using the fact that the variables $x_i$ are all binary to parametrize the messages with a single real number. We define

$$\zeta_{ia} \equiv m_{i \to a}(x_i = J_{ai}) \in [0, 1],$$

$$\hat{\zeta}_{ai} \equiv \hat{m}_{a \to i}(x_i = J_{ai}) \in [0, 1].$$

From the normalization of the messages, it follows that $m_{i \to a}(x_i = 1 - J_{ai}) = 1 - \zeta_{ia}$ and $\hat{m}_{a \to i}(x_i = 1 - J_{ai}) = 1 - \hat{\zeta}_{ai}$. The variables $\zeta_{ai}$ and $\hat{\zeta}_{ia}$ can be interpreted as the message associated with the wrong direction of $x_i$. In terms of $\zeta_{ai}$ and $\hat{\zeta}_{ia}$, the BP equations (4.32)–(4.33) read

$$\hat{\zeta}_{ai} = \frac{1 - \prod_{j \in \partial a \setminus i} \zeta_{ja}}{1 + \left(1 - \prod_{j \in \partial a \setminus i} \zeta_{ja}\right)}, \tag{4.34}$$

$$\zeta_{ia} = \frac{\left[\prod_{b \in \mathcal{S}_{ia}} \hat{\zeta}_{bi}\right] \left[\prod_{b \in \mathcal{U}_{ia}} (1 - \hat{\zeta}_{bi})\right]}{\left[\prod_{b \in \mathcal{S}_{ia}} \hat{\zeta}_{bi}\right] \left[\prod_{b \in \mathcal{U}_{ia}} (1 - \hat{\zeta}_{bi})\right] + \left[\prod_{b \in \mathcal{S}_{ia}} (1 - \hat{\zeta}_{bi})\right] \left[\prod_{b \in \mathcal{U}_{ia}} \hat{\zeta}_{bi}\right]}, \tag{4.35}$$

where we use the convention that a product of zero factors is 1. The numbers of operations required to evaluate the right-hand sides of these two equations are $O(|\partial a|)$ and $O(|\partial i|)$, respectively, where $|A|$ is the cardinality of $A$. To solve (4.34), we update the messages until a fixed point is reached, after which we can obtain the marginals.

### Statistical analysis

We can go further and use the equations to derive the overall distribution of the messages. The idea is to draw a random edge $(i, a)$ in the factor graph and consider the corresponding fixed point of the messages $\zeta_{ia}, \hat{\zeta}_{ai}$ as random variables. Within the RS assumption, and when $N \to \infty$, these variables converge in distribution to edge-independent random variables $\zeta, \hat{\zeta}$, with distribution

$$\hat{\zeta} \overset{d}{=} \frac{1 - \zeta_1 \cdots \zeta_{K-1}}{2 - \zeta_1 \cdots \zeta_{K-1}}, \tag{4.36}$$

$$\zeta \overset{d}{=} \frac{\hat{\zeta}_1 \cdots \hat{\zeta}_p (1 - \hat{\zeta}_{p+1}) \cdots (1 - \hat{\zeta}_{p+q})}{\hat{\zeta}_1 \cdots \hat{\zeta}_p (1 - \hat{\zeta}_{p+1}) \cdots (1 - \hat{\zeta}_{p+q}) + (1 - \hat{\zeta}_1) \cdots (1 - \hat{\zeta}_p) \hat{\zeta}_{p+1} \cdots \hat{\zeta}_{p+q}}, \tag{4.37}$$

where $\overset{d}{=}$ means "equal in distribution." The numbers $p$ and $q$ are two independent and identically distributed (i.i.d.) Poisson random variables with mean $K\alpha/2$, and correspond to the random number of unnegated and negated edges in a variable node—namely, the numbers $|\partial_0 i|$ and $|\partial_1 i|$. The variables $\zeta_1, \ldots, \zeta_{K-1}$ are i.i.d. copies of $\zeta$, and $\hat{\zeta}_1, \ldots, \hat{\zeta}_{p+q}$ are i.i.d. copies of $\hat{\zeta}$. The probability density functions for $\zeta$ and $\hat{\zeta}$ defined by (4.36)–(4.37) are to be understood as

$$p(\hat{\zeta}) = \int \prod_{i=1}^{K-1} \{ d\zeta_i \, p(\zeta_i) \} \, \delta\left( \frac{1 - \zeta_1 \cdots \zeta_{K-1}}{2 - \zeta_1 \cdots \zeta_{K-1}} \right), \tag{4.38}$$

$$p(\zeta) = \sum_{r=0}^{\infty} \sum_{s=0}^{\infty} P(r) P(s) \int \prod_{i=1}^{K-1} \{ d\hat{\zeta}_i \, p(\hat{\zeta}_i) \}$$

$$\times \, \delta\left( \frac{\displaystyle\prod_{a=1}^{r} \hat{\zeta}_a \prod_{b=r+1}^{r+s} (1 - \hat{\zeta}_b)}{\displaystyle\prod_{a=1}^{r} \hat{\zeta}_a \prod_{b=r+1}^{r+s} (1 - \hat{\zeta}_b) + \prod_{a=1}^{r} (1 - \hat{\zeta}_a) \prod_{b=r+1}^{r+s} \hat{\zeta}_b} \right), \tag{4.39}$$

where $P(r)$ is the probability distribution of a Poisson random variable $X$, $\Pr(X = r) = e^{-\lambda} \lambda^r / r!$, with mean $\lambda = K\alpha/2$.

The generic way to solve the set of coupled equations (4.36)–(4.37) is by using *population dynamics* (see page 111). In this numerical method, one approximates the distribution of $\zeta$ (or $\hat{\zeta}$) through a sample of $N$ i.i.d. copies of the variable and exploits the property that, in the limit of large $N$, the empirical distribution of the sample converges to the actual distribution.

### 4.3.2   Free entropy

Recall from Section 4.2 that the free entropy informs us about the number of solutions, and it is a function of the messages of the factor graph. We now evaluate the free entropy for a $K$-SAT problem. If $E$ denotes the set of edges in the graph, there are $2|E|$ messages, which we collectively denote by $m \equiv \{ m_{i \to a}(\cdot), \hat{m}_{a \to i}(\cdot) \}$. The free entropy then reads

$$F(m) = \sum_{a \in F} F_a(m) + \sum_{i \in V} F_i(m) - \sum_{(ai) \in E} F_{ai}(m),$$

where $F$ is the set of factor nodes, $V$ is the set of variable nodes, and

$$F_a(m) = \log\left[ \sum_{x_{\partial a}} \psi_a(x_{\partial a}) \prod_{i \in \partial a} m_{i \to a}(x_i) \right], \tag{4.40}$$

$$F_i(m) = \log\left[ \sum_{x_i} \prod_{b \in \partial i} \hat{m}_{b \to i}(x_i) \right], \tag{4.41}$$

$$F_{ai}(m) = \log\left[ \sum_{x_i} m_{i \to a}(x_i) \, \hat{m}_{a \to i}(x_i) \right]. \tag{4.42}$$

In $F_a(m)$, the sum $\sum_{x_{\partial a}}$ is over all the possible configurations of the variable nodes adjacent to $a$. In terms of $\zeta \equiv \{\zeta_{ia}, \hat{\zeta}_{ai}\}$, (4.40)–(4.42) read

$$F_a(\zeta) = \log\Big[1 - \prod_{i\in\partial a} \zeta_{ia}\Big], \tag{4.43}$$

$$F_i(\zeta) = \log\Bigg[\prod_{a\in\partial_0 i} \hat{\zeta}_{ai} \prod_{b\in\partial_1 i} (1 - \hat{\zeta}_{bi}) + \prod_{a\in\partial_0 i} (1 - \hat{\zeta}_{ai}) \prod_{b\in\partial_1 i} \hat{\zeta}_{bi}\Bigg], \tag{4.44}$$

$$F_{ai}(\zeta) = \log\Big[\zeta_{ia}\hat{\zeta}_{ai} + (1 - \zeta_{ia})(1 - \hat{\zeta}_{ai})\Big]. \tag{4.45}$$

In Section 4.2.4 we saw that under RS assumptions, the Bethe free entropy density in the thermodynamic limit is

$$\lim_{N\to\infty} \frac{F}{N} = f^{\mathrm{RS}} = f_v^{\mathrm{RS}} + \alpha f_c^{\mathrm{RS}} - K\alpha f_e^{\mathrm{RS}}, \tag{4.46}$$

where

$$f_v^{\mathrm{RS}} = \mathbb{E}\log\Bigg[\prod_{a=1}^{p} \hat{\zeta}_a \prod_{b=p+1}^{p+q} (1 - \hat{\zeta}_b) + \prod_{a=1}^{p}(1 - \hat{\zeta}_a) \prod_{b=p+1}^{p+q} \hat{\zeta}_b\Bigg],$$

$$f_c^{\mathrm{RS}} = \mathbb{E}\log[1 - \zeta_1 \cdots \zeta_{K-1}],$$

$$f_e^{\mathrm{RS}} = \mathbb{E}\log\Big[(1 - \zeta_1)(1 - \hat{\zeta}_1) + \zeta_1\hat{\zeta}_1\Big].$$

Here $\mathbb{E}$ denotes expectation with respect to the variables $\zeta_1, \ldots, \zeta_K$ (the i.i.d. copies of $\zeta$), $\hat{\zeta}_1, \ldots, \hat{\zeta}_{p+q}$ (the i.i.d. copies of $\hat{\zeta}$), and the Poisson random variables $p$ and $q$. We can use population dynamics to estimate the distributions of $\zeta$ and $\hat{\zeta}$, and then use the resulting samples to estimate the free entropy density (4.46). The outcome of this procedure, repeated for several values of $\alpha$, is summarized in Fig. 4.9. The entropy

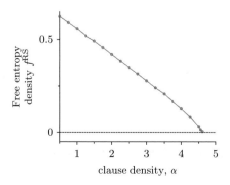

**Fig. 4.9** Estimate of the Bethe free entropy density as a function of the clause density, for 3-SAT and assuming replica symmetry. The curve reaches 0 entropy at around $\alpha_*(3) \approx 4.677$.

density is strictly positive and decreasing for $\alpha \leq \alpha_*(K)$, with $\alpha_*(3) \approx 4.677$. The value $\alpha_*(K)$ is the RS prediction for the SAT–UNSAT threshold $\alpha_s(K)$, where $K$-SAT instances cease to be satisfiable.

Unfortunately, this result is inconsistent with the upper bound $\alpha_{UB}(3) \approx 4.666$, derived rigorously from the first-moment method (see Chapter 2). The reason for this contradiction is that the RS assumption is expected to be correct only up to the condensation transition $\alpha_c(3) \approx 3.86$, where pure states start to proliferate (see Section 4.2).

### BP-guided decimation

Another way to realize that the RS assumption cannot be valid close to the SAT–UNSAT threshold is by using the BP iteration. We can just pick a random $K$-SAT instance, initialize the messages with uniform random numbers, and then iterate the BP equations (4.34)–(4.35) until no message changes by more than some prescribed small number $\delta$. If we fix a large time $t_{max}$, we can estimate the probability of convergence within $t_{max}$ by repeating the same experiment many times. Figure 4.10 summarizes such an experiment for $K = 3$ and 4. The estimated probability curves show a sharp decrease around a critical value of $\alpha$, which we denote by $\alpha_{BP}$ and which turns out to be robust to variations of $\delta$ and $t_{max}$.

We can go further and find a SAT assignment based on the messages obtained after convergence of the BP iteration. The method is called BP-guided decimation and is as follows. Given the BP estimate of the marginal of $x_i$, we compute the *bias* $\pi_i \equiv P_i(1) - P_i(1)$ for each variable, and then pick the variable with highest $|\pi_i|$. This variable is fixed to its favored value (i.e., $x_i$ is set to 0 if $\pi_i > 0$, or to 1 otherwise), and the SAT formula is reduced (decimated) using this individual assignment. The method is repeated until all the variables are assigned, or until the BP fails to converge.

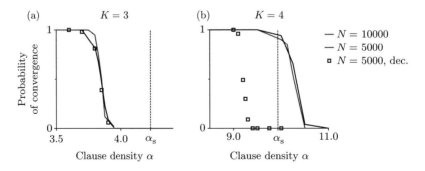

**Fig. 4.10** Empirical probability that the BP algorithm converges to a fixed point, as a function of the the clause density, for 3-SAT (a) and 4-SAT (b). The estimate is based on 100 instances with the number of variables indicated in the legend. Squares indicate empirical probabilities that BP-guided decimation finds a SAT assignment, using 100 instances with 5000 variables each. The vertical dashed line is located at the SAT–UNSAT threshold $\alpha_s$. Parameters of the decimation: $\delta = 10^{-2}$, $t_{max} = 10^3$. Figures adapted from Mézard and Montanari (2009).

The probability that BP-guided decimation results in a SAT assignment is shown in Fig. 4.10, for several values of $\alpha$ and for $K = 3, 4$. Note that for 3-SAT the decimation method returns a SAT assignment almost every time the BP iteration converges (i.e., for $\alpha \lesssim 3.85$). In contrast, for 4-SAT, BP-guided decimation finds SAT assignments for $\alpha \lesssim 9.25$, while BP converges most of the time for $\alpha \lesssim 10.3$ (a value that is larger than the conjectured SAT–UNSAT threshold, $\alpha_s(4) \approx 10.93$).

This numerical experiment shows that something goes wrong when $\alpha$ is large enough. It also shows that 4-SAT is qualitatively different from 3-SAT; what makes BP fail at large $\alpha$ differs depending on the $K$ we consider. For $K = 3$, the BP fixed point becomes unstable at around $\alpha_{st} \approx 3.86$, which leads to errors in decimations. For $K = 4$, in contrast, the BP fixed point remains stable but does not lead to the correct marginals, because the 1RSB condensation threshold $\alpha_c$ is crossed.

### The 1RSB cavity method

We could proceed with the strategy outlined in Section 4.2.2, using the BP approximation in the auxiliary model to estimate the complexity function $\Sigma(f)$. This can be done, but it gets complicated because we need to operate on probability functions (the Bethe measures) rather than on simple real numbers. If we just want to compute the entropy to find whether or not there exist solutions, we can take a shortcut, based on the min–sum algorithm.

Instead of computing the marginals of the distribution in (4.30), we consider the problem of minimizing the following cost (energy) function:

$$E(x) = \sum_{a=1}^{M} E_a(x_{\partial a}), \tag{4.47}$$

where $E_a(x_{\partial a}) = 0$ if clause $a$ is satisfied by the assignment $x = (x_1, \ldots, x_N)$, while $E_a(x_{\partial a}) = 1$ otherwise. The two problems are mapped onto each other through $\psi_a(x_{\partial a}) = e^{-\beta E_a(x_{\partial a})}$, with $\beta > 0$. The particular choice of the factor $\psi_a$ as the indicator function of clause $C_a$, (4.31), corresponds to the zero-temperature limit $\beta \to \infty$.

In this formulation, the SAT–UNSAT threshold $\alpha_s(K)$ is identified as the value of $\alpha$ above which the probability of having a configuration with ground state energy, $E(x) = 0$, vanishes. We will estimate the ground state density with the cavity method. For this, we need to adapt the message-passing rules (4.32)–(4.33) in two steps. First, we need to compute max-marginals, rather than marginals. This is a straightforward step that consists of replacing sums with maximizations, and leads to the so-called *max–product* update rules

$$\hat{m}_{a \to i}^{(t)}(x_i) \cong \max_{x_{\partial a \setminus i}} \left\{ \psi_a(x_{\partial a}) \prod_{k \in \partial a \setminus i} m_{k \to a}^{(t)}(x_k) \right\}, \tag{4.48}$$

$$m_{i \to a}^{(t+1)}(x_i) \cong \prod_{b \in \partial i \setminus a} \hat{m}_{b \to i}^{(t)}(x_i). \tag{4.49}$$

Second, we express these update rules in terms of the energy $E(x)$, which amounts to taking the logarithms of 4.48–(4.49). The resulting algorithm is the so-called *min–sum* algorithm:

$$\hat{E}_{a \to i}^{(t)}(x_i) = \min_{x_{\partial a \backslash i}} \left\{ E_a(x_{\partial a}) + \sum_{k \in \partial a \backslash i} E_{k \to a}^{(t)}(x_k) \right\} + \hat{C}_{a \to i}^{(t)}, \tag{4.50}$$

$$E_{i \to a}^{(t+1)}(x_i) = \sum_{b \in \partial i \backslash a} \hat{E}_{b \to i}^{(t)}(x_i) + C_{i \to a}^{(t)}. \tag{4.51}$$

The fixed points of these equations are known as the *energetic cavity equations*. In the same way that the max–product marginals are defined up to a multiplicative constant, min–sum messages are defined up to an overall additive constant. We set the constants $C_{i \to a}^{(t)}$ and $\hat{C}_{a \to i}^{(t)}$ so that $\min_{x_i} E_{i \to a}^{(t+1)}(x_i) = 0$ and $\min_{x_i} \hat{E}_{i \to a}^{(t)}(x_i) = 0$. With this arrangement, all energies are relative to the ground state energy.

### Warning propagation

The fact that the energy function (4.47) counts the number of violated constraints allows us to simplify the min–sum updates given by (4.50)–(4.51). It can be shown that if messages are initialized so that $\hat{E}_{a \to i}^{(0)}$ are either 0 or 1, the subsequent values of $\hat{E}^{(t)}$ obtained from the min–sum updates will also be either 0 or 1 (see Mézard and Montanari, 2009). As a consequence of this property, instead of keeping track of the variable-to-node messages $E_{i \to a}(\cdot)$, we will only bother to use the projections on $\{0, 1\}$:

$$\mathcal{E}_{i \to a}(x_i) = \min\{1, E_{i \to a}(x_i)\}.$$

The update rules become

$$\hat{E}_{a \to i}^{(t)}(x_i) = \min_{x_{\partial a \backslash i}} \left\{ E_a(x_{\partial a}) + \sum_{k \in \partial a \backslash i} \mathcal{E}_{k \to a}^{(t)}(x_k) \right\} + \hat{C}_{a \to i}^{(t)}, \tag{4.52}$$

$$\mathcal{E}_{i \to a}^{(t+1)}(x_i) = \min \left\{ 1, \sum_{b \in \partial i \backslash a} \hat{E}_{b \to i}^{(t)}(x_i) + C_{i \to a}^{(t)} \right\}. \tag{4.53}$$

This simplified min–sum algorithm with update equations (4.52)–(4.53) is called the *warning propagation* algorithm. The name stems from the interpretation of $\mathcal{E}_{i \to a}$ as a warning: $\mathcal{E}_{i \to a} = 1$ means that according to the set of constraints $b \in \partial i \backslash a$, the $i$th variable should not take the value $x_i$; analogously, $\mathcal{E}_{i \to a} = 0$ means that according to the set of constraints $b \in \partial i \backslash a$, the $i$th variable has a green light to take the value $x_i$. The main advantage of warning propagation is that messages are are either 0 or 1, rather than distributions.

Because our problem involves binary variables and hard constraints, the messages of the 1RSB cavity equations are triples: $(Q_{ia}(0), Q_{ia}(1), Q_{ia}(*))$ for variable-to-function messages and $(\hat{Q}_{ai}(0), \hat{Q}_{ai}(1), \hat{Q}_{ai}(*))$ for function-to-variable messages. In the case of $K$-satisfiability, these messages can be simplified further: if $J_{ai} = 0$, then $\hat{Q}_{ai}(1)$ is necessarily 0; if $J_{ai} = 1$, then $\hat{Q}_{ai}(0)$ must be 0. This is because a "0" message means that the constraint $a$ forces $x_i$ to take the value 0 in order to minimize the system's energy. In $K$-SAT, this can happen only if $J_{ai} = 0$, because $x_i = 0$ is the value that satisfies $a$. An analogous argument applies for the "1" message. The bottom line is that function-to-variable messages can be parametrized by a single real number. We take this number to be $\hat{Q}_{ai}(0)$ if $J_{ai} = 0$ and $\hat{Q}_{ai}(1)$ if $J_{ai} = 1$, and denote it simply by $\hat{Q}_{ai}$.

Similarly, we can use a parametrization for the variable-to-function message $Q_{ia}(\cdot)$ that takes into account the value of $J_{ai}$. We denote by $Q_{ia}(0)$, $Q_{ia}(*)$, and $Q_{ia}(1)$ the three possible type of messages: $m(1) > m(0) = 0$, $m(0) = m(1) = 0$, and $m(0) > m(1) = 0$, respectively. We then define, if $J_{ai} = 0$, $Q_{ia}^S \equiv Q_{ia}(0)$, $Q_{ia}^* \equiv Q_{ia}(*)$, and $Q_{ia}^U \equiv Q_{ia}(1)$. Conversely, if $J_{ai} = 1$, we have $Q_{ia}^S \equiv Q_{ia}(1)$, $Q_{ia}^* \equiv Q_{ia}(*)$, and $Q_{ia}^U \equiv Q_{ia}(0)$. The interpretation of the newly defined variables is as follows:

$$Q_{ia}^S = \Pr(x_i \text{ is forced to satisfy } a \text{ by } b \in \mathcal{S}_{ia}),$$
$$Q_{ia}^U = \Pr(x_i \text{ is forced to violate } a \text{ by } b \in \mathcal{U}_{ia}),$$
$$Q_{ia}^* = \Pr(x_i \text{ is not forced by } b \in \mathcal{S}_{ia} \cup \mathcal{U}_{ia}),$$
$$\hat{Q}_{ai} = \Pr(x_i \text{ is forced by clause } a \text{ to satisfy } a).$$

At this point, we could derive the explicit 1RSB equations in terms of the messages $Q^S$, $Q^U$, $Q^*$, and $\hat{Q}$. Another option is to use the above interpretation of the messages to guess the 1RSB cavity equations. Note first that clause $a$ forces variable $x_i$ to satisfy $a$ only when all the other variables involved in $a$ are forced (by some other clause) not to satisfy $a$. This can be stated as

$$\hat{Q}_{ai} = \prod_{j \in \partial a \setminus i} Q_{ja}^U.$$

Let us define $\Omega^S$ and $\Omega^U$ as, respectively, the subsets of clauses $\mathcal{S}_{ia}$ and $\mathcal{U}_{ia}$ that send a warning. For concreteness, let us pick the variable node $i$ and assume that $J_{ia} = 0$ (the opposite case leads to identical equations). In that case, $\mathcal{S}_{ia}$ is the subset $b \in \partial i \setminus a$ for which $J_{ib} = 0$, while $\mathcal{U}_{ia}$ is the remaining set of neighbors except $a$ for which $J_{ib} = 1$. Let us also assume that the clauses $\Omega^S \subseteq \mathcal{S}_{ia}$ and $\Omega^U \subseteq \mathcal{U}_{ia}$ force the variable node $i$ to take the value $x_i$ that satisfies them. It follows that $x_i$ is forced to satisfy $a$ if $|\Omega^S| > |\Omega^U|$, and it is forced to violate $a$ if $|\Omega^S| < |\Omega^U|$; $x_i$ is not forced if $|\Omega^S| = |\Omega^U|$. The energy shift equals the number of "forcing" clauses in $\partial i \setminus a$ that are violated when $x_i$ is set to satisfy the largest number of clauses. This leads to $\min(|\Omega^S|, |\Omega^U|)$ violated clauses. The resulting 1RSB message-passing algorithm, also known as the survey propagation equations, reads

$$Q_{ia}^U \cong \sum_{|\Omega^U| > |\Omega^S|} e^{-y|\Omega^S|} \prod_{b \in \Omega^U \cup \Omega^S} \hat{Q}_{bi} \prod_{b \notin \Omega^U \cup \Omega^S} (1 - \hat{Q}_{bi}), \qquad (4.54)$$

$$Q_{ia}^S \cong \sum_{|\Omega^S| > |\Omega^U|} e^{-y|\Omega^U|} \prod_{b \in \Omega^U \cup \Omega^S} \hat{Q}_{bi} \prod_{b \notin \Omega^U \cup \Omega^S} (1 - \hat{Q}_{bi}), \qquad (4.55)$$

$$Q_{ia}^* \cong \sum_{|\Omega^U| = |\Omega^S|} e^{-y|\Omega^U|} \prod_{b \in \Omega^U \cup \Omega^S} \hat{Q}_{bi} \prod_{b \notin \Omega^U \cup \Omega^S} (1 - \hat{Q}_{bi}). \qquad (4.56)$$

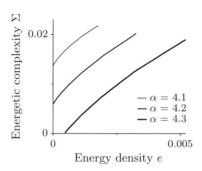

**Fig. 4.11** Energetic complexity density versus energy density for the 3-SAT problem, and for three different clause densities, indicated in the legend.

The overall normalization is fixed by the condition $Q^U_{ia}+Q^S_{ia}+Q^*_{ia} = 1$. These equations can be solved by iteration as easily as those of BP. As in the BP equations, we can use (4.54)–(4.56) to find the fixed point of the messages $\{\hat{Q}_{ai}, Q_{ia}\}$ for a given instance, or, rather, we can do a statistical analysis. In the latter case, we can compute with population dynamics the probabilities $P(\hat{Q}_{ai})$ and $P(Q^U_{ia}, Q^S_{ia}, Q^*_{ia})$. We can then compute the Bethe free energy, and then the Legendre transform of the resulting formula, from which we obtain the complexity as a function of the energy. We get Fig. 4.11, from which we can see that $\alpha = 4.3$ leads to a certain number of contradictions (given by the finite energy at $\Sigma = 0$, i.e., the intersection with the abscissa). The number of contradictions decreases as we reduce $\alpha$, until they vanish. This happens when the value of $\alpha$ is such that the curve crosses the origin of the $\Sigma$ versus energy curve, which is approximately $\alpha \approx 4.2667$. This is the prediction for the SAT–UNSAT threshold. An analogous derivation for the 4-SAT problem leads to the estimate $\alpha \approx 9.667$.

## References

de Almeida, J. R. L. and Thouless, David J. (1978). Stability of the Sherringten–Kirkpatrick solution of a spin glass model. *J. Phys. A*, **11**(5), 983.

Edwards, S. F. and Anderson, P. W. (1975). Theory of spin glasses. *J. Phys. F*, **5**(5), 965.

Mézard, M. and Montanari, A. (2009). *Information, Physics, and Computation*. Oxford University Press.

Mézard, M. and Mora, T. (2009). Constraint satisfaction problems and neural networks: a statistical physics perspective. *J. Physiol. Paris*, **103**(1), 107–113.

Mézard, M. and Parisi, G. (2001). The Bethe lattice spin glass revisited. *Euro. Phys. J. B*, **233**, 217–233.

Mézard, M. and Parisi, G. (2003). The cavity method at zero temperature. *J. Stat. Phys.*, **111**(April), 1–34.

Mézard, M., Parisi, G., and Virasoro, M. A. (1986). SK model: the replica solution without replicas. *Europhys. Lett.*, **1**(2), 77–82.

Parisi, G. (1979). Infinite number of order parameters for spin-glasses. *Phys. Rev. Lett.*, **43**(23), 1754.

Parisi, G. (1980). The order parameter for spin glasses: a function on the interval 0–1. *J. Phys. A*, **13**(3), 1101.

Sherrington, D. and Kirkpatrick, S. (1975, December). Solvable model of a spin-glass. *Phys. Rev. Lett.*, **35**, 1792–1796.

Thouless, D. J., Anderson, P. W., and Palmer, R. G. (1977). Solution of "solvable model of a spin glass". *Philos. Mag.*, **35**(3), 593–601.

Touchette, H. (2009). The large deviation approach to statistical mechanics. *Phys. Rep.*, **478**(1–3), 1–69.

# 5
# Statistical estimation: from denoising to sparse regression and hidden cliques

## Andrea MONTANARI

Department of Electrical Engineering
and Department of Statistics,
Stanford University,
Stanford, CA 94304

*Lecture notes taken by*
Eric W. Tramel, École Normale Supérieure, France
Santhosh Kumar Vanaparthy, Texas A&M University, Texas
Andrei Giurgiu, EPFL, Switzerland

*Statistical Physics, Optimization, Inference, and Message-Passing Algorithms.* First Edition.
F. Krzakala et al. © Oxford University Press 2016. Published in 2016 by Oxford University Press.

# Chapter Contents

These lectures provide a gentle introduction to some modern topics in high-dimensional statistics, statistical learning, and signal processing for an audience without any previous background in these areas. The point of view we take is to connect recent advances to the basic background in statistics (estimation, regression, and the bias–variance trade-off), and to classical—although non-elementary—developments (sparse estimation and wavelet denoising).

The first three sections cover these basic and classical topics. We then cover more recent research, and discuss sparse linear regression in Section 5.4, and its analysis for random designs in Section 5.5. Finally, in Section 5.6, we discuss an intriguing example of a class of problems in which sparse and low-rank structures have to be exploited simultaneously.

Needless to say, the selection of topics presented here is very partial. The reader interested in a deeper understanding can choose from a number of excellent options for further study. Very readable introductions to the fundamentals of statistical estimation can be found in the books by Wasserman (2004, 2006). More advanced references (with a focus on high-dimensional and nonparametric settings) are the monographs by Johnstone (2011) and Tsybakov and Zaiats (2009). The recent book by Bühlmann and Van De Geer (2011) provides a useful survey of recent research in high-dimensional statistics. For the last part of this chapter, dedicated to most recent research topics, we will provide references to specific papers.

## 5.1 Statistical estimation and linear models

### 5.1.1 Statistical estimation

The general problem of statistical estimation is that of estimating an unknown object from noisy observations. To be concrete, we can consider the model

$$y = f(\theta; \text{noise}), \tag{5.1}$$

where $y$ is a set of observations, $\theta$ is the unknown object, for instance a vector, a set of parameters, or a function. Finally, $f(\cdot; \text{noise})$ is an observation model that links together the observations and the unknown parameters that we wish to estimate. Observations are corrupted by random noise according to this model. The objective is to produce an estimation $\widehat{\theta} = \widehat{\theta}(y)$ that is accurate under some metric. The estimation of $\theta$ from $y$ is commonly aided by some hypothesis about the structure, or behavior, of $\theta$. Several examples are described below.

Statistical estimation can be regarded as a subfield of statistics, and lies at the core of a number of areas of science and engineering, including data mining, signal processing, and inverse problems. Each of these disciplines provides some information on how to model data acquisition, on computation, and on how best to exploit the hidden structure of the model of interest. Numerous techniques and algorithms have been developed over a long period of time, and they often differ in the assumptions and in the objectives that they try to achieve. As an example, the following are a few major distinctions to keep in mind:

**Parametric versus nonparametric.** In parametric estimation, stringent assumptions are made about the unknown object, hence reducing $\theta$ to be determined by a small set of parameters. In contrast, nonparametric estimation strives to make minimal modeling assumptions, resulting in $\theta$ being a high-dimensional or infinite-dimensional object (e.g., a function).

**Bayesian versus frequentist.** The Bayesian approach assumes $\theta$ to be a random variable as well, whose "prior" distribution plays an obviously important role. From a frequentist point of view, $\theta$ is instead an arbitrary point in a set of possibilities. In this chapter, we shall mainly follow the frequentist point of view, but we stress that the two are in fact closely related.

**Statistical efficiency versus computational efficiency.** Within classical estimation theory, a specific estimator $\widehat{\theta}$ is mainly evaluated in terms of its accuracy: How close (or far) is $\widehat{\theta}(y)$ to $\theta$ for typical realizations of the noise? We can broadly refer to this figure of merit as "statistical efficiency."

Within modern applications, computational efficiency has arisen as a second central concern. Indeed, $\theta$ is often high-dimensional: it is not uncommon to fit models with millions of parameters. The number of observations has grown in parallel. It has therefore become crucial to devise estimators whose complexity scales gently with the dimensions and with the amount of data.

We next discuss informally a few motivating examples.

### *Example 1: Exploration seismology*

Large-scale statistical estimation plays a key role in the field of exploration seismology. This technique uses seismic measurements on the earth's surface to reconstruct geological structures and the composition and density field of a geological substrate (Herrmann *et al.*, 2012). Measurements are generally acquired by sending a known seismic wave through the ground, perhaps by a controlled detonation, and measuring the response at multiple spatially dispersed sensors.

The following is a simple dictionary that points at the various elements of the model (5.1) in this example.:

Exploration Seismology

| | |
|---|---|
| $y$ | seismographic measurements |
| $\theta$ | earth density field |
| *Hypothesis* | smooth density field |

The function $f(\cdot)$ in (5.1) expresses the outcome of the seismographic measurements, given a certain density field $\theta$ and a certain source of seismic waves (left implicit since it is known). While this relation is of course complex, and ultimately determined by the physics of wave propagation, it is in principle perfectly known.

Because of the desired resolution of the recovered earth density field (Fig. 5.1), this statistical estimation problem is often ill-posed, since sampling is severely limited by the cost of generating the source signal and the distribution and set-up of the receivers. Resolution can be substantially improved by using some structural insights into the nature of the earth density field. For instance, one can exploit the fact that this is mostly smooth with the exception of some surfaces of discontinuity.

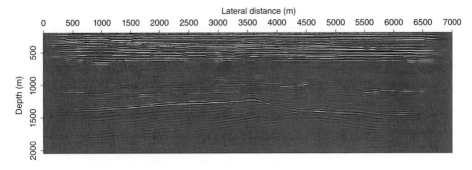

**Fig. 5.1** A recovered earth density field. From Herrmann *et al.* (2012). ©2012 IEEE. Reprinted, with permission, from IEEE.

### Example 2: Hidden structure in networks

Many modern data sets are relational, i.e., they express pairwise relations within a set of objects. This is the case in social networks, communication networks, unsupervised learning, and so on.

In the simplest case, for each pair of nodes in a network, we know whether they are connected or not. Finding a hidden structure in such a network is a recurring problem with these datasets. A highly idealized but nevertheless very interesting problem requires one to find a highly connected subgraph in an otherwise random graph:

Hidden network structure

| | |
|---|---|
| $y$ | large network |
| $\theta$ | hidden subset of nodes |
| *Hypothesis* | hidden network is highly connected |

From Fig. 5.2, it is apparent that the discovery of such networks can be a difficult task.

### Example 3: Collaborative filtering

Recommendation systems are ubiquitous in e-commerce and web services. They aim at personalizing each user's experience through an analysis of her past behavior, and—crucially—the past behavior of similar users. The algorithmic and statistical techniques that allow the exploitation of this information are referred to as "collaborative filtering." Amazon, Netflix, and YouTube all make intensive use of collaborative filtering technologies.

In a idealized model for collaborative filtering, each user of a e-commerce site is associated with a row of a matrix, and each product with a column. Entry $\theta_{i,j}$ in this matrix corresponds to the evaluation that user $i$ gives of product $j$. A small subset of the entries is observed because of feedback provided by the users (reviews, ratings, purchasing behavior). In this setting, collaborative filtering aims at estimating the whole matrix, on the basis of noisy observations of relatively few of its entries.

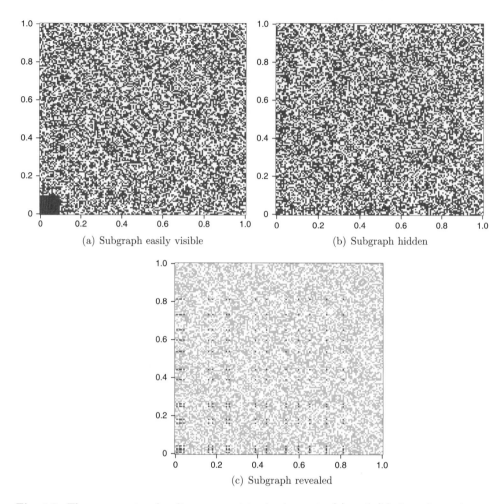

(a) Subgraph easily visible

(b) Subgraph hidden

(c) Subgraph revealed

**Fig. 5.2** The same network adjacency matrix, is shown in (a) and (b), but the nodes are permuted in (b). In (c), the hidden subgraph is revealed.

While this task is generally hopeless, it is observed empirically that such data matrices are often well approximated by low-rank matrices. This corresponds to the intuition that a small number of factors (corresponding to the approximate rank) explain the opinions of many users concerning many items. The problem is then modeled as that of estimating a low-rank matrix from noisy observations of some of its entries:

Collaborative filtering

| | |
|---|---|
| $y$ | small set of entries in a large matrix |
| $\theta$ | unknown entries of matrix |
| *Hypothesis* | matrix has a low-rank representation |

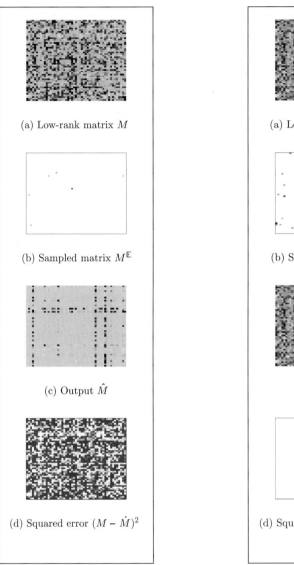

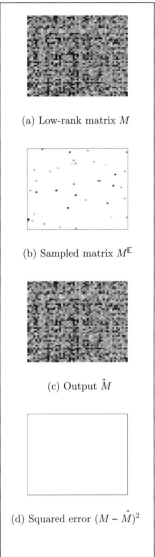

**Fig. 5.3** Recovering a 2000 × 2000 rank-8 matrix from 0.25% of its values.

**Fig. 5.4** Recovering a 2000 × 2000 rank-8 matrix from 1.75% of its values.

A toy example of this problem is demonstrated in Figs. 5.3 and 5.4. It can be observed that accurate estimation of the original matrix is possible even when very few of its coefficients are known.

### 5.1.2 Denoising

We will begin by considering in greater depth a specific statistical estimation problem, known as "denoising." On the one hand, denoising is interesting, since it is a very common signal processing task: in essence, it seeks to restore a signal that has been corrupted by some random process, for instance additive white noise. On the other, it will allow us to introduce some basic concepts that will play an important role throughout this chapter. Finally, recent research by Donoho *et al.* (2013*b*) has unveiled a deep and somewhat surprising connection between denoising and the rapidly developing field of compressed sensing.

To define the problem formally, we assume that the signal to be estimated is a function $t \mapsto f(t)$ Without loss of generality, we will restrict the domain of $f(t)$, $f : [0, 1] \to \mathbb{R}$. We measure $n$ uniformly spaced samples over the domain of $f$,

$$y_i = f(i/n) + w_i, \tag{5.2}$$

where $i \in \{1, 2, \ldots, n\}$ is the sample index, and $w_i \sim \mathcal{N}(0, \sigma^2)$ is the additive noise term. Each of $y_1, \ldots y_n$ is a sample, as seen in Fig. 5.5.

For the denoising problem, we desire to calculate the original function from the noise-corrupted observables $y_i$. How might we go about doing this?

### 5.1.3 Least squares estimation

One of the most direct approaches to signal estimation in the presence of noise is the **least squares (LS) method**, which dates back to Gauss and Legendre (1823). The method starts with the natural idea to parametrize the function $f$. For instance, we can tentatively assume that it is a polynomial of degree $p - 1$:

$$f(t) = \sum_{j=1}^{p} \theta_j \, t^{j-1}. \tag{5.3}$$

Each monomial is weighted according to the coefficients $\theta_j$ for $j \in \{1, 2, \ldots, p\}$, and we will collect these coefficients in a vector $\theta = (\theta_1, \theta_2, \ldots, \theta_p) \in \mathbb{R}^p$. Thus, the problem of recovering $f(t)$ boils down to the recovery of the $p$ coefficients $\theta_j$ from the set of observables $y_i$. We therefore seek to find the set of coefficients that generate a function that most closely matches the observed samples.

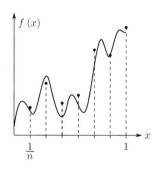

**Fig. 5.5** Depiction of a discrete-time sampling of the continuous-time function $f(x)$. Notice that the additive noise term $w_i$ prevents exact knowledge of $f(i/n)$.

It is natural to set this up as an optimization problem (here RSS stands for "residual sum of squares"),

$$\widehat{\theta}^{LS}(y) \equiv \operatorname*{argmin}_{\theta} RSS(\theta), \tag{5.4}$$

$$RSS(\theta) \equiv \sum_{i=1}^{n} \left( y_i - \sum_{j=1}^{p} \theta_j \left( \frac{i}{n} \right)^{j-1} \right)^2. \tag{5.5}$$

Fitting a low-degree polynomial to a dataset by LS is a very common practice, and the reader has probably tried this exercise at least once. A moment's reflection reveals that nothing is special about the polynomials used in this procedure. In general, we can consider a set of functions $\{\varphi_1, \varphi_2, \ldots, \varphi_p\}$, where

$$\varphi_j : [0, 1] \to \mathbb{R}. \tag{5.6}$$

Of course, the quality of our estimate depends on how well the functions $\{\varphi_j\}$ capture the behavior of the signal $f$. Assuming that $f$ can be represented as a linear combination of these functions, we can rewrite our model as

$$y_i = \sum_{j=1}^{p} \theta_{0j} \varphi_j (i/n) + w_i. \tag{5.7}$$

Equivalently, if we define $\varphi : [0, 1] \to \mathbb{R}^p$ by letting $\varphi(x) = (\varphi_1(x), \varphi_2(x), \ldots, \varphi_p(x))$, $\theta_0 = (\theta_{0,1}, \theta_{0,2}, \ldots, \theta_{0,p})$, and denoting by $\langle a, b \rangle \equiv \sum_{i=1}^{m} a_i b_i$ the usual scalar product in $\mathbb{R}^m$, we have

$$y_i = \langle \theta_0, \varphi(i/n) \rangle + w_i. \tag{5.8}$$

Before continuing further, it is convenient to pass to matrix notation. Let us define a matrix $\mathbf{X}(X_{ij}) \in \mathbb{R}^{n \times p}$ whose entry $i, j$ is given by

$$X_{ij} = \varphi_j \left( \frac{i}{n} \right), \tag{5.9}$$

Using this notation, and letting $y = (y_1, y_2, \ldots, y_n)$, $w = (w_1, w_2, \ldots, w_n)$, our model reads

$$y = X\theta_0 + w, \tag{5.10}$$

$w \sim \mathcal{N}\left(0, \sigma^2 I_n\right)$ (here and below, $I_n$ denotes the identity matrix in $n$ dimensions: the subscript will be dropped if it is clear from the context).

From (5.10), we see that the vector of observations $y$ is approximated as a linear combination of the columns of $\mathbf{X}$, each column corresponding to one of the functions $\varphi_1, \varphi_2, \ldots, \varphi_p$, evaluated on the sampling points.

This is a prototype of a very general idea in statistical learning, data mining, and signal processing. Each data point $x$ (or each point in a complicated space, e.g., a space of images) is represented by a vector in $\mathbb{R}^p$. This vector is constructed by evaluating $p$ functions at $x$, hence yielding the vector $(\varphi_1(x), \varphi_2(x), \ldots, \varphi_p(x))$. Of course, the choice of suitable functions $\{\varphi_j\}$ is very important and domain-specific.

The functions $\{\varphi_j\}$ (or, correspondingly, the columns of the matrix $\mathbf{X}$) have a variety of names. They are known as "covariates" and "predictors" in statistics, and as "features" in the context of machine learning and pattern recognition. The set of features $\{\varphi_j\}$ is sometimes called a "dictionary," and the matrix $\mathbf{X}$ is also referred to as the "design matrix." Finding an appropriate set of features, i.e., "featurizing," is a problem in and of itself. The observed $y_i$ are commonly referred to as the "responses" or "labels" within statistics and machine-learning, respectively. The act of finding the true set of coefficients $\theta_0$ is known as both "regression" and "supervised learning."

So, how do we calculate the coefficients $\theta_0$ from $y$? Going back to LS estimation, we desire to find a set of coefficients $\widehat{\theta}$ that best match our observations. Specifically, in matrix notation, (5.4) reads

$$\widehat{\theta}^{\mathrm{LS}} = \underset{\theta \in \mathbb{R}^p}{\operatorname{argmin}} \, \mathcal{L}(\theta), \tag{5.11}$$

where

$$\mathcal{L}(\theta) = \frac{1}{2n}\|y - \mathbf{X}\theta\|_2^2$$

$$= \frac{1}{2n}\sum_{i=1}^{n}(y_i - \langle x_i, \theta \rangle)^2, \tag{5.12}$$

with $x_i$ the $i$th row of $\mathbf{X}$. Here and below, $\|a\|_2$ denotes the $\ell_2$ norm of the vector $a$: $\|a\|_2^2 = \sum_i a_i^2$. The minimizer can be found by noting that

$$\nabla\mathcal{L}(\theta) = -\frac{1}{n}\mathbf{X}^\mathsf{T}(y - \mathbf{X}\theta), \tag{5.13}$$

and therefore $\quad \widehat{\theta}^{\mathrm{LS}} = \left(\mathbf{X}^\mathsf{T}\mathbf{X}\right)^{-1}\mathbf{X}^\mathsf{T}y. \tag{5.14}$

Looking at (5.14), we note that an important role is played by the sample covariance matrix

$$\widehat{\Sigma} = \frac{1}{n}\mathbf{X}^\mathsf{T}\mathbf{X}. \tag{5.15}$$

This is the matrix of correlations of the predictors $\{\varphi_1, \ldots, \varphi_p\}$. The most immediate remark is that for $\widehat{\theta}^{\mathrm{LS}}$ to be well defined, $\widehat{\Sigma}$ needs to be invertible, which is equivalent to requiring $\operatorname{rank}(\mathbf{X}) = p$. This of course can only happen if the number of parameters is no larger than the number of observations: $n \leq p$. If $\widehat{\Sigma}$ is invertible, but nearly singular, $\widehat{\theta}$ is unstable and hence a poor estimator. A natural way to quantify the "goodness" of $\widehat{\Sigma}$ is through its condition number $\kappa(\widehat{\Sigma})$, which is the ratio of its largest to its smallest eigenvalue: $\kappa(\widehat{\Sigma}) = \lambda_{\max}(\widehat{\Sigma})/\lambda_{\min}(\widehat{\Sigma})$. From this point of view, an optimal design has minimal condition number $\kappa(\widehat{\Sigma}) = 1$, which corresponds to $\mathbf{X}$ being proportional to an orthogonal matrix. In this case, $\mathbf{X}$ is called an "orthogonal design" and we shall fix normalizations by assuming $\widehat{\Sigma} = (\mathbf{X}^\mathsf{T}\mathbf{X}/n) = \mathrm{I}_p$

In functional terms, we see that the LS estimator is calculated according to the correlations between $y$ and the predictors,

$$\widehat{\Sigma}_{jl} = \frac{1}{n} \sum_{i=1}^{n} \varphi_j\left(i/n\right) \varphi_l\left(i/n\right), \tag{5.16}$$

and therefore
$$\widehat{\theta}_l^{\mathrm{LS}} = \sum_{j=1}^{p} \left(\widehat{\Sigma}^{-1}\right)_{lj} \left(\frac{1}{n} \sum_{i=1}^{n} \varphi_j(i/n)y_i\right). \tag{5.17}$$

### 5.1.4 Evaluating the estimator

Now that we have calculated the LS estimator for our problem, a natural question arises: Is this indeed the best estimator we could use? In order to answer this question, we need a way of comparing one estimator with another.

This is normally done by considering the *risk function* associated with the estimator. If the model depends on a set of parameters $\theta \in \mathbb{R}^p$, the risk function is a function $R : \mathbb{R}^p \to \mathbb{R}$, defined by

$$R(\theta) = \mathbb{E}\left\{\|\widehat{\theta}(y) - \theta\|_2^2\right\},$$

$$= \sum_{j=1}^{p} \mathbb{E}\left\{(\widehat{\theta}_j(y) - \theta_j)^2\right\}. \tag{5.18}$$

Here expectation is taken with respect to $y$, distributed according to the model (5.10) with $\theta_0 = \theta$. Note that the $\ell_2$ distance is used to measure the estimation error.

Other measures (called "loss functions") could be used as well, but we will focus on this for the sake of concreteness. We can also calculate risk over the function space and not just over the parameter space. This is also known as the "prediction error":

$$R_{\mathrm{p}}(\theta) = \frac{1}{n} \sum_{i=1}^{n} \mathbb{E}\left\{\left(\widehat{f}\left(i/n\right) - f\left(i/n\right)\right)^2\right\},$$

$$= \frac{1}{n} \sum_{i=1}^{n} \mathbb{E}\left\{\left[\sum_{j=1}^{p} \mathbf{X}_{ij}\left(\widehat{\theta}_j - \theta_j\right)\right]^2\right\},$$

$$= \frac{1}{n} \mathbb{E}\left\{\|\mathbf{X}(\widehat{\theta} - \theta)\|_2^2\right\}. \tag{5.19}$$

In particular, for $\mathbf{X}$ an orthogonal design, $R_{\mathrm{p}}(\theta) = c\,R(\theta)$.

Let us apply this definition of risk to the LS estimator $\widehat{\theta}^{\mathrm{LS}}$. Returning to the signal sampling model,

$$y = \mathbf{X}\theta_0 + w, \tag{5.20}$$

and therefore
$$\widehat{\theta}^{\mathrm{LS}} = \left(\mathbf{X}^{\mathsf{T}}\mathbf{X}\right)^{-1}\mathbf{X}^{\mathsf{T}}y,$$

$$= \theta_0 + (\mathbf{X}^{\mathsf{T}}\mathbf{X})^{-1}\mathbf{X}^{\mathsf{T}}w, \tag{5.21}$$

which shows that the LS estimator will return the true parameters $\theta_0$, perturbed by some amount due to noise. Now, we will calculate the risk function

$$R(\theta) = \mathbb{E}\left\{ \|\hat{\theta}^{\mathrm{LS}}(y) - \theta_0\|_2^2 \right\},$$

$$= \mathbb{E}\left\{ \| \left( \mathbf{X}^{\mathsf{T}}\mathbf{X} \right)^{-1} \mathbf{X}^{\mathsf{T}} w \|_2^2 \right\},$$

$$= \mathbb{E}\left\{ w^{\mathsf{T}}\mathbf{X} \left( \mathbf{X}^{\mathsf{T}}\mathbf{X} \right)^{-2} \mathbf{X}^{\mathsf{T}} w \right\},$$

$$= \sigma^2 \mathrm{Tr}\left( \mathbf{X} \left( \mathbf{X}; \mathsf{T} \right) \mathbf{X} \right)^{-2} \mathbf{X}^{\mathsf{T}},$$

$$= \sigma^2 \mathrm{Tr}\left( \left( \mathbf{X}^{\mathsf{T}}\mathbf{X} \right)^{-1} \right),$$

$$= \frac{\sigma^2 p}{n} \left[ \frac{\mathrm{Tr}\left( \widehat{\Sigma}^{-1} \right)}{p} \right], \tag{5.22}$$

where we add the $p$ term to the final result because we expect $(1/p)\mathrm{Tr}(\widehat{\Sigma}^{-1})$ to be of order one, under the assumption of near-orthonormal predictors.

To further illustrate this point, let us consider the case in which the functions $\{\varphi_j\}$ are orthonormal (more precisely, they are an orthonormal set in $L^2([0,1])$). This means that

$$\int_0^1 \varphi_i(x)\varphi_j(x)\,\mathrm{d}x = \delta_{ij}, \tag{5.23}$$

where $\delta_{ij}$ is 1 when $i = j$ and 0 for all $i \neq j$. For $n$ large, this implies

$$\widehat{\Sigma}_{jl} = \frac{1}{n}\sum_{i=1}^n \varphi_j\left(i/n\right)\varphi_l\left(i/n\right) \approx \delta_{jl}, \tag{5.24}$$

where we have assumed that the sum can be approximated by an integral. In other words, if the functions $\{\varphi_j\}$ are orthonormal, the design is nearly orthogonal, and this approximation improves as the number of samples increases. Thus, under such good conditions, $\mathrm{Tr}(\widehat{\Sigma}^{-1}) \approx p$. Under these conditions, we can simplify the risk function for the LS estimator in the case of orthonormal or near-orthonormal predictors to be

$$R(\theta) \approx \frac{p\sigma^2}{n}. \tag{5.25}$$

This result has several interesting properties:

- The risk is proportional to the noise variance. This makes sense: the larger the noise, the less well can we estimate the function.
- It is inversely proportional to the number of samples $n$: the larger the number of observations, the better can we estimate $f$.
- The risk is proportional to the number of parameters, $p$. This can be interpreted as an over-fitting phenomenon. If we choose a large $p$, then our estimator will be more sensitive to noise. Conversely, we are effectively searching the right parameters

in a higher-dimensional space, and a larger number of samples is required to determine it with the same accuracy.

- The risk $R(\theta)$ is independent of $\theta$. This is closely related to the linearity of the LS estimator.

At this point, two questions naturally arise:

Q1. Is this the best that we can do? Is it possible to use a different estimator and decrease risk?

Q2. What happens if the function to be estimated is not *exactly* given by a linear combinations of the predictors? Indeed in general, we cannot expect to have a perfect model for the signal, and any set of predictors is only approximate:

$$f(t) \neq \sum_{j=1}^{p} \theta_j \varphi_j(t). \tag{5.26}$$

To discuss these two issues, let us first change the notation for the risk function, by making explicit the dependence on the estimator $\widehat{\theta}$:

$$R(\theta; \widehat{\theta}), \quad \text{where} \quad \theta \in \mathbb{R}^p, \tag{5.27}$$

$$\widehat{\theta} : \mathbb{R}^n \to \mathbb{R}^p.$$

Note that estimators $\widehat{\theta}$ are functions of $y \in \mathbb{R}^n$. For two estimators $\widehat{\theta}^1$ and $\widehat{\theta}^2$, we can compare $R(\theta; \widehat{\theta}^1)$ and $R(\theta; \widehat{\theta}^2)$. This leads to the next, crucial question: how do we compare these two curves? For instance, in Fig 5.6, we sketch two cartoon risk functions $R(\theta; \widehat{\theta}^1)$ and $R(\theta; \widehat{\theta}^2)$. Which is the better? The way this question is answered has important consequences.

Note that, naively, one could hope to find an estimator that is *simultaneously* the best at all points $\theta$. Denoting this ideal estimator by $\widehat{\theta}^{\mathrm{opt}}$, we would get

$$R(\theta; \widehat{\theta}^{\mathrm{opt}}) \leq R(\theta; \widehat{\theta}) \quad \forall\, \theta, \widehat{\theta}. \tag{5.28}$$

However, assuming the existence of such an ideal estimator leads to a contradiction. To see this, we will let the predictors be, for simplicity,

$$\mathbf{X} = \sqrt{n}\mathbf{I}_n, \qquad p = n, \tag{5.29}$$

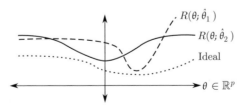

**Fig. 5.6** Comparing the risk functions of two different estimators $\widehat{\theta}^1$, $\widehat{\theta}^2$ over the space of possible parameters $\theta$. Also shown is a risk function for some estimator that is *ideal* in the sense that it is below both of the known estimators for all $\theta$.

which means our regression problem is now

$$y = \theta + \frac{w}{\sqrt{n}}, \qquad w \sim \mathcal{N}(0, \sigma^2 I_n). \tag{5.30}$$

Note that in this case, the LS estimator is simply $\widehat{\theta}^{\mathrm{LS}}(y) = y$.

Next, fix $\xi \in \mathbb{R}^p$, and consider the oblivious estimator that always returns $\xi$:

$$\widehat{\theta}^{\xi}(y) = \xi. \tag{5.31}$$

This has the risk function

$$R(\theta; \widehat{\theta}^{\xi}) = \|\xi - \theta\|_2^2, \tag{5.32}$$

as seen in Fig. 5.7. If an "ideal" estimator $\widehat{\theta}^{\mathrm{opt}}$ as above existed, it would beat $\widehat{\theta}^{\xi}$, which implies in particular

$$R(\xi; \widehat{\theta}^{\mathrm{opt}}) = 0. \tag{5.33}$$

Since $\xi$ is arbitrary, this would imply that the ideal estimator has risk everywhere equal to 0; i.e., it always reconstructs the true signal perfectly, independently of the noise. This is of course impossible.

One approach to overcome this problem is to evaluate for each risk function the corresponding Bayes risk. This amounts to averaging $R(\theta; \widehat{\theta})$ over $\theta$, using a certain prior distribution of the parameters $P(\theta)$. Namely,

$$R_{\mathrm{B}}(\mathrm{P}; \widehat{\theta}) \equiv \int R(\theta; \widehat{\theta}) \, \mathrm{P}(\mathrm{d}\theta). \tag{5.34}$$

However, it is not clear in every case how one might determine this prior. Further, the choice of $P(\theta)$ can completely skew the comparison between $\widehat{\theta}^1$ and $\widehat{\theta}^2$. If $P(\theta)$ is concentrated in a region in which, say, $\widehat{\theta}^1$ is superior to $\widehat{\theta}^2$, then $\widehat{\theta}^1$ will obviously win the comparison, and vice versa.

In the next section, we shall discuss the minimax approach to comparing estimators.

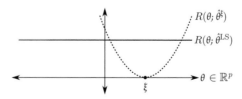

**Fig. 5.7** The risk associated with a trivial estimator as compared with the LS risk.

## 5.2   Nonlinear denoising and sparsity

### 5.2.1   Minimax risk

The previous section discussed estimating a set of parameters $\theta$, given the linear model

$$y = \mathbf{X}\theta + w. \tag{5.35}$$

In this discussion, we stated that there exists no estimator that dominates all other possible estimators in terms of risk, $R(\theta; \widehat{\theta})$ (Fig. 5.8). Still the question remains of how to compare two different estimators $\widehat{\theta}^1$ and $\widehat{\theta}^2$.

A fruitful approach to this question is to consider the worst-case risk over some region $\Omega \subseteq \mathbb{R}^p$. Formally, we define the *minimax risk* of $\widehat{\theta}$ over $\Omega$ as

$$R_*(\Omega; \widehat{\theta}) = \sup_{\theta \in \Omega} \ R(\theta; \widehat{\theta}). \tag{5.36}$$

Such a definition of risk is useful if we have some knowledge a priori about the region in which the true parameters live. The minimax risk allows us to compare the maximal risk of a given estimator over the set $\Omega$ to find an estimator with minimal worst-case risk. The minimax risk is also connected to the Bayes risk, which we defined earlier:

$$R_{\mathrm{B}}(\mathrm{P}; \widehat{\theta}) \equiv \int R(\theta; \widehat{\theta}) \, \mathrm{P}(\mathrm{d}\theta) \,, \tag{5.37}$$

$$R_*(\Omega; \widehat{\theta}) = \sup_{\mathrm{supp}(\mathrm{P}) \subseteq \Omega} R_{\mathrm{B}}(\mathrm{P}; \widehat{\theta}). \tag{5.38}$$

With this definition of minimax risk, it is easy to compute the minimax risk of LS:

$$R_*(\mathbb{R}^p; \widehat{\theta}^{\mathrm{LS}}) = \frac{p\sigma^2}{n} \left[ \frac{\mathrm{Tr}(\widehat{\Sigma}^{-1})}{p} \right]. \tag{5.39}$$

The LS estimator is optimal in the minimax sense:

**Theorem 5.1** *The LS estimator is minimax-optimal over $\mathbb{R}^p$. Namely, any estimator $\widehat{\theta}$ has minimax risk $R_*(\mathbb{R}^p; \widehat{\theta}) \geq R_*(\mathbb{R}^p; \widehat{\theta}^{LS})$.*

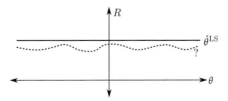

**Fig. 5.8**  The risk of an unknown estimator that dominates $\widehat{\theta}^{\mathrm{LS}}$ for all $\theta$.

**Proof** The proof of this result relies on the connection with Bayes risk. Consider for the sake of simplicity the case of orthogonal designs, $\widehat{\Sigma} = I$. It is not hard to show that if P is Gaussian with mean 0 and covariance $c^2 I_p$, then

$$\inf_{\widehat{\theta}} R_B(P; \widehat{\theta}) = \frac{c^2 \sigma^2}{c^2 + \sigma^2}. \tag{5.40}$$

Hence, for any estimator $\widehat{\theta}$,

$$\sup_{\theta \in \mathbb{R}^p} R(\theta; \widehat{\theta}) \geq R_B(P; \widehat{\theta}) \geq \frac{c^2 \sigma^2}{c^2 + \sigma^2}. \tag{5.41}$$

Since $c$ is arbitrary, we can let $c \to \infty$, whence

$$R_*(\mathbb{R}^p; \widehat{\theta}) \geq \sigma^2. \tag{5.42}$$

A full treatment of a more general result can be found, for instance, in Wasserman (2006, Chapter 7). $\square$

A last caveat. One might suspect—on the grounds of Theorem 5.1—that LS estimation is optimal "everywhere" in $\mathbb{R}^p$. This was indeed common belief among statisticians until the surprising discovery of the "Stein phenomenon" in the early 1960s (James and Stein, 1961). In a nutshell, for $p \geq 3$, there exist estimators that have risk $R(\theta; \widehat{\theta}) < R(\theta; \widehat{\theta}^{LS})$ strictly for every $\theta \in \mathbb{R}^p$! (The gap vanishes as $\theta \to \infty$.) We refer to Wasserman (2006, Chapter 7) for further background on this.

### 5.2.2 Approximation error and the bias-variance tradeoff

Until now we have assumed that the unknown function $f(t)$ could be exactly represented by the set of predictors, corresponding to columns of $\mathbf{X}$. How is our ability to estimate the parameters set $\theta$, and thus $f(t)$, affected when this assumption is violated?

In order to study this case, we assume that we are given an infinite sequence of predictors $\{\varphi_j\}_{j \geq 1}$, and use only the first $J$ to estimate $f$. For any fixed $J$, $f(t)$ can be approximated as a linear combination of the first $J$ predictors, plus an error term that is dependent upon $J$:

$$f(t) = \sum_{j=1}^{J} \theta_j \varphi_j(t) + \Delta_J(t). \tag{5.43}$$

For a complete set $\{\varphi_j\}$, we can ensure $\lim_{J \to \infty} \|\Delta_J\| = 0$ in a suitable norm. This can be formalized by assuming $\{\varphi_j\}_{j \geq 1}$ to be a orthonormal basis in the Hilbert space $L^2([0,1])$ and the above to be the orthonormal decomposition. In particular, the remainder will be orthogonal to the expansion:

$$\int_0^1 \Delta_J(t) \varphi_j(t) \, dt = 0 \quad \forall j \in \{1, \dots, J\}. \tag{5.44}$$

Alternatively, we can require orthogonality with respect to the sampled points (the resulting expansions are very similar for $n$ large):

$$\frac{1}{n} \sum_{i=1}^{n} \Delta_J\left(i/n\right) \varphi_j\left(i/n\right) = 0 \quad \forall j \in \{1, 2, \ldots, J\}. \tag{5.45}$$

With the remainder $\Delta_J$, our regression model becomes

$$y = \mathbf{X}\theta_0 + \Delta_J + w, \tag{5.46}$$

where $\mathbf{X} \in \mathbb{R}n \times J$, $w \sim \mathcal{N}\left(0, \sigma^2 I_n\right)$, and $\mathbf{X}^\mathsf{T} \Delta_J = 0$. Recall that the LS estimator is given by

$$\widehat{\theta} = (\mathbf{X}^\mathsf{T}\mathbf{X})^{-1}\mathbf{X}^\mathsf{T} y,$$

$$= \theta_0 + (\mathbf{X}^\mathsf{T}\mathbf{X})^{-1}\mathbf{X}^\mathsf{T} w. \tag{5.47}$$

We can compute the prediction risk as follows:

$$R_p(f) = \frac{1}{n} \mathbb{E}\left\{ \sum_{i=1}^{n} \left( \widehat{f}\left(i/n\right) - f\left(i/n\right) \right)^2 \right\},$$

$$= \frac{1}{n} \mathbb{E}\left\{ \|\mathbf{X}\widehat{\theta} - \mathbf{X}\theta_0 - \Delta_J\|_2^2 \right\},$$

$$= \frac{1}{n} \mathbb{E}\left\{ \|\mathbf{X}(\theta - \widehat{\theta}_0)\|_2^2 \right\} + \frac{1}{n}\|\Delta_J\|_2^2,$$

$$= \frac{1}{n} \mathbb{E}\left\{ \|\mathbf{X}(\mathbf{X}^\mathsf{T}\mathbf{X})^{-1}\mathbf{X}^\mathsf{T} w\|_2^2 \right\} + \frac{1}{n}\|\Delta_J\|_2^2,$$

$$= \frac{1}{n} \|\Delta_J\|_2^2 + \frac{\sigma^2}{n} \mathrm{Tr}\left( \mathbf{X}(\mathbf{X}^\mathsf{T}\mathbf{X})^{-1}\mathbf{X}^\mathsf{T} \right). \tag{5.48}$$

Finally, note that $\mathbf{X}(\mathbf{X}^\mathsf{T}\mathbf{X})^{-1}\mathbf{X}^\mathsf{T} \in \mathbb{R}^{n \times n}$ is the orthogonal projector to the space spanned by the columns of $\mathbf{X}$. Hence, its trace is always equal to $J$. This gives us the final form of the estimation risk at $f(\,\cdot\,)$, as a function of $J$:

$$R_p(f) = \frac{\|\Delta_J\|_2^2}{n} + J\frac{\sigma^2}{n}. \tag{5.49}$$

In other words, the estimation risk associated with $f(\,\cdot\,)$ is a sum of two terms, both of which are dependent upon the choice of $J$:

- The first term is associated with the approximation error induced by the choice of the predictor $\{\varphi_j\}$. It is independent of the noise variance $\sigma^2$, and decreases with $J$. We interpret it therefore as a *bias* term.
- The second term depends on the noise level, and is related to the fluctuations that the noise induces in $\widehat{\theta}$. It increases with the number of predictors $J$, as the fit becomes more unstable. We interpret it therefore as a *variance* term.

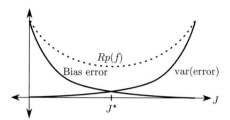

**Fig. 5.9**  The effect of bias and variance on estimation risk.

Therefore, the optimal number of predictors to use, $J^*$, is the one that minimizes the risk by striking a balance between bias and variance (Fig. 5.9). In other words, we want to find the optimal point between under- and over-fitting the model. Note that how we choose the predictors themselves determines the rate at which the bias term goes to zero as $J$ increases.

### *Example: The Fourier basis*

In this example, we select our set of predictors to be a Fourier basis

$$\varphi_j(t) = \sqrt{2}\cos((j-1)\pi t), \tag{5.50}$$

for $t \in [0, 1]$. If $f(\,\cdot\,)$ is square-integrable, then it can be represented as an infinite series (converging in $L^2([0, 1])$, i.e., in mean square error)

$$f(t) = \sum_{j=1}^{\infty} \theta_{0j}\varphi_j(t). \tag{5.51}$$

However, if only $J$ sinusoids are used, the remainder $\Delta_J$ is

$$\Delta_J(t) = f(t) - \sum_{j=1}^{J} \theta_{0j}\varphi_j(t). \tag{5.52}$$

And, finally, the squared norm of $\Delta_J$ is, by orthogonality,

$$\|\Delta_J\|_2^2 = \sum_{i=1}^{n} \Delta_J(i/n)^2$$

$$\approx n \int_0^2 \Delta_J(t)^2 \, \mathrm{d}t = n \sum_{j=J+1}^{\infty} \theta_{0j}^2. \tag{5.53}$$

Here we have replaced the sum by an integral, an approximation that is accurate for large $n$.

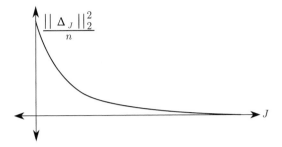

**Fig. 5.10** Depiction of the rate of decay of the bias term as a function of $J$.

Note that the decay of the bias term mirrors the decay of the Fourier coefficients of $f$, by (5.53). In particular, if $f$ is smooth, its Fourier coefficients decay faster, and hence the bias decays rapidly with $J$ (Fig. 5.10). In this case, the Fourier basis is a good set of features/predictors (a good dictionary) for our problem.

We now look at the case of Fourier predictors for a specific class of smooth functions, namely, functions whose second derivative is square-integrable. Formally, we define

$$W(C) \equiv \left\{ \int_0^1 \left(f''(t)\right)^2 \, dt \leq C^2 \right\},  \tag{5.54}$$

and we will consider estimation over $\Omega = W(C)$. This space is known in functional analysis as the "Sobolev ball of radius $C$ and order 2."

In terms of Fourier coefficients, this set of smooth functions can be characterized as

$$\int_0^1 \left( \sum_{j=1}^\infty \pi^2 (j-1)^2 \theta_{0j} \varphi_j(t) \right)^2 \, dt \leq C,  \tag{5.55}$$

or, equivalently,

$$\sum_{j=1}^\infty \pi^4 (j-1)^4 \theta_{0j}^2 \leq C.  \tag{5.56}$$

Now, for this to happen, we must have (5.56) satisfied,

$$\sum_{j=J+1}^\infty \theta_{0j}^2 \lesssim \frac{C'}{J^4},  \tag{5.57}$$

and hence we can estimate a bound on the squared norm of the remainder term:

$$\|\Delta_J\|_2^2 \approx n \sum_{j=J+1}^\infty \theta_{0j}^2 \lesssim \frac{nC'}{J^4}.  \tag{5.58}$$

Therefore, the prediction risk for the set of functions $\Omega = W(C)$ is upper-bounded:

$$R_{p,*}(\Omega; \widehat{\theta}) \lesssim \frac{C}{J^4} + \frac{\sigma^2 J}{n}. \tag{5.59}$$

The optimum value of $J$ is achieved when the two terms are of the same order, or by setting equal to 0 the derivative with respect to $J$:

$$\frac{\partial}{\partial J} \left\{ \frac{C}{J^4} + \frac{\sigma^2 J}{n} \right\} = \frac{\sigma^2}{n} - \frac{4C}{J^5}, \tag{5.60}$$

$$\text{and therefore} \quad J_* \sim \left( \frac{n}{\sigma^2} \right)^{1/5}. \tag{5.61}$$

Finally, with the optimal choice of $J$, we obtain the upper bound for the prediction risk, in general:

$$R_{p,*}(\Omega; \widehat{\theta}) \lesssim \left( \frac{\sigma^2}{n} \right)^{4/5}. \tag{5.62}$$

As in the standard parametric case, see (5.25), the risk depends on the ratio of the noise variance to the number of samples. However the decay with the number of samples is slower: $n^{-4/5}$ instead of $n^{-1}$. This is the price paid for not knowing in advance the $p$-dimensional space to which $f$ belongs. It can be proved that the exponent derived here is optimal.

### 5.2.3   Wavelet expansions

As we have emphasized several times, the quality of our function estimation procedure is depends strongly on the choice of the features $\{\varphi_j\}$. More precisely, it depends on the ability to represent the signal of interest with a few elements of this dictionary. While the Fourier basis works well for smooth signals, it is not an adequate dictionary for many signals of interest. For instance, the Fourier expansion does not work very well for images.

Why is this the case? Let us reconsider what are the LS estimates for the Fourier coefficients. Using orthonormality of the Fourier basis, we have

$$\widehat{\theta}_j^{\mathrm{LS}} \approx \frac{1}{n} \sum_{i=1}^{n} \varphi_j (i/n) \, y_i = \theta_{0,j} + \widehat{w}_j, \tag{5.63}$$

where

$$\widehat{w}_i = n^{-1} \sum_{i=1}^{n} \varphi_j (i/n) \, y_i.$$

Figure 5.11 shows a cartoon of these coefficients.

In other words, each estimated coefficients is a sum of two contributions: the true Fourier coefficients $\theta_{0,j}$ and the noise contribution $\widehat{w}_j$. Since the noise is white, its

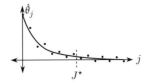

**Fig. 5.11**  Cutting off coefficients at $J^*$.

energy is equally spread across all Fourier modes. On the other hand, if the signal is smooth, its energy is concentrated on low-frequency modes. By selecting a cut-off at $J^*$, we are sacrificing some true signal information, in order to get rid of most of the noise. For frequencies higher than $J^*$, the noise energy surpasses any additional information these coefficients contain about the original signal we wish to estimate.

In other words, by selecting $J = J_*$, we are filtering out high frequencies in our measurements. In the "time" domain, this is essentially equivalent to averaging the observations over a sliding window of size of order $J_*^{-1}$. Formally, this is done by convolving the observations $y$ with some smooth kernel $K(\cdot)$:

$$\hat{f}(t) = \frac{1}{n} \sum_{i=1}^{n} K\left(\frac{i}{n} - t\right) y_i. \tag{5.64}$$

The full details of the kernel $K(\cdot)$ can be worked out, but what is important here is that it is $K(s)$ significantly different from 0 if and only if $|s| \le J_*^{-1}$. This point of view gives a different perspective on the bias–variance trade-off:

- For small $J_*$, we are averaging over a large window, and hence reducing the variance of our estimates. On the other hand, we are introducing a large bias in favor of smooth signals.
- For large $J_*$, we are averaging over a small window. The estimate is less biased, but has a lot of variance.

For truly smooth signals, this approach to denoising is adequate. However, for many signals, the degree of smoothness changes dramatically from one point of the signal to another. For instance, an image is mostly smooth, because of homogeneous surfaces corresponding to the same object or degree of illumination. However, they contain a lot of important discontinuities (e.g. edges) as well. Missing or smoothing out edges has a dramatic impact on the quality of reconstruction.

Smoothing with a kernel with uniform width produces a very bad reconstruction on such signals. If the width is large, the image becomes blurred across these edges. If the width is small, it will not filter out noise efficiently in the smooth regions. A different predictor set must be used that adapts to different levels of smoothness at different points of the image.

Wavelets are one such basis. A wavelet expansion of a function allows for localization of frequency terms, which means high-frequency coefficients can be localized to edges, while smoother content of the image can be more concisely described with just a few low-frequency coefficients. Wavelets, as in our previous example of the Fourier basis, are an orthonormal basis of $[0, 1]$. The expansion is formed via two functions: the *father-wavelet*, or *scaling*, function $\varphi(\cdot)$ and the *mother-wavelet* function $\psi_{jk}(\cdot)$. The mother-wavelet function is used to generate set of self-similar functions that are composed of scaled and shifted versions of the mother-wavelet:

$$\psi_{jk}(t) = 2^{j/2}\psi(2^j t - k), \quad \text{where} \quad j \in \{0, 1, 2, \dots\}, \tag{5.65}$$

$$k \in \{0, 1, \dots, 2^{j-1}\}.$$

Hence $j$ is an index related to frequency, and $k$ is related to position. The full wavelet expansion is then

$$f(t) = \theta_0 \varphi(t) + \sum_{j=0}^{\infty} \sum_{k=0}^{2^{j-1}} \theta_{0jk} \psi_{jk}(t). \tag{5.66}$$

There exist many families of wavelet functions, but the simplest among them is the Haar wavelet family. For the Haar wavelet, the wavelet functions are defined as

$$\varphi(t) = \begin{cases} 1 & \text{if } 0 \le t < 1, \\ 0 & \text{otherwise,} \end{cases}$$

and

$$\psi(t) = \begin{cases} -1 & \text{if } 0 \le t < \dfrac{1}{2}, \\ 1 & \text{if } \dfrac{1}{2} \le t < 1, \\ 0 & \text{otherwise.} \end{cases}$$

In Fig. 5.12, we see an example wavelet expansion of a piecewise-continuous function. Larger-magnitude wavelet coefficients will be located with the discontinuities in the original function across all scales.

Two problems arise naturally:

1. Unlike the Fourier basis, wavelet coefficients have no natural ordering of "importance," since each wavelet coefficient describes the function at a certain length scale, and in a certain position. Hence, the simple idea of fitting all coefficients up to a certain maximum index $J_*$ cannot be applied. If we select all coefficients corresponding to all positions up to a certain maximum frequency, we will not exploit the spatial adaptivity property of the wavelet basis.
2. Any linear estimation procedure that is also translation-invariant can be represented as a convolution, cf. (5.64), and thus incurs the problems outlined above. In order to treat edges differently from smooth regions in an image, a nonlinear procedure must be used.

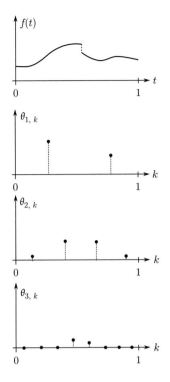

**Fig. 5.12** Wavelet coefficients of a piecewise-continuous function for increasing scale levels.

The simplest approach that overcomes these problems is the wavelet denoising method that was developed in a sequence of seminal papers by David Donoho and Iain Johnstone (Donoho and Johnstone, 1994*a,b*, 1995, 1998; Johnstone, 2011). The basic idea is to truncate, not according to the wavelet index, but according to the magnitude of the measured wavelet coefficient. In the simplest implementation, we proceed in two steps. First, we perform LS estimation of each coefficient. In the case of the orthogonal designs considered here, this yields

$$\tilde{y} = \frac{1}{n}\mathbf{X}^{\mathsf{T}}y = \theta + \tilde{w}, \tag{5.67}$$

Here $\tilde{w} = \mathbf{X}^{\mathsf{T}}w/n$ is again white noise $\tilde{w} \sim \mathcal{N}\left(0, (\sigma^2/n)\mathbf{I}_n\right)$. After this, coefficients are *thresholded*, independently:

$$\widehat{\theta}_i = \begin{cases} \tilde{y}_i, & \text{if } |\tilde{y}_i| \geq \lambda, \\ 0, & \text{otherwise}, . \end{cases}$$

The overall effect of this thresholding is to preserve large-magnitude wavelet coefficients while zeroing those that are "below the noise level." Since larger coefficients correspond to edges in the image, this approach seek to estimate higher frequencies near edges, only retaining low frequencies in smooth regions. This allows denoising without blurring across edges.

## 5.3   Denoising with thresholding

In the last section, we briefly described a denoising method, wavelet thresholding, that can adapt to a degree of smoothness that varies across a signal (e.g. an image). In this section, we work out some basic properties of this method, under a simple signal model. Apart from being interesting per se, this analysis provides key insights for generalizing the same method to high-dimensional statistical estimation problems beyond denoising. For an in-depth treatment, we refer, for instance, to Johnstone (2011) and Donoho and Johnstone (1994a).

To recall the our set-up, we are considering the model

$$y = \mathbf{X}\theta + w, \tag{5.68}$$

where $y \in \mathbb{R}^n$, $\mathbf{X} \in \mathbb{R}^{n \times p}$ are observed, and we want to estimate the vector of coefficients $\theta \in \mathbb{R}^p$. The vector $w$ is noise $w \sim \mathcal{N}\left(0, \sigma^2 \mathbf{I}_n\right)$. We are focusing on orthogonal designs, i.e., on the case $n = p$ with $\mathbf{X}^\mathsf{T}\mathbf{X} = n, \mathbf{I}_{n \times n}$.

There is no loss of generality in carrying out LS as a first step, which in this case reduces to

$$\tilde{y} = \frac{1}{n}\mathbf{X}^\mathsf{T}y = \theta + \tilde{w}, \quad \tilde{w} \sim \mathcal{N}\left(0, \frac{\sigma^2}{n}\mathbf{I}_{n \times n}\right). \tag{5.69}$$

In other words, in the case of orthogonal designs, we can equivalently assume that the unknown object $\theta$ has been observed directly, with additive Gaussian noise.

Since we expect $\theta$ to be sparse, it is natural to return a sparse estimate $\widehat{\theta}$. In particular, if $\tilde{y}_i$ is of the same order as the noise standard deviation $\sigma$, it is natural to guess that $\theta_i$ is actually very small or vanishing, and hence set $\widehat{\theta}_i = 0$. Two simple ways to implement this idea are "hard thresholding" and "soft thresholding."

Under *hard thresholding*, the estimate $\widehat{\theta} = (\widehat{\theta}_1, \ldots, \widehat{\theta}_p)$ of $\theta$ is given by

$$\widehat{\theta}_i = \begin{cases} \tilde{y}_i & \text{if } |\tilde{y}_i| \geq \lambda, \\ 0 & \text{else.} \end{cases} \tag{5.70}$$

Under *soft thresholding*, the estimate $\widehat{\theta}$ is given by

$$\widehat{\theta}_i = \begin{cases} \tilde{y}_i - \lambda & \text{if } \tilde{y}_i \geq \lambda, \\ 0 & \text{if } |\tilde{y}_i| \leq \lambda, \\ \tilde{y}_i + \lambda & \text{if } \tilde{y}_i \leq -\lambda. \end{cases} \tag{5.71}$$

These hard and soft thresholding functions are plotted in Fig. 5.13. While the two approaches have comparable properties (in particular, similar risk over sparse vectors),

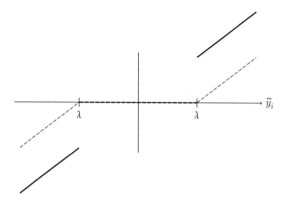

**Fig. 5.13** Soft thresholding (dashed lines) and hard thresholding (full lines).

we shall focus here on soft thresholding since it is most easily generalizable to other estimation problems.

Note that both hard and soft thresholding depend on a threshold parameter that we have denoted by $\lambda$. Entries below $\lambda$ are set to zero: to achieve minimal risk, it is of course crucial to select an appropriate $\lambda$. Ideally, the threshold should cut off the coefficients resulting from the noise, and hence we expect $\lambda$ to be proportional to the noise standard deviation $\sigma$. To determine the optimal choice of $\lambda$, let us first consider the case $\theta = 0$. Note that, when $\theta = 0$, $\tilde{y} \sim \mathcal{N}(0, (\sigma^2/n)\mathrm{I}_n)$ is a vector with i.i.d. Gaussian entries. We claim that, in this case,

$$\max_{i \in 1, \cdots, p} |\tilde{y}_i| \approx \sigma \sqrt{\frac{2 \log p}{n}}, \tag{5.72}$$

with probability very close to one.[1]

To see why this is the case, let $N(z) = \mathbb{E}\#\{i \in [p] : |\tilde{y}_i| \geq |z|\}$ be the expected number of coordinates in the vector $\tilde{y}$ that are above level $|z|$ or below $-|z|$. By linearity of expectation, we have

$$N(z) = 2p\,\Phi\left(-\frac{n|z|}{\sigma}\right), \tag{5.73}$$

where $\Phi(x) = \int_{-\infty}^{x} e^{-t^2/2} \mathrm{d}t / \sqrt{2\pi}$ is the Gaussian distribution function. Using the inequality $\Phi(-x) \leq e^{-x^2/2}/2$, valid for $x \geq 0$, we obtain $N(z) \leq p \exp(-nz^2/2\sigma^2)$. In particular, for any $\delta > 0$,

$$\mathbb{P}\left\{\max_{i \in 1, \cdots, p} |\tilde{y}_i| \geq \sigma \sqrt{\frac{2(1+\delta)\log p}{n}}\right\} \leq N\left(\sigma \sqrt{\frac{2(1+\delta)\log p}{n}}\right) \leq p^{-\delta}, \tag{5.74}$$

---

[1] In this derivation, we will be by choice somewhat imprecise, so as to increase readability. The reader is welcome to fill in the details, or to consult, for instance, Johnstone (2011) and Donoho and Johnstone (1994*a*).

which vanishes as $p \to \infty$. Roughly speaking, this proves that

$$\max_{i \in 1, \cdots, p} |\tilde{y}_i| \lesssim \sigma \sqrt{\frac{2 \log p}{n}} \qquad \text{with high probability.}$$

A matching lower bound can be proved by a second-moment argument, and we leave it to the reader to do this (or to refer to the literature).

Figure 5.14 reproduces the behavior of $\log N(z)$. The reader with a background in statistical physics has probably noticed the similarity between the present analysis and Derrida's treatment of the "random-energy model" (Derrida, 1981). In fact, the two models are identical and there is a close relationship between the problem addressed within statistical physics and estimation theory.

Figure 5.15 is a cartoon of the vector of observations $\tilde{y}$ in the case in which the signal vanishes: $\theta = 0$. All the coordinates of $\tilde{y}$ lie between $-\sigma \sqrt{(2 \log p)/n}$ and $+\sigma \sqrt{(2 \log p)/n}$. This suggests that the threshold $\lambda$ should be set such that all the entries that are pure noise are zeroed. This leads to the so-called following thresholding rule, proposed in Donoho and Johnstone (1994c):

$$\lambda = \sigma \sqrt{\frac{2 \log p}{n}} . \tag{5.75}$$

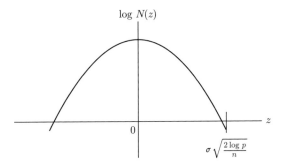

**Fig. 5.14**  Sketch of $\log N(z)$ (logarithm of the number of coordinates with noise level $z$).

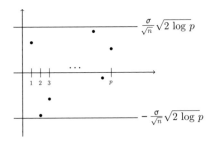

**Fig. 5.15**  Cartoon of the vector of observations $\tilde{y}$ when $\theta = 0$.

We now turn to evaluating the risk for such an estimator, when $\theta \neq 0$ is a sparse signal:

$$R(\theta; \widehat{\theta}) = \mathbb{E}\{\|\theta - \widehat{\theta}\|^2\} = \sum_{i=1}^{p} \mathbb{E}\{(\theta_i - \widehat{\theta}_i)^2\}. \tag{5.76}$$

We can decompose this risk as

$$R = R_0 + R_{\neq 0},$$

where $R_0$ (respectively, $R_{\neq 0}$) is the risk from entries $\theta_i$ that are zero (respectively, nonzero). The two contributions depend differently on $\lambda$: the contribution of zeros decreases with $\lambda$ since for large $\lambda$ more entries are set to 0. The contribution of nonzero entries instead increases with $\lambda$ since large $\lambda$ produces a larger bias; see Fig. 5.16 for a cartoon.

Under universal thresholding (Fig. 5.17), since $\max_{i:\theta_i=0} |\tilde{y}_i| \lesssim \sigma\sqrt{(2\log p)/n} = \lambda$, we have $R_0 \approx 0$. To evaluate the contribution of nonzero entries, we assume that $\theta$ is $s_0$ sparse; i.e., letting $\mathrm{supp}(\theta) \equiv \{i \in [p] : \theta_i \neq 0\}$, we have $|\mathrm{supp}(\theta)| \leq s_0$. Note that soft thresholding introduces a bias of size $\lambda$ on these entries, as soon as they are sufficiently less than $\lambda$. This gives an error per coordinate proportional to $\lambda^2$ (the variance contribution is negligible on these entries). This gives

$$R(\theta; \widehat{\theta}) \approx R_{\neq 0} \approx s_0 \lambda^2 = \frac{s_0 \sigma^2}{n}(2 \log p). \tag{5.77}$$

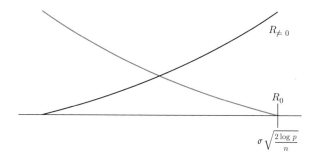

**Fig. 5.16** Risk $R_{\neq 0}$ in comparison with $R_0$.

**Fig. 5.17** Illustration of universal thresholding. The solid dots represent $\tilde{\theta}$ and the bars represent $\tilde{y}$.

We can now step back and compare this result with the risk of LS estimation (5.25). Neglecting the factor $2 \log p$, which is small even for very high dimension, our formula for sparse vectors (5.77) is the same as for LS, except that the dimension $p$ is replaced by the number of nonzero entries $s_0$. In other words, we basically achieve the same risk *as if* we knew a priori $\mathrm{supp}(\theta)$ and ran LS on that support! The extra factor $2 \log p$ is the price we pay for not knowing where the support is. For sparse vectors, we achieve an impressive improvement over LS.

Notice that this improvement is achieved simultaneously over all possible sparsity levels $s_0$, and the estimator does not need to know a priori $s_0$.

### 5.3.1   An equivalent analysis: Estimating a random scalar

There is a different, and essentially equivalent, way to analyze soft thresholding denoising. We will quickly sketch this approach because it provides an alternative point of view and, most importantly, because we will use some of its results in the following sections. We will omit spelling out the correspondence with the analysis in the last section.

We state this analysis in terms of a different—but essentially equivalent—problem. A source of information produces a random variable $\Theta$ in $\mathbb{R}$ with distribution $p_\Theta$, and we observe it corrupted by Gaussian noise. Namely, we observe $Y$ given by

$$Y = \Theta + \tau Z , \tag{5.78}$$

where $Z \sim \mathcal{N}(0,1)$ independent of $\Theta$, and $\tau$ is the noise standard deviation. We want to estimate $\Theta$ from $Y$. The following is a block diagram of this process:

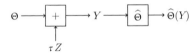

(Hint: the correspondence with the problem in the previous section is obtained by setting $\tau = \sigma/\sqrt{n}$ and $p_\Theta = (1/p) \sum_{i=1}^{P} \delta_{\theta_i}$.)

We saw in the previous sections that sparse vectors can be used to model natural signals (e.g., images in the wavelet domain). In the present framework, this can be modeled by the set of probability distributions that attribute mass at least $1 - \varepsilon$ to 0:

$$\mathcal{F}_\varepsilon = \left\{ p_\Theta \in \mathcal{P} \mid p_\Theta(\{0\}) \geq 1 - \varepsilon \right\}, \tag{5.79}$$

where $\mathcal{P}$ is the set of all probability distributions over the real line $\mathbb{R}$. Equivalently, $\mathcal{F}_\varepsilon$ is the class of probability distributions that can be written as $p_\Theta = (1 - \varepsilon)\delta_0 + \varepsilon Q$, where $\delta_0$ is the Dirac measure at 0 and $Q$ is an arbitrary probability distribution.

The Bayes risk of an estimator $\widehat{\Theta}$ is given by

$$R_{\mathrm{B}}(p_\Theta; \widehat{\Theta}) = \mathbb{E}\left\{ [\widehat{\Theta} - \Theta]^2 \right\}. \tag{5.80}$$

In view of the interesting properties of soft thresholding, unveiled in the previous section, we will assume that $\Theta$ is obtained by soft thresholding $Y$. It is convenient at this point to introduce some notation for soft thresholding:

$$\eta(z;\lambda) = \begin{cases} z - \lambda & \text{if } z \geq \lambda, \\ 0 & \text{if } |z| \leq \lambda, \\ z + \lambda & \text{if } z \leq -\lambda. \end{cases} \tag{5.81}$$

With an abuse of notation, we write $R_B(p_\Theta;\lambda) = R_B(p_\Theta;\eta(\,\cdot\,;\lambda))$ for the Bayes risk of soft thresholding with threshold $\lambda$. Explicitly,

$$R_B(p_\Theta;\lambda) = \mathbb{E}\big\{[\eta(Y;\lambda) - \Theta]^2\big\}. \tag{5.82}$$

We are interested in bounding the risk $R_B(p_\Theta;\lambda)$ for all sparse signals, i.e., in the present framework, for all the probability distributions $p_\Theta \in \mathcal{F}_\varepsilon$. We then consider the minimax risk:

$$R_*(\varepsilon;\tau^2) = \inf_\lambda \sup_{p_\Theta \in \mathcal{F}_\varepsilon} R_B(p_\Theta;\lambda). \tag{5.83}$$

First note that the class $\mathcal{F}_\varepsilon$ is scale-invariant. If $p_\Theta \in \mathcal{F}_\varepsilon$, also the probability distribution that is obtained by "stretching" $p_\Theta$ by any positive factor $s$ is in $\mathcal{F}_\varepsilon$. Hence the only scale in the problem is the noise variance $\tau^2$. It follows that

$$R_*(\varepsilon;\tau^2) = M(\varepsilon)\,\tau^2 \tag{5.84}$$

for some function $M(\varepsilon)$. Explicit formulae for $M(\varepsilon)$ can be found, for instance, in Donoho *et al.* (2009, Supplementary Material) or Donoho *et al.* (2011). A sketch is shown in Fig. 5.18: in particular, $M(\varepsilon) \approx 2\varepsilon \log(1/\varepsilon)$ as $\varepsilon \to 0$. By the same scaling argument as above, the optimal threshold $\lambda$ takes the form

$$\lambda^* = \tau \ell(\varepsilon), \tag{5.85}$$

where the function $\ell(\varepsilon)$ can also be computed and behaves as $\ell(\varepsilon) \approx \sqrt{2\log(1/\varepsilon)}$ for small $\varepsilon$. Finally, the worst-case signal distribution is

$$p_\Theta^* = (1 - \varepsilon)\delta_0 + \varepsilon\delta_\infty. \tag{5.86}$$

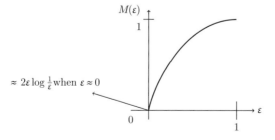

**Fig. 5.18**   Sketch of the minimax risk of soft thresholding $M(\varepsilon)$.

Note that the small-$\varepsilon$ behavior matches—as expected—the very sparse limit for vector denoising derived in the previous section. The correspondence is obtained by substituting $\epsilon = s_0/p$ for the fraction of nonzero entries and noting that the vector risk is

$$R = pR_*(\varepsilon; \tau) = pM(\varepsilon)\tau^2. \tag{5.87}$$

When $s_0 \ll p$, we have $\epsilon \approx 0$ and

$$R = pM(\varepsilon)\tau^2 \approx p\, 2\varepsilon \log \frac{1}{\varepsilon} \tau^2$$

$$= p\, 2\frac{s_0}{p} \log \frac{p}{s_0} \frac{\sigma^2}{n}$$

$$= \frac{2s_0\sigma^2}{n} \log \frac{p}{s_0}, \tag{5.88}$$

which matches the behavior derived earlier.

## 5.4   Sparse regression

Up to now, we have focused on estimating $\theta \in \mathbb{R}^p$ from observations of the form

$$y = \mathbf{X}\theta + w, \tag{5.89}$$

where $y \in \mathbb{R}^n$, $\mathbf{X} \in \mathbb{R}^{n \times p}$ are known and $w \in \mathbb{R}^n$ is an unknown noise vector. We focused in the previous case on orthogonal designs $n \geq p$ and $\mathbf{X}^{\mathsf{T}}\mathbf{X} = n\, \mathrm{I}_{p \times p}$.

Over the last decade, there has been a lot of interest in the underdetermined case $n \ll p$, with as general a $\mathbf{X}$ as possible, which naturally emerges in many applications. It turns out that good estimation is possible provided $\theta$ is highly structured, and in particular when it is very sparse. Throughout, we let $\|\theta\|_0$ denote the "$\ell_0$ norm" of $\theta$, i.e., the number of nonzero entries in $\theta$ (note that this is not really a norm). The main outcome of the work in this area is that the number of measurements needs to scale with the number of nonzero $\|\theta\|_0$ instead of the ambient dimension $p \gg \|\theta\|_0$. This setup is the so-called *sparse regression* or *high-dimensonal regression* problem.

### 5.4.1   Motivation

It is useful to overview a few scenarios where the above framework applies and, in particular, the high-dimensional regime $n \ll p$ plays a crucial role.

**Signal processing.** An image can be modeled, for instance, by a function $f : [0, 1] \times [0, 1] \rightarrow \mathbb{R}$ if it is grayscale. Color images require three scalars at each point, three-dimensional imaging requires a domain $[0, 1] \times [0, 1] \times [0, 1]$, and so on. Many imaging devices can be modeled (to first order) as linear operators $A$ creating a vector $y \in \mathbb{R}^n$ corrupted by noise $w$:

$$y = Af + w, \tag{5.90}$$

As a useful example to be kept in mind, $A$ can be the partial Fourier matrix, i.e., the operator computing a subset of Fourier coefficients. As emphasized in the previous sections, the image $f$ is often sparse in some domain, say the wavelet transform domain. That is, $f = T\theta$, where $\theta$ is sparse and $T$ is the wavelet transform, or some other sparsifying transform. This gives rise to the model

$$y = Af + w = (AT)\theta + w = \mathbf{X}\theta + w, \tag{5.91}$$

where $\mathbf{X} = AT$. Here $n$ corresponds to the number of measurements, while $p$ scales with the number of wavelet coefficients and hence with the resolution that we want to achieve. The high-dimensional regime $n \ll p$ is therefore very useful, since it corresponds to simpler measurements and higher resolution.

**Machine learning.** In web services, we often want to predict an unknown property of a user, on the basis of a large amount of known data about her. For instance, an online social network such as Facebook might want to estimate the income of its users, in order to display targeted advertisements. For each user $i$, we can construct a feature vector $x_i \in \mathbb{R}^p$, where, for example,

$$x_i = (\text{age, location, number of friends, number of posts,}$$
$$\text{time of first post in a day, ...}). \tag{5.92}$$

In a linear model, we assume

$$\underbrace{y_i}_{\text{income}} = \langle x_i, \theta \rangle + w_i, \tag{5.93}$$

Combining all users, we have

$$y = \mathbf{X}\theta + w, \tag{5.94}$$

where $y = (y_1, y_2, \ldots, y_n)$ is a vector comprising all response variables (e.g., the customers' income), and $\mathbf{X}$ is a matrix whose $i$th row is the feature vector $x_i$ of the $i$th customer. Typically, one constructs feature vectors with tens of thousands of attributes, hence giving rise to $p = 10^4$ to $10^5$. On the other hand, in order to fit such a model, the response variable (income) $y_i$ needs to be known for a set of users, and this is often possible only for $n \ll p$ users.

Luckily, only a small subset of features is actually relevant to predict income, and hence we are led to use sparse estimation techniques.

## 5.4.2  The LASSO

The LASSO (least absolute shrinkage and selection operator) presented in Tibshirani (1996), also known as basis pursuit denoising (BPDN) (Chen and Donoho, 1995; Chen *et al.*, 1998) is arguably the most successful method for sparse regression. The LASSO estimator is defined in terms of an optimization problem

$$\widehat{\theta} = \underset{\theta \in \mathbb{R}^p}{\operatorname{argmin}} \left\{ \underbrace{\frac{1}{2n} \|y - \mathbf{X}\theta\|_2^2}_{\text{Residual sum of squares}} + \underbrace{\lambda \|\theta\|_1}_{\text{Regularizer}} \right\} \tag{LASSO}$$

The term $\mathcal{L}(\theta) = (1/2n)\|y - \mathbf{X}\theta\|_2^2$ is the ordinary LS cost function, and the regularizer $\lambda\|\theta\|_1$ promotes sparse vectors by penalizing coefficients different from 0. Note that the optimization problem is convex and hence can be solved efficiently: we will discuss a simple algorithm in the following.

To gain insight as to why the LASSO is well suited for sparse regression, let us start by revisiting the case of orthogonal designs, namely, $n \geq p$ and

$$\mathbf{X}^\mathsf{T}\mathbf{X} = n\,\mathrm{I}_{n \times n}. \tag{5.95}$$

Rewriting $\mathcal{L}(\theta)$,

$$\begin{aligned}
\mathcal{L}(\theta) &= \frac{1}{2n}\,\langle y - \mathbf{X}\theta, y - \mathbf{X}\theta\rangle \\
&= \frac{1}{2n}\,\left\langle y - \mathbf{X}\theta, \frac{1}{n}\mathbf{X}\mathbf{X}^\mathsf{T}(y - \mathbf{X}\theta)\right\rangle \\
&= \frac{1}{2n^2}\,\left\langle \mathbf{X}^\mathsf{T}y - n\theta, \mathbf{X}^\mathsf{T}y - n\theta\right\rangle \\
&= \frac{1}{2}\,\left\|\theta - \frac{1}{n}\mathbf{X}^\mathsf{T}y\right\|^2 \\
&= \frac{1}{2}\,\|\theta - \tilde{y}\|^2, \qquad \text{where } \tilde{y} = \frac{1}{n}\mathbf{X}^\mathsf{T}y. \tag{5.96}
\end{aligned}$$

Thus, in this case, the LASSO problem is equivalent to

$$\text{minimize } \sum_{i=1}^{p}\left\{\frac{1}{2}\left|\tilde{y}_i - \theta_i\right|^2 + \lambda|\theta_i|\right\}. \tag{5.97}$$

This is a "separable" cost function, and we can minimize each coordinate separately. Let $F(\theta_i) = \frac{1}{2}(\tilde{y} - \theta_i)^2 + \lambda|\theta_i|$. Now,

$$\frac{\partial F}{\partial \theta_i} = \theta_i - \tilde{y}_i + \lambda\,\mathrm{sign}(\theta_i), \tag{5.98}$$

where $\mathrm{sign}(\cdot)$ denotes the sign function shown in Fig. 5.19. Note that $|\theta_i|$ is nondifferentiable at $\theta_i = 0$. How should we interpret its derivative $\mathrm{sign}(\theta_i)$ in this case? For convex functions (which is the case here), the derivative can be safely replaced by the "subdifferential," i.e., the set of all possible slopes of tangent lines at $\theta_i$ that stay below the graph of the function to be differentiated. The subdifferential coincides with the usual derivative when the function is differentiable. For the function $\theta_i \to |\theta_i|$, it is an easy exercise to check that the subdifferential at $\theta_i = 0$ is given by the interval $[-1, 1]$. In other words, we can think of Fig. 5.19 as the correct graph of the subdifferential of $|\theta_i|$ if we interpret its value at 0 as given by the whole interval $[-1, 1]$.

The minimizer of $F(\theta_i)$ must satisfy

$$\tilde{y}_i = \theta_i + \lambda\,\mathrm{sign}(\theta_i). \tag{5.99}$$

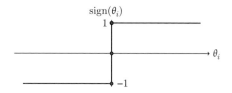

**Fig. 5.19** The sign function.

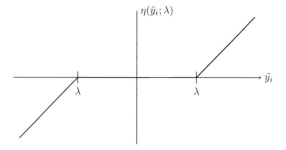

**Fig. 5.20** The soft thresholding function.

Hence we can obtain the minimizer as a function of $\tilde{y}_i$ by adding $\theta_i$ to the graph in Fig. 5.19 and flipping the axis. The result is plotted in Fig. 5.20. The reader will recognize the soft thresholding function $\eta(\,\cdot\,;\lambda)$ already encountered in the previous section. Summarizing, in the case of orthogonal designs, the LASSO estimator admits the explicit representation

$$\widehat{\theta} = \eta\Big(\frac{1}{n}\mathbf{X}^\mathsf{T} y; \lambda\Big), \tag{5.100}$$

where it is implicitly understood that the soft thresholding function is applied componentwise to the vector $(1/n)\mathbf{X}^\mathsf{T} y$. As we saw in the previous section, componentwise soft thresholding has nearly optimal performances on this problem, and hence the same holds for the LASSO.

In the high-dimensional setting, $p \gg n$ and $\mathbf{X}$ is obviously not orthogonal, and the LASSO estimator is non-explicit. Nevertheless it can be computed efficiently, and we will discuss next a simple algorithm that is guaranteed to converge. It is an example of a generic method for convex optimization known as a "subgradient" or "projected gradient" approach. The important advantage of these algorithms (and more generally of "first-order methods") is that their complexity per iteration scales only linearly in the dimensions $p$ of the problem, and hence they are well suited for high-dimensional applications (Juditsky and Nemirovski, 2011). They do not converge so rapidly as, for instance, Newton's method, but this is often not crucial. For statistical problems, a "low-precision" solution is often as good as a "high-precision" one, since in any case there is an unavoidable statistical error to deal with.

We want to minimize the cost function

$$F(\theta) = \frac{1}{2n}\|y - \mathbf{X}\theta\|^2 + \lambda\|\theta\|_1 . \tag{5.101}$$

At each iteration, the algorithm constructs an approximation $\theta^{(t)}$ of the minimizer $\widehat{\theta}$. In order to update this state, the idea is to construct an upper bound to $F(\theta)$ that is easy to minimize and is a good approximation of $F(\theta)$ close to $\theta^{(t)}$ (Fig. 5.21). Rewriting $\mathcal{L}(\theta)$,

$$\begin{aligned}
\mathcal{L}(\theta) &= \frac{1}{2n}\|y - \mathbf{X}\theta\|_2^2 \\
&= \frac{1}{2n}\|y - \mathbf{X}\theta^{(t)} - \mathbf{X}(\theta - \theta^{(t)})\|_2^2 \\
&= \frac{1}{2n}\|y - \mathbf{X}\theta^{(t)}\|_2^2 - \frac{1}{n}\left\langle \mathbf{X}(\theta - \theta^{(t)}), y - \mathbf{X}\theta^{(t)} \right\rangle + \frac{1}{2n}\|\mathbf{X}(\theta - \theta^{(t)})\|_2^2. \tag{5.102}
\end{aligned}$$

Note that the first two terms are "simple" in that they are linear in $\theta$. The last term is "small" for $\theta$ close to $\theta^{(t)}$) (quadratic in $(\theta - \theta^{(t)})$). We will upper- bound the last term. Suppose the largest eigenvalue of $(1/n)\mathbf{X}^{\mathsf{T}}\mathbf{X}$ is bounded by $L$,

$$\lambda_{\max}\left(\frac{1}{n}\mathbf{X}^{\mathsf{T}}\mathbf{X}\right) \le L, \tag{5.103}$$

and let $v = (1/n)\mathbf{X}^{\mathsf{T}}(y - \mathbf{X}\theta^{(0)})$. Then

$$\begin{aligned}
\mathcal{L}(\theta) &= \frac{1}{2n}\|y - \mathbf{X}\theta^{(t)}\|_2^2 - \left\langle v, \theta - \theta^{(t)} \right\rangle + \frac{1}{2}\left\langle \theta - \theta^{(t)}, \frac{1}{n}\mathbf{X}^{\mathsf{T}}\mathbf{X}(\theta - \theta^{(t)}) \right\rangle \\
&\le \frac{1}{2n}\|y - \mathbf{X}\theta^{(t)}\|_2^2 - \left\langle v, \theta - \theta^{(t)} \right\rangle + \frac{L}{2}\|\theta - \theta^{(t)}\|_2^2 \\
&= \underbrace{\frac{1}{2n}\|y - \mathbf{X}\theta^{(t)}\|_2^2 - \frac{1}{2L}\|v\|_2^2}_{\triangleq C} + \frac{L}{2}\left\|\theta - \theta^{(t)} - \frac{1}{L}v\right\|_2^2 . \tag{5.104}
\end{aligned}$$

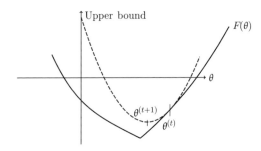

**Fig. 5.21**  Subgradient approach: construct a convenient upper bound to $F(\theta)$.

We therefore obtain the following upper bound, where $C$ is a constant independent of $\theta$:

$$F(\theta) \leq C + \lambda\|\theta\|_1 + \frac{L}{2}\left\|\theta - \theta^{(0)} - \frac{v}{L}\right\|_2^2. \tag{5.105}$$

We compute the next iterate $\theta^{(t+1)}$ by minimizing the above upper bound:

$$\text{minimize} \quad \frac{\lambda}{L}\|\theta\|_1 + \frac{1}{2}\|\theta - \tilde{\theta}^{(0)}\|^2, \tag{5.106}$$

$$\tilde{\theta}^{(0)} = \theta^{(0)} + \frac{1}{nL}\mathbf{X}^{\mathsf{T}}(y - \mathbf{X}\theta^{(0)}). \tag{5.107}$$

We have already solved this problem when discussing the case of orthogonal designs. The solution is given by the soft thresholding operator:

$$\theta^{(t+1)} = \eta\left(\theta^{(t)} + \frac{1}{nL}\mathbf{X}^{\mathsf{T}}(y - \mathbf{X}\theta^{(t)}); \frac{\lambda}{L}\right). \tag{5.108}$$

This yields an iterative procedure known as iterative soft thresholding that can be initialized arbitrarily, for example, with $\theta^{(0)} = 0$. This algorithm is guaranteed to always converge, as shown in Daubechies *et al.* (2004) and Beck and Teboulle (2009), in the sense that

$$F(\theta^{(t)}) - F(\widehat{\theta}) \leq \frac{\text{constant}}{t}. \tag{5.109}$$

Note that this is much slower that the rate achieved by Newton's method. However, it can be proved that no first-order method (i.e., no method using only gradient information) can achieve a global convergence rate faster than $1/t^2$ for any problem in the class of the LASSO. We refer to Juditsky and Nemirovski (2011) for a recent introduction to first-order methods.

To conclude this section, it is instructive to quickly consider two special cases: $\lambda \to \infty$ and $\lambda \to 0$. We rewrite the minimization problem as

$$\widehat{\theta} = \operatorname*{argmin}_{\theta \in \mathbb{R}^p} F(\theta), \tag{5.110}$$

$$\frac{1}{\lambda}F(\theta) = \frac{1}{2n\lambda}\|y - \mathbf{X}\theta\|_2^2 + \|\theta\|_1. \tag{5.111}$$

When $\lambda \to \infty$, the first term vanishes and $\widehat{\theta} \to \operatorname{argmin}_\theta \|\theta\|_1 = 0$. In fact, $\widehat{\theta} = 0$ for all $\lambda \geq \lambda_*$ for some critical $\lambda_*$. When $\lambda \to 0$, the weight in front of the first term goes to infinity, and hence the equality $y = \mathbf{X}\theta$ is enforced strictly. In the high-dimensional regime $p > n$, this linear system is underdetermined and has multiple solutions. The most relevant is selected by minimizing the $\ell_1$ norm. In other words, as $\lambda \to 0$, the LASSO estimator $\widehat{\theta}$ converges to the solution of the following problem (known as "basis pursuit"):

$$\text{minimize} \quad \|\theta\|_1, $$

$$\text{subject to} \quad y = \mathbf{X}\theta. \tag{5.112}$$

### 5.4.3   Behavior of the LASSO under the restricted isometry property

A significant amount of theory has been developed to understand and generalize the remarkable properties of the LASSO estimator and its empirical success. The theory establishes certain optimality properties under suitable assumptions on the design matrix $\mathbf{X}$. The most popular of these assumptions goes under the name of the restricted isometry property (RIP) and was introduced in the groundbreaking work of Candés, Tao, and collaborators (Candés and Tao, 2005, 2007). Several refinements of this condition have been developed in recent years (the restricted eigenvalue condition of Bickel *et al.* (2009), the compatibility condition of Bühlmann and Van De Geer (2011), and so on).

In order to motivate the RIP, we notice that the LASSO estimator performs well when the columns of the matrix $\mathbf{X}$ are orthogonal, $\mathbf{X}^\mathsf{T}\mathbf{X} = n\,\mathrm{I}_{n \times n}$. Indeed, this is the case of orthogonal designs explored above. The orthogonality condition $\mathbf{X}^\mathsf{T}\mathbf{X} = n\,\mathrm{I}_{n \times n}$ is equivalent to

$$\|\mathbf{X}v\|^2 = n\|v\|^2 \quad \text{for all } v \in \mathbb{R}^p \tag{5.113}$$

This is of course impossible in the high-dimensional regime $p > n$ (indeed, the null space of $\mathbf{X}$ has dimension at least $p - n$). The idea is to relax this condition, by requiring that $\mathbf{X}$ is "almost orthogonal" instead of orthogonal, and only when it acts on sparse vectors. Explicitly, we say that $\mathbf{X}$ satisfies the condition $\mathrm{RIP}(k, \delta)$ for some integer $k$ and $\delta \in (0, 1)$ if

$$(1 - \delta)\|v\|^2 \le \frac{1}{n}\|\mathbf{X}v\|^2 \le (1 + \delta)\|v\|^2$$

$$\text{for all } v \in \mathbb{R}^p \text{ with } \|v\|_0 \le k \quad \text{(RIP property).} \tag{5.114}$$

It is possible to show that this definition is non-empty and indeed—in a certain sense—most matrices satisfy it. For instance, if $\mathbf{X}$ has i.i.d. entries $\mathbf{X}_{ij} \sim \mathrm{Unif}\{+1, -1\}$ or $\mathbf{X}_{ij} \sim \mathcal{N}(0, 1)$, then, with high probability, $\mathbf{X}$ satisfies $\mathrm{RIP}(k, \delta)$ for a fixed $\delta$ and $n \ge Ck\log(p/k)$. The RIP property has been established for a large number of matrix constructions. For instance, partial Fourier matrices[2] satisfy RIP with high probability for $n \ge C_1\,k(\log p)^4$, as shown in Rudelson and Vershynin (2008).

The following theorem illustrates the utility of RIP matrices for sparse estimation. It is a simplified version of stronger results established in Bickel *et al.* (2009) (without any attempt at reproducing optimal constants, or the explicit dependence of all the quantities). Results of the same nature were proved earlier in Candés and Tao (2007) for a closely related estimator, known as the "Dantzig selector."

**Theorem 5.2 (Candés and Tao (2007); Bickel *et al.* (2009))** *If $\theta$ is $s_0$-sparse and $\mathbf{X}$ satisfies $\mathrm{RIP}(10\,s_0, 0.1)$, then, by choosing $\lambda = \sigma\sqrt{(5\log p)/n}$, we have, with high probability for a suitable constant $C > 0$,*

$$\|\widehat{\theta} - \theta\|_2^2 \le \frac{Cs_0\sigma^2}{n}\log p\,. \tag{5.115}$$

---

[2] That is, matrices obtained by subsampling randomly the rows of the $p \times p$ discrete Fourier transform.

A few observations are in order. It is—once again—instructive to compare this bound with the risk of ordinary LS, cf. (5.25). Apart from the $\log p$ factor, the error scales as if $\theta_0$ were $s_0$-dimensional. As in the case of orthogonal designs discussed above, we obtain roughly the same scaling as if the support of $\theta$ were known. Also, the choice of $\lambda$ scales as in the case where $\mathbf{X}$ is orthogonal.

Finally, as $\sigma \to 0$, we have $\widehat{\theta} \to \theta$ provided the RIP condition is satisfied. As mentioned above, this happens for random design matrices if $n \geq Cs_0 \log p$. In other words, we can reconstruct exactly an $s_0$-sparse vector from about $s_0 \log p$ random linear observations.

## *Modeling the design matrix* **X**

The restricted isometry property and its refinements/generalizations allow one to develop a powerful theory of high-dimensional statistical estimation (both in the context of linear regression and beyond). This approach has a number of strengths:

(a) Given a matrix $\mathbf{X}$, we can characterize it in terms of its RIP constant, and hence obtain a bound on the resulting estimation error. The bound holds uniformly over all signals $\theta$.

(b) The resulting bound is often nearly optimal.

(c) Many classes of random matrices of interest have been proved to possess RIP.

(d) RIP allows one to decouple the analysis of the statistical error, for example, the risk of the LASSO estimator $\widehat{\theta}$ (which is the main object of interest to statisticians) from the development of algorithms to compute $\widehat{\theta}$ (which is the focus within the optimization community).

The RIP theory also has some weaknesses. It is useful to understand them since this exercise leads to several interesting research directions that are—to a large extent—still open:

(a) In practice, given a matrix $\mathbf{X}$, it is NP-hard to determine whether it has the RIP. Hence, one has often to rely on the intuition provided by random matrix constructions.

(b) The resulting bounds are typically optimal *up to a constant factor*, which can be quite large. This makes it difficult to compare different estimators for the same problem. If the estimator $\widehat{\theta}^{(1)}$ has a risk that is, say, twice as large as that of $\widehat{\theta}^{(2)}$, this is often not captured by this theory.

(c) As a special case of the last point, RIP theory provides little guidance for the practically important problem of selecting the right amount of regularization $\lambda$. It is observed in practice that changing $\lambda$ by a modest amount has important effects on the quality of estimation, but this is hardly captured by RIP theory.

(d) Since RIP theory aims at bounding the risk uniformly over all (sparse) vectors $\theta$, it is typically driven by the "worst-case" vectors, and is overly conservative for most $\theta$'s.

Complementary information on the LASSO, and other high-dimensional estimation methods, can be gathered by studying simple random models for the design matrix $\mathbf{X}$. This will be the object of the next section.

## 5.5  Random designs and approximate message passing

In this section, we revisit the linear model (5.89) and the LASSO estimator, while assuming a very simple probabilistic model for the design matrix $\mathbf{X}$. Before proceeding, we should therefore ask: Is there any application for which probabilistic design matrices are well suited? Two type of examples come to mind:

- In statistics and machine learning, we are given pairs (response variable, covariate vector), $(y_1, x_1)$, ..., $(y_n, x_n)$, and postulate a relationship as for instance in (5.93). These pairs can often be thought as samples from a larger "population"; for example, customers of an e-commerce site are samples of a population of potential customers.
  One way to model this is to assume that the covariate vectors $x_i$, i.e., the rows of $\mathbf{X}$, are i.i.d. samples from a distribution.
- In compressed sensing, the matrix $\mathbf{X}$ models a sensing or sampling device that is designed within some physical constraints. Probabilistic constructions have been proposed and implemented by several authors (see, e.g., Tropp *et al.*, 2010). A cartoon example of these constructions is obtained by sampling i.i.d. random rows from the discrete $p \times p$ Fourier transform.

In other words, random design matrices $\mathbf{X}$ with i.i.d. rows can be used to model several applications. Most work, however, has focused on the special case in which the rows are i.i.d. with distribution $\mathcal{N}(0, \mathrm{I}_{p \times p})$. Equivalently, the matrix $\mathbf{X}$ has i.i.d. entries $\mathbf{X}_{i,j} \sim \mathcal{N}(0, 1)$. Despite its simplicity, this model has been an important playground for the development of many ideas in compressed sensing, starting with the pioneering work of Donoho (2006) and Donoho and Tanner (2005*a,b*). Recent years have witnessed an explosion of contributions thanks to the convergence of powerful ideas from high-dimensional convex geometry and Gaussian processes (see, e.g., Chandrasekaran *et al.*, 2012; Candés and Recht, 2013; Stojnic, 2013; Oymak and Hassibi, 2013; Amelunxen *et al.*, 2013). Non-rigorous ideas from statistical physics have also been used (Kabashima *et al.*, 2009; Rangan *et al.*, 2009; Guo *et al.*, 2009; Krzakala *et al.*, 2012).

Here we follow a rigorous approach that builds upon ideas from statistical physics, information theory, and graphical models, and is based on the analysis of a highly efficient reconstruction algorithm. We will sketch the main ideas, referring to Donoho *et al.* (2009) for the original idea, to Donoho *et al.* (2011) and Bayati and Montanari (2011, 2012) for the analysis of the LASSO, and to Donoho *et al.* (2013*a,b*) and Javanmard and Montanari (2013) for extensions. This approach was also used in Bayati *et al.* (2015) to establish universality of the compressed sensing phase transition for non-Gaussian i.i.d. entries $\mathbf{X}_{i,j}$.

### 5.5.1  Message-passing algorithms

The plan of our analysis is as follows:

1. We define an approximate message-passing (AMP) algorithm to solve the LASSO optimization problem. The derivation presented here starts from the subgradient

method described in Section 5.4.2 and obtain a slight—but crucial—modification of the same algorithm. Also, in this case, the algorithm is iterative and computes a sequence of iterates $\{\theta^{(t)}\}$.

An alternative approach (susceptible of generalization—for instance, to Bayesian estimation) is presented in Donoho *et al.* (2010).

2. We derive an exact asymptotic characterization of the same algorithm as $n, p \to \infty$, *for t fixed*. The characterization is given in terms of the so-called state evolution method developed rigorously in Bayati and Montanari (2011) (with generalizations in Javanmard and Montanari (2013) and Bayati *et al.* (2015)).

3. We prove that AMP converges rapidly to the optimized $\widehat{\theta}$; that is, with high probability as $n, p \to \infty$, we have $\|\theta^{(t)} - \widehat{\theta}\|_2^2 / p \leq c_1, \mathrm{e}^{-c_2 t}$, with $c_1, c_2$ two dimension-independent constants. A full proof of this step can be found in Bayati and Montanari (2012).

4. We select $t$ a large enough constant and use the last two results to deduce properties of the optimizer $\widehat{\theta}$.

We next provide a sketch of the above steps. We start by considering iterative soft thresholding with $L = 1$:

$$
\begin{cases}
\theta^{(t+1)} = \eta \left( \theta^{(t)} + \dfrac{1}{n} \mathbf{X}^{\mathsf{T}} r^{(t)}; \gamma_t \right), \\[2mm]
r^{(t)} = y - \mathbf{X}\theta^{(t)},
\end{cases}
\tag{5.116}
$$

where have we introduced the additional freedom of an iteration-dependent threshold $\gamma_t$ (instead of $\lambda$). Componentwise, the iteration takes the form

$$
\begin{cases}
\theta_i^{(t+1)} = \eta \left( \theta_i^{(t)} + \dfrac{1}{n} \displaystyle\sum_{a=1}^{n} \mathbf{X}_{ai} r_a^{(t)}; \gamma_t \right), \\[2mm]
r_a^{(t)} = y_a - \displaystyle\sum_{i=1}^{p} \mathbf{X}_{ai} \theta_i^{(t)}.
\end{cases}
\tag{5.117}
$$

We next derive a message-passing version of this iteration[3] (we refer for instance to Richardson and Urbanke (2008) and Mezard and Montanari (2009) for background). The motivation for this modification is that message-passing algorithms have appealing statistical properties. For instance, they admit an exact asymptotic analysis on locally tree-like graphs. While—in the present case—the underlying graph structure is not locally tree-like, the conclusion (exact asymptotic characterization) continues to hold.

In order to define the message-passing version, we need to associate a factor graph with the LASSO cost function:

$$
F(\theta) = \frac{1}{2n} \sum_{a=1}^{n} \left( y_a - \langle x_a, \theta \rangle \right)^2 + \lambda \sum_{i=1}^{p} |\theta_i|.
\tag{5.118}
$$

---

[3] We use the expression "message passing" in the same sense attributed in information theory and graphical models.

Following a general prescription from Mezard and Montanari (2009), we associate a factor node with each term $(y_a - \langle x_a, \theta \rangle)^2/(2n)$ in the cost function indexed by $a \in \{1, 2, \ldots, n\}$ (we do not need to represent the singletons $|\theta_i|$ by factor nodes), and we associate a variable node with each variable, indexed by $i \in \{1, 2, \ldots, p\}$. We connect the factor node $a$ and the variable node $i$ by an edge $(a, i)$ if and only if the term $a$ depends on the variable $\theta_i$, i.e., if $\mathbf{X}_{ai} \neq 0$. Note that for Gaussian design matrices, all the entries $\mathbf{X}_{ai}$ are nonzero with probability one. Hence, the resulting factor graph is a complete bipartite graph with $n$ factor nodes and $p$ variable nodes.

The message-passing version of the iteration (5.117) has iteration variables (messages) associated with directed edges of the factor graph. Namely, for each edge $(a, i)$, we introduce a message $r_{a \to i}^{(t)}$ and a message $\theta_{i \to a}^{(t)}$. We replace the update rule (5.117) by the following:

$$
\begin{cases}
\theta_{i \to a}^{(t+1)} = \eta \left( \dfrac{1}{n} \sum_{b \in [n] \setminus a} \mathbf{X}_{bi} r_{b \to i}^{(t)}; \gamma_t \right), \\[2em]
r_{a \to i}^{(t)} = y_a - \displaystyle\sum_{j \in [p] \setminus i} \mathbf{X}_{aj} \theta_{j \to a}^{(t)}.
\end{cases}
\tag{5.119}
$$

The key property of this iteration is that an outgoing message from node $\alpha$ is updated by evaluating a function of all messages incoming in the same node $\alpha$, except the one along the same edge. An alternative derivation of this iteration follows by considering the standard belief propagation algorithm (in its sum–product or min–sum forms), and using a second-order approximation of the messages as in Donoho *et al.* (2010).

Note that with respect to standard iterative soft thresholding, cf. (5.116), the algorithm (5.119) has higher complexity, since it requires tracking of $2np$ messages, as opposed to the $n + p$ variables in (5.116). Also, there is an obvious interpretation of the fixed points of the iteration (5.119).

It turns out that a simpler algorithm can be defined, whose state has dimension $n + p$ as for iterative soft thresholding, but tracks closely the iteration (5.119). This builds on the remark that the messages $\theta_{i \to a}^{(t)}$ issued from node $i$ do not differ too much, since their definitions in (5.119) only differ in one out of $n$ terms. A similar argument applies to the messages $r_{a \to i}^{(t)}$ issued by node $a$/ We then write $\theta_{i \to a}^{(t)} = \theta_i^{(t)} + \delta\theta_{i \to a}^{(t)}$, $r_{a \to i}^{(t)} = r_a^{(t)} + \delta r_{a \to i}^{(t)}$ and linearize the iteration (5.119) in $\{\delta\theta_{i \to a}^{(t)}\}$, $\{\delta r_{a \to i}^{(t)}\}$. After eliminating these quantities (Donoho *et al.*, 2010), the resulting iteration takes the form, known as approximate message passing (AMP),

$$
\begin{cases}
\theta^{(t+1)} = \eta \left( \theta^{(t)} + \dfrac{1}{n} \mathbf{X}^{\mathsf{T}} r^{(t)}; \gamma_t \right), \\[2em]
r^{(t)} = y - \mathbf{X}\theta^{(t)} + \mathsf{b}_t r^{(t-1)},
\end{cases}
\tag{AMP}
$$

where $\mathsf{b}_t \equiv \|\theta^{(t)}\|_0/n$ is a scalar. In other words, we have recovered iterative soft thresholding except for the memory term $\mathsf{b}_t r^{(t-1)}$, which is straightforward to evaluate.

In the context of statistical physics, a similar correction is known as the Onsager term. Remarkably, this memory term changes the statistical behavior of the algorithm.

It is an instructive exercise (left to the reader) to prove that fixed points of the AMP algorithm (with $\gamma_t = \gamma_*$ fixed) are minimizers of the LASSO. In particular, for Gaussian sensing matrices, such a minimizer is unique with probability one.

We notice in passing that there is nothing special about the LS objective, or the $\ell_1$ regularization in our derivation. Indeed, similar ideas have been developed and applied to a large number of problems; for a very incomplete list of examples, see Som and Schniter (2012), Rangan (2011), Donoho *et al.* (2013*b*), Donoho and Montanari (2013), Metzler *et al.* (2014), Tan *et al.* (2014), and Barbier and Krzakala (2014).

### 5.5.2 Analysis of AMP and the LASSO

We next carry out a heuristic analysis of AMP, referring to Bayati and Montanari (2011) for a rigorous treatment that uses ideas developed by Bolthausen (2014) in the context of mean-field spin glasses.

We use the message-passing version of the algorithm, cf. (5.119), and we will make the assumption that the pairs $\{(r_{a\to i}^{(t)}, \mathbf{X}_{ai})\}_{a\in[n]}$ are "as if" independent, and likewise for $\{(\theta_{a\to i}^{(t)}, \mathbf{X}_{ai})\}_{i\in[p]}$. This assumption is only approximately correct, but leads to the right asymptotic conclusions.

Consider the first equation in (5.119), and further assume (this assumption will be verified inductively)

$$\mathbb{E}(r_{a\to i}^{(t)}) = \mathbf{X}_{ai}\theta_i, \quad \text{Var}(r_{a\to i}^{(t)}) = \tau_t^2. \tag{5.120}$$

Letting $\tilde{r}_{a\to i}^{(t)} \equiv r_{a\to i}^{(t)} - \mathbb{E}(r_{a\to i}^{(t)})$, the argument of $\eta(\,\cdot\,;\gamma_t)$ in (5.119) can be written as

$$\frac{1}{n}\sum_{b\in[n]\backslash a}\mathbf{X}_{bi}r_{b\to i}^{(t)} = \frac{1}{n}\sum_{b\in[n]\backslash a}\mathbf{X}_{bi}^2\theta_i + \frac{1}{n}\sum_{b\in[n]\backslash a}\mathbf{X}_{bi}\tilde{r}_{b\to i}^{(t)} \approx \theta_i + \frac{\tau_t}{\sqrt{n}}Z_{i\to a}^{(t)}, \tag{5.121}$$

where, by the central limit theorem, $Z_{i\to a}^{(t)}$ is approximately distributed as $\mathcal{N}(0,1)$.

Rewriting the first equation in (5.119) , we obtain

$$\theta_{i\to a}^{(t+1)} = \eta\left(\theta_i + \frac{\tau_t}{\sqrt{n}}Z_{i\to a}^{(t)};\gamma_t\right). \tag{5.122}$$

In the second message equation, we substitute $y_a = w_a + \sum_{j=1}^p \mathbf{X}_{aj}\theta_j$, thus obtaining

$$r_{a\to i}^{(t+1)} = w_a + \mathbf{X}_{ai}\theta_i - \sum_{j\in[p]\backslash i}\mathbf{X}_{aj}(\theta_{j\to a}^{(t+1)} - \theta_j). \tag{5.123}$$

The first and last terms have zero mean, thus confirming the induction hypothesis $\mathbb{E}(r_{a\to i}^{(t+1)}) = \mathbf{X}_{ai}\theta_i$. The variance of $r_{a\to i}^{t+1}$ is given by (neglecting sublinear terms)

$$\tau_{t+1}^2 = \sigma^2 + \sum_{j=1}^p\left[\eta(\theta_j + \frac{\tau_t}{\sqrt{n}}Z_j;\gamma_t) - \theta_j\right]^2. \tag{5.124}$$

It is more convenient to work with the rescaled quantities $\tilde{\theta}_i = \theta_i \sqrt{n}$ and $\tilde{\gamma}_t = \gamma_t \sqrt{n}$ (this allows us to focus on the most interesting regime, whereby $\theta_i$ is of the same order as the noise level $\tau_t / \sqrt{n}$). Using the scaling property of the thresholding function $\eta(ax, a\gamma) = a\eta(x, \gamma)$, the last equation becomes

$$\tau_{t+1}^2 = \sigma^2 + \frac{1}{n} \sum_{j=1}^{p} \left[ \eta(\tilde{\theta}_j + \tau_t Z_j; \tilde{\gamma}_t) - \tilde{\theta}_j \right]^2. \tag{5.125}$$

We now define the probability measure $p_\Theta$ as the asymptotic empirical distribution of $\tilde{\theta}$, $p^{-1} \sum_{j=1}^{p} \delta_{\tilde{\theta}_j}$ (formally, we assume that $p^{-1} \sum_{j=1}^{p} \delta_{\tilde{\theta}_j}$ converges weakly to $p_\Theta$, and that low-order moments converge as well). We also let $\delta = \lim_{n \to \infty} (n/p)$ be the asymptotic aspect ratio of $\mathbf{X}$. We then obtain

$$\tau_{t+1}^2 = \sigma^2 + \frac{1}{\delta} \mathbb{E} \left\{ (\eta(\Theta + \tau_t Z; \tilde{\gamma}_t) - \Theta)^2 \right\}, \tag{5.126}$$

where expectation is with respect to $\Theta \sim p_\Theta$ independent of $Z \sim \mathcal{N}(0, 1)$. The last equation is known as *state evolution*: despite the many unjustified assumptions in our derivation, it can be proved to correctly describe the $n, p \to \infty$ asymptotics of the message-passing algorithm (5.119) as well as of the AMP algorithm.

Reconsidering the above derivation, we can derive asymptotically exact expressions for the risk at $\theta$ for of the AMP estimator $\theta^{(t+1)}$. Namely, we define the asymptotic risk

$$R_\infty(\theta; \theta^{(t+1)}) = \lim_{n,p \to \infty} \mathbb{E} \left\{ \| \theta - \theta^{(t)} \|^2 \right\}, \tag{5.127}$$

the limit being taken along sequences of vectors $\theta$ with converging empirical distribution. Then we claim that the limit exists and is given by

$$R_\infty(\theta; \theta^{(t+1)}) = \frac{1}{\delta} \mathbb{E} \left\{ (\eta(\Theta + \tau_t Z; \tilde{\gamma}_t) - \Theta)^2 \right\}, \tag{5.128}$$

or, equivalently,

$$R_\infty(\theta; \theta^{(t+1)}) = \tau_{t+1}^2 - \sigma^2. \tag{5.129}$$

Thus, apart from an additive constant, $\tau_t^2$ coincides with risk, and the latter can be tracked using state evolution.

In Bayati and Montanari (2012), it is proved that the AMP iterates $\theta^{(t)}$ converge rapidly to the LASSO estimator $\hat{\theta}$. We are therefore led to consider the large-$t$ behavior of $\tau_t$, which yields the risk of the LASSO, or—equivalently—the risk of AMP after a sufficiently large (constant in $n, p$) number of iterations. Before addressing this question, we need to set the values of $\tilde{\gamma}_t$. A reasonable choice is to fix $\tilde{\gamma}_t = \kappa \tau_t$, for some constant $\kappa$, since $\tau_t$ can be thought as the "effective noise level" at iteration $t$. There is a one-to-one correspondence between $\kappa$ and the regularization parameter $\lambda$ in the LASSO (Donoho *et al.*, 2011). We thus define the function

$$G(\tau^2; \sigma^2) \equiv \sigma^2 + \frac{1}{\delta} \mathbb{E} \left\{ (\eta(\Theta + \tau Z; \kappa \tau) - \Theta)^2 \right\}, \tag{5.130}$$

which of course depends implicitly on $p_\Theta$, $\kappa$, and $\delta$. State evolution is then the one-dimensional recursion $\tau_{t+1}^2 = G(\tau_t^2; \sigma^2)$. For the sequence $\tau_t$ to stay bounded, we assume $\lim_{\tau^2 \to \infty} G(\tau^2; \sigma^2)/\tau^2 < 1$, which can always be ensured by taking $\kappa$ sufficiently large.

Let us first consider the noiseless case $\sigma = 0$. Since $G(0; 0) = 0$, we know that $\tau = 0$ is always a fixed point. It is not hard to show (Donoho *et al.*, 2009) that indeed $\lim_{t \to \infty} \tau_t^2 = 0$ if and only if this is the unique non-negative fixed point (see Fig. 5.22). If this condition is satisfied, AMP reconstructs exactly the signal $\theta$, and also, owing to the correspondence with the LASSO, basis pursuit (the LASSO with $\lambda \to 0$) reconstructs exactly $\theta$. Notice that this condition is sharp: if it is not satisfied, then AMP and the LASSO fail to reconstruct $\theta$, despite vanishing noise. In order to derive the phase transition location, remember that by the definition of minimax risk of soft thresholding (cf. Section 5.3.1), we have, assuming $\kappa = \ell(\varepsilon)$ to be set in the optimal way,

$$G(\tau^2; 0) = \frac{1}{\delta}\mathbb{E}\left\{(\eta(\Theta + Z; \kappa\tau) - \Theta)^2\right\} \le \frac{M(\varepsilon)}{\delta}\tau^2. \tag{5.131}$$

Hence $\tau_{t+1}^2 \le (M(\varepsilon)/\delta)\tau_t^2$ and if

$$\delta > M(\varepsilon), \tag{5.132}$$

then $\tau_t^2 \to 0$ and AMP (LASSO) reconstructs $\theta$ with vanishing error. This bound is in fact tight: for $\delta < M(\varepsilon)$, any probability distribution $p_\Theta$ with $p_\Theta(\{0\}) = 1 - \varepsilon$, and any threshold parameter $\kappa$, the mean square error remains bounded away from zero.

Recalling the definition of $\delta = n/p$, the condition $\delta > M(\varepsilon)$ corresponds to requiring a sufficient number of samples, as compared with the sparsity. It is interesting to recover the very sparse regime from this point of view. Recall from previous sections that $M(\varepsilon) \approx 2\varepsilon \log(1/\varepsilon)$ for small $\epsilon$. The condition $\delta > M(\varepsilon)$ then translates to $\delta \gtrsim 2\varepsilon \log(1/\varepsilon)$ or, in other words, $(n/p) \gtrsim 2(s_0/p)\log(p/s_0)$. Thus, we obtain the

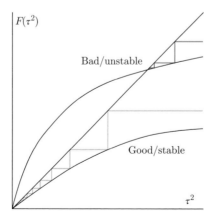

**Fig. 5.22** State evolutions for successful and unsuccessful recoveries in the noiseless setting.

condition—already discussed—that the number of samples must be as large as the number of nonzero coefficients, times a logarithmic factor. Reconstruction is possible if and only if

$$n \gtrsim 2s_0 \log \frac{p}{s_0}, \tag{5.133}$$

a condition that we have seen in previous sections.

In the noisy case, we cannot hope to achieve perfect reconstruction. In this case, we say that estimation is stable if there is a constant $C$ such that for any $\theta \in \mathbb{R}^p$, $R(\theta; \widehat{\theta}) \leq C\sigma^2$. This setting is sketched in Fig. 5.23. Exact reconstruction at $\sigma = 0$ translate into a fixed point $\tau_*^2 = O(\sigma^2)$ and hence stability. Inexact reconstruction corresponds to a fixed point of order 1 and hence lack of stability.

Again by choosing a suitable threshold value $\kappa$, we can ensure that the minimax bound (5.131) is valid and hence

$$\tau_{t+1}^2 \leq \sigma^2 + \frac{M(\varepsilon)}{\delta} \tau_t^2 . \tag{5.134}$$

Taking the limit $t \to \infty$, in the case that $\delta > M(\varepsilon)$, we have that

$$\tau_*^2 \leq \frac{\sigma^2}{1 - (M(\varepsilon)/\delta)}. \tag{5.135}$$

This establishes that the following is an upper bound on the asymptotic mean square error of AMP, and hence (by the equivalence discussed above) of the LASSO:

$$R_\infty(\theta; \widehat{\theta}) = \begin{cases} \dfrac{M(\epsilon)}{\delta - M(\varepsilon)} \sigma^2 & \text{if } M(\varepsilon) < \delta, \\ \infty & \text{otherwise.} \end{cases} \tag{5.136}$$

As proved in Donoho *et al.* (2011) and Bayati and Montanari (2012), this result indeed holds with equality (note that these papers have slightly different normalizations of the noise variance $\sigma^2$).

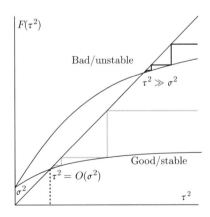

**Fig. 5.23**  State evolutions for successful and unsuccessful recoveries in a noisy setting.

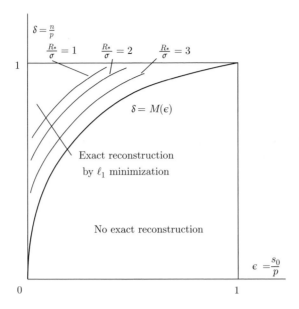

**Fig. 5.24** Phase diagram of signal recovery regions over $(\varepsilon, \delta)$.

A qualitative sketch of the resulting phase diagram in $\varepsilon$ and $\delta$ is shown in Fig. 5.24. As anticipated above, if $\delta > M(\varepsilon)$, i.e., in the regime in which exact reconstruction is feasible through basis pursuit in zero noise, reconstruction is also stable with respect to noise.

Again, let us consider the sparse regime $\varepsilon \to 0$. Assuming $M(\varepsilon) \ll \delta$, and substituting $M(\varepsilon) \approx 2\varepsilon \log(1/\varepsilon)$ together with the definitions of $\varepsilon$ and $\delta$, we get

$$R_\infty(\theta; \widehat{\theta}) = \sigma^2 \frac{M(\varepsilon)}{\delta - M(\varepsilon)} \approx \frac{\sigma^2}{\delta} 2\varepsilon \log \frac{1}{\varepsilon} = \frac{s_0 \sigma^2}{n} 2 \log \frac{p}{s_0} . \qquad (5.137)$$

We have therefore rederived the same behavior already established in the previous section under the RIP assumption. Apart from the factor $2 \log(p/s_0)$, the risk is the same "as if" we knew the support of $\theta$.

## 5.6 The hidden clique problem

One of the most surprising facts about sparse regression is that we can achieve ideal estimation error using a low-complexity algorithm, namely, by solving a convex optimization problem such as the LASSO. Indeed—at first sight—one might have suspected it necessary to search over possible supports of size $s_0$, a task that requires at least $\binom{p}{s_0}$ operations and is therefore nonpolynomial. Unfortunately, this is not always the case. There are problems in which a huge gap exists between the statistical limits of estimation (i.e., the minimax risk achieved by an arbitrary estimator) and the computational limits (i.e., the minimax risk achieved by any estimator computable in polynomial time). The hidden clique (or hidden submatrix) problem, an example of

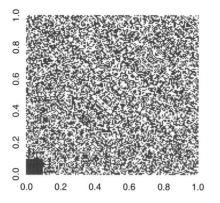

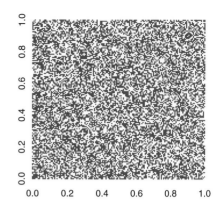

(a) Arando graph with a planted clique.

(b) The same graph, but with the vertices shuffled.

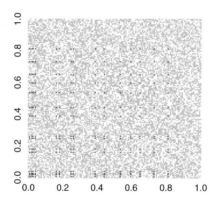

(c) Retrieving the clique in the shuffled graph

**Fig. 5.25**  Example of a planted hidden clique problem.

which is given in Fig. 5.25, is a prototypical example of this class of computationally hard estimation problems. Recently, reductions to this problem have been used to prove that other estimation problems are hard as well (Berthet and Rigollet, 2013).

We next define the problem. Let $Q_0$ and $Q_1$ be two given probability distributions on $\mathbb{R}$. For a set $S \subseteq \{1, 2, \ldots, n\}$, we let $\mathbf{W} \in \mathbb{R}^{n \times n}$ be a symmetric random matrix with entries $(\mathbf{W}_{ij})_{i \leq j}$ independent, with distribution

$$\mathbf{W}_{ij} \sim Q_1, \ \text{if } i, j \in S \,, \tag{5.138}$$

$$\mathbf{W}_{ij} \sim Q_0, \ \text{otherwise} \,. \tag{5.139}$$

The problem is to find the set $S$ given one realization of $\mathbf{W}$.

**Example 1.** Suppose $Q_0 = \mathcal{N}(0, 1)$ and $Q_1 = \mathcal{N}(\mu, 1)$ to be two Gaussian distributions with different means and the same known variance (which we set—without loss of generality—equal to one). The model is then equivalent to the following:

$$\mathbf{W} = \mu \, u_S u_S^\mathsf{T} + \mathbf{Z}, \tag{5.140}$$

where $u_S$ is the indicator vector of the set $S$, namely, $(u_S)_i = 1$ if $i \in S$ and $(u_S)_i = 0$ otherwise.

**Example 2.** This is the original setting of the hidden clique problem from Jerrum (1992). Both $Q_0$ and $Q_1$ are Bernoulli distributions:

$$Q_0 = \frac{1}{2}\delta_{-1} + \frac{1}{2}\delta_{+1}, \tag{5.141}$$

$$Q_1 = \delta_{+1}. \tag{5.142}$$

There is a straightforward way to interpret this as a graph problem. Let $G$ be the random graph on $n$ vertices $\{1, \ldots, n\}$ whereby two vertices $i, j$ are joined by an edge if and only if $\mathbf{W}_{ij} = +1$. Then $G$ is an Erdős–Rényi random graph (with edge density $\frac{1}{2}$) to which a clique has been added with support on $S$.

For simplicity of exposition, we will focus for the rest of this lecture on the Bernoulli case, i.e. on the second example. We will use interchangeably the language of random graphs and that of random matrices. All of our results can in fact be generalized to arbitrary probability distributions $Q_0$, $Q_1$ under suitable tail conditions, as shown in Deshpande and Montanari (2013).

We will denote by $k = |S|$ the size of the hidden set. It is not hard to see that the problem is easy for $k$ large (from both the statistical and computational points of view) and hard for $k$ small (both computationally and statistically). Indeed, for $k$ sufficiently large, a simple degree-based heuristics is successful. This is based on the remark that vertices in the clique have a slightly higher degree than others. Hence, sorting the vertices by degree, the first $k$ vertices should provide a good estimate of $S$.

**Proposition 5.3** *Let $\widehat{S}$ be the set of $k$ vertices with large degree in $G$. If*

$$k \geq \sqrt{(2 + \varepsilon)n \log n},$$

*then, with high probability, $\widehat{S} = S$.*

**Proof** Let $D_i$ denote the degree of vertex $i$. If $i \notin S$, then $D_i \sim \text{Binom}(n - 1, \frac{1}{2})$. In particular, standard concentration bounds on independent random variables yield $\mathbb{P}\{D_i \geq \mathbb{E}D_i + t\} \leq \exp(-2t^2/n)$. By a union bound (the same already used to analyze denoising in Section 5.3), and using $\mathbb{E}(D_i) = (n - 1)/2$, we have, for any $\varepsilon' > 0$, with probability converging to 1 as $n \to \infty$,

$$\max_{i \notin S} D_i \leq \frac{n}{2} + \sqrt{(1 + \varepsilon')\frac{n \log n}{2}}. \tag{5.143}$$

On the other hand, if $i \in S$, then $D_i \sim k - 1 + \text{Binom}(n - k, 1/2)$. Hence, by a similar union bound,

$$\min_{i \in S} D_i \geq \frac{n + k}{2} - \sqrt{(1 + \varepsilon') \frac{n \log k}{2}}. \tag{5.144}$$

The claim follows by using together the above and selecting a suitable value $\varepsilon'$.  □

For $k$ too small, the problem becomes statistically intractable because the planted clique is not the unique clique of size $k$. Hence no estimator can distinguish between the set $S$ and another set $S'$ that supports a different (purely random) clique. The following theorem characterizes this statistical threshold.

**Proposition 5.4** *Let $\varepsilon > 0$ be fixed. Then, for $k < 2(1 - \varepsilon) \log_2 n$, any estimator $\widehat{S}$ is such that $\widehat{S} \neq S$ with probability converging to 1 as $n \to \infty$.*
*Vice versa, for $k < 2(1 - \varepsilon) \log_2 n$, there exists an estimator $\widehat{S}$ such that $\widehat{S} = S$ with probability converging to 1 as $n \to \infty$.*

**Proof**  We will not present a complete proof but only sketch the fundamental reason for a threshold $k \approx 2 \log_2 n$ and leave to the reader the task of filling in the details.

The basic observation is that the largest "purely random" clique is of size approximately $2 \log_2 n$. As a consequence, for $k$ larger than this threshold, searching for a clique of size $k$ returns the planted clique.

More precisely, let $\mathcal{G}(n, \frac{1}{2})$ be an Erdős–Rényi random graph with edge density $\frac{1}{2}$ (i.e. a random graph where each edge is present independently with probability $\frac{1}{2}$). We will show that the largest clique in $\mathcal{G}(n, \frac{1}{2})$ is, with high probability, of size between $2(1 - \varepsilon) \log_2 n$ and $2(1 + \varepsilon) \log_2 n$.

This claim can be proved by a moment's calculation. In particular, for proving that the largest clique cannot be much larger than $2 \log_2 n$, it is sufficient to compute the expected number of cliques of size $\ell$. Letting $N(\ell; n)$ denote the number of cliques of size $\ell$ in $\mathcal{G}(n, \frac{1}{2})$, we have

$$\mathbb{E}\, N(\ell; n) = \binom{n}{\ell} 2^{-\binom{\ell}{2}} \approx n^\ell 2^{-\ell^2/2} = 2^{\ell \log_2 n - \ell^2/2}. \tag{5.145}$$

For $\ell > 2(1 + \varepsilon) \log_2 n$, the exponent is negative and the expectation vanishes as $n \to \infty$. In fact $\sum_{\ell \geq 2(1+\varepsilon) \log_2 n} \mathbb{E} N(\ell; n)$ vanishes as well. By the Markov inequality, it follows that—with high probability—no clique has size larger than $2(1 + \varepsilon) \log_2 n$.  □

The catch with the last proposition is that the estimator might not be computable in polynomial time. Indeed, the estimator implicitly assumed in the proof requires searching over all subsets of $k$ vertices, which takes time at least $\binom{n}{k} \approx n^k$. For $k$ above the threshold, this is $\exp\{c(\log n)^2\}$, i.e., super-polynomial.

To summarize, with unlimited computational resources, we can find planted cliques as soon as their size is larger than $c \log_2 n$ for any $c > 2$. This is the fundamental statistical barrier to estimating the set $S$. On the other hand, the naive degree-based heuristic described above correctly identifies the clique if $k \geq \sqrt{cn\log n}$. There is a huge gap between the fundamental statistical limit and what is achieved by a simple polynomial-time algorithm. This begs the question as to whether this gap can be filled by more advanced algorithmic ideas.

A key observation, due to Alon, Krivelevich, and Sudakov (Alon *et al.*, 1998) is that the matrix $\mathbf{W}$ is—in expectation—a rank-one matrix. That is,

$$\mathbb{E}\{\mathbf{W}\} = u_S u_S^\mathsf{T}, \tag{5.146}$$

and therefore $S$ can be reconstructed from the eigenvalue decomposition of $\mathbb{E}\{\mathbf{W}\}$. Of course, $\mathbb{E}\{\mathbf{W}\}$ is not available, but one can hope that the random part of $\mathbf{W}$ does not perturb the leading eigenvector too much. In other words, one can compute the principal eigenvector $v_1(\mathbf{W})$, i.e., the eigenvector of $\mathbf{W}$ with largest eigenvalue, and use its largest entries to estimate the clique. For instance, one can take the $k$ vertices corresponding to the entries of $v_1(\mathbf{W})$ with largest absolute value.

This spectral approach allows us to reduce the minimum detectable clique size by a factor $\sqrt{\log n}$, with respect to the degree heuristics of Proposition 5.3.

**Theorem 5.5 (Alon *et al.* (1998))** *There exists an algorithm that returns an estimate $\widehat{S}$ of the set $S$, with the same complexity as computing the principal eigenvector of $\mathbf{W}$, and such that if $k > 100\sqrt{n}$, then $\widehat{S} = S$ with probability converging to 1 as $n \to \infty$.*

**Proof [Proof sketch]** Again, we will limit ourselves to explaining the basic argument. The actual proof requires some additional steps.

The matrix $\mathbf{W}$ has the form

$$\mathbf{W} = u_S u_S^\mathsf{T} + \mathbf{Z} - \mathbf{Z}_{SS}, \tag{5.147}$$

where $\mathbf{Z}$ is a Wigner matrix, i.e., a matrix with i.i.d. zero-mean entries $(\mathbf{Z}_{ij})_{i \leq j}$, and $\mathbf{Z}_{S,S}$ is the restriction of $\mathbf{Z}$ to indices in $S$. In the present case, the distribution of entries is Bernoullian:

$$\mathbf{Z}_{ij} = \begin{cases} +1, & \text{with probability } \dfrac{1}{2}, \\[2ex] -1, & \text{with probability } \dfrac{1}{2}. \end{cases} \tag{5.148}$$

By the celebrated Füredi–Komlos theorem (Füredi and Komlós, 1981), the operator norm of this matrix (i.e., the maximum of the largest eigenvalue of $\mathbf{Z}$ and the largest eigenvalue of $-\mathbf{Z}$) is upper-bounded as $\|\mathbf{Z}\|_2 \leq (2+\varepsilon)\sqrt{n}$, with high probability. By the same argument $\|\mathbf{Z}_{SS}\|_2 \leq (2+\varepsilon)\sqrt{k}$, which is much smaller than $\|\mathbf{Z}\|_2$.

We view $\mathbf{W}$ as a perturbation of the matrix $u_S u_S^\mathsf{T}$ (whose principal, normalized, eigenvector is $u_S/\sqrt{k}$). Matrix perturbation theory implies that the largest eigenvector is perturbed by an amount proportional to the norm of the perturbation and inversely proportional to the gap between the top eigenvalue and the second eigenvalue of the perturbed matrix. More precisely, the Davis–Kahan "sine theta" theorem yields (for $v_1 = v_1(\mathbf{W})$)

$$\sin\theta(v_1, u_S) \leq \frac{\|\mathbf{Z} - \mathbf{Z}_{S,S}\|_2}{\lambda_1(u_S u_S^\mathsf{T}) - \lambda_2(\mathbf{W})}, \tag{5.149}$$

where $\lambda_\ell(\mathbf{A})$ denotes the $\ell$th largest eigenvalue of the matrix $\mathbf{A}$, and $\theta(a, b)$ is the angle between the vectors $a$ and $b$. We have, of course, $\lambda_1(u_S u_S^\mathsf{T}) = \|u_S\|_2^2 = k$, and $\lambda_2(\mathbf{W}) \leq \lambda_2(u_S u_S^\mathsf{T}) + \|\mathbf{Z} - \mathbf{Z}_{S,S}\|_2$. Therefore,

$$\sin\theta(v_1, u_S) \leq \frac{\|\mathbf{Z} - \mathbf{Z}_{S,S}\|_2}{k - \|\mathbf{Z} - \mathbf{Z}_{S,S}\|_2}$$

$$\leq \frac{2.1\sqrt{n}}{k - 2.1\sqrt{n}} \leq \frac{1}{45}, \tag{5.150}$$

where the last inequality holds with high probability by the Füredi–Komlos theorem. Using standard trigonometry, this bound can be immediately converted into a bound on the $\ell_2$ distance between $v_1(\mathbf{W})$ and the unperturbed eigenvector:

$$\left\| v_1 - \frac{u_S}{\sqrt{k}} \right\|_2 \leq \frac{1}{40}. \tag{5.151}$$

We can then select the set $B$ of $k$ vertices that correspond to the $k$ entries of $v_1$ with largest absolute value. The last bound does not guarantee that $B$ coincides with $S$, but it implies that $B$ must have a substantial overlap with $S$. The estimator $\widehat{S}$ is constructed by selecting the $k$ vertices in $\{1, 2, \ldots, p\}$ that have the largest number of neighbors in $B$. $\qquad\square$

It is useful to pause for a few remarks on this result:

**Remark 5.6** *The complexity of the above algorithm is the same as that of computing the principal eigenvector $v_1(\mathbf{W})$. Under the assumptions of the theorem, this is nondegenerate and, in fact, there is a large gap between the first eigenvalue and the second, say, $\max(\lambda_2(\mathbf{W}), |\lambda_n(\mathbf{W})|) \leq \frac{1}{2}\lambda_1(\mathbf{W})$.*

*Hence, $v_1(\mathbf{W})$ can be computed efficiently through power iteration, i.e., by computing the sequence of vectors $v^{(t+1)} = \mathbf{W}v^{(t)}$. Each operation takes at most $n^2$ operations, and, owing to the rapid convergence, $O(\log n)$ iterations are sufficient for implementing the above algorithm. We will revisit power iteration in the following.*

**Remark 5.7** *The eigenvalues and eigenvectors of a random matrix of the form (5.147) have been studied in detail in statistics, under the name of the "spiked model," and probability theory, as "low-rank perturbation of Wigner matrices" (see, e.g., Féral and Péché, 2007; Capitaine et al., 2009, 2012). These works have unveiled a* *phase transition phenomenon that, in the present application, can be stated as follows. Assume $k, n \to \infty$ with $k/\sqrt{n} = \kappa \in (0, \infty)$. Then*

$$\lim_{n \to \infty} |\langle v_1(\mathbf{W}), u_S/\sqrt{k}\rangle| = \begin{cases} 0 & \text{if } \kappa \leq 1, \\ \sqrt{1 - \kappa^{-2}} & \text{otherwise.} \end{cases} \tag{5.152}$$

*In other words, for $k \leq (1 - \varepsilon)\sqrt{n}$, the principal eigenvector of $\mathbf{W}$ is essentially uncorrelated with the hidden set $S$. The barrier at $k$ of order $\sqrt{n}$ is not an artifact of the proof, but instead a fundamental limit related to this phase transition.*

*On the other hand, a more careful analysis of the spectral method can possibly show that it succeeds for all $k \geq (1+\varepsilon)\sqrt{n}$. (Here and above, $\varepsilon > 0$ is an arbitrary constant.)*

**Remark 5.8** *A clever trick by Alon et al. (1998), allows us to find cliques of size $k \geq \delta\sqrt{n}$ for any fixed constant $\delta > 0$ in polynomial time. The price to pay is that the computational complexity increases rapidly as $\delta$ gets smaller. More precisely, we can identify sets of size $k \geq \delta\sqrt{n}$ for an algorithm with time complexity of order $n^{O(\log(1/\delta))}$.*

*To see this, we use the spectral method as a routine that is able to find the clique with high probability provided $k \geq c\sqrt{n}$ for some constant $c$. First, we assume that an oracle gives us one node in the clique. We can solve the problem with $k \gtrsim c\sqrt{n/2}$. Indeed, we can focus our attention on the set of neighbors of the node provided by the oracle. There are about $n/2$ such neighbors, and they contain a clique of size $k - 1$; hence the spectral method will succeed under the stated condition.*

*We then observe that we do not need any such approach: we can search for the vertex that the oracle would tell us by blowing up the runtime by a factor at most $n$ (indeed, a $\sqrt{n}$ factor is sufficient, since one of every $\sqrt{n}$ vertices is in the clique). In this way, we can trade a factor of $\sqrt{2}$ in $k$ by an $n$-fold increase in runtime. This construction can be repeated $O(\log(1/\delta))$ times to achieve the trade-off mentioned above.*

### 5.6.1   An iterative thresholding approach

Throughout this section, we shall normalize the data and work with the matrix $\mathbf{A} = \mathbf{W}/\sqrt{n}$. As we saw in the previous section, the principal eigenvector of $\mathbf{A}$ carries important information about the set $S$, and in particular it is correlated with the indicator vector $u_S$ if the hidden set $S$ is large enough. Also, an efficient way to compute the principal eigenvector is through power iteration:

$$v^{(t+1)} = \mathbf{A}\, v^{(t)}. \tag{5.153}$$

Note that the resulting vector $v^{(t)}$ will not, in general, be sparse—if not a posteriori, then because of the correlation with $u_S$. It is therefore a natural idea to modify the power iteration by introducing a nonlinearity that enforces sparsity:

$$\theta^{(t+1)} = \mathbf{A}\, f_t(\theta^{(t)}), \tag{5.154}$$

where $\theta \in \mathbb{R}^n$ and $f_t : \mathbb{R}^n \to \mathbb{R}^n$ is a nonlinear function that enforces sparsity. To be definite, we will assume throughout that the initialization is $\theta^{(0)} = (1, 1, \dots, 1)$, the "all-ones" vector.

For ease of exposition, we shall focus on separable functions and denote by $f_t$ the action of this function on each component. In other words, with a slight abuse of notation, we will write $f_t(v) = (f_t(v_1), f_t(v_2), \dots, f_t(v_n))$ when $v = (v_1, v_2, \dots, v_n)$. Examples of such a function might be the following:

- *Positive soft thresholding:* $f_t(x) = (x - \lambda_t)_+$ for some iteration-dependent threshold $\lambda_t$. The threshold can be chosen so that, on average, $f_t(\theta^{(t)})$ has a number of nonzeros of order $k$.
- *Positive hard thresholding:* $f_t(x) = x \, \mathbb{I}(x \geq \lambda_t)$ (here $\mathbb{I}$ is the indicator function: $\mathbb{I}(B) = 1$ if $B$ is true and $0$ otherwise). Again $\lambda_t$ is a threshold.
- *Logistic nonlinearity:*

$$f_t(x) = \frac{1}{1 + \exp(-a_t(x - \lambda_t))}, \tag{5.155}$$

where $\lambda_t$ plays the role of a "soft threshold."

Which function should we choose? Which thresholds? Will this approach beat the simple power iteration (i.e., $f_t(x) = x$)?

To address these questions, we will carry out a simple heuristic analysis of the above nonlinear power iteration. Remarkably, we will see in the next section that this analysis yields the correct answer for a modified version of the same algorithm—a message-passing algorithm. Our discussion is based on Deshpande and Montanari (2013), and we refer to that paper for all omitted details, formal statements, and derivations.

The heuristic analysis requires that we consider separately vertices in $S$ and those outside $S$:

1. For $i \notin S$, the nonlinear power iteration (5.154) reads

$$\theta_i^{(t+1)} = \sum_{j=1}^{n} \mathbf{A}_{ij} f_t(\theta_j^{(t)}). \tag{5.156}$$

Since in this case the variables $\{\mathbf{A}_{ij}\}_{j \in [n]}$ are i.i.d. with mean zero and variance $1/n$, it is natural to guess—by the central limit theorem—that $\theta_i^{(t+1)}$ is approximately normal with mean $0$ and variance $(1/n) \sum_{j=1}^{n} f_t^2(\theta_j^{(t)})$. Repeating this argument inductively, we conclude that that $\theta_i^t \sim \mathcal{N}\left(0, \sigma_t^2\right)$, where, by the law of large numbers applied to $(1/n) \sum_{j=1}^{n} f_t^2(\theta_j^{(t)})$,

$$\sigma_{t+1}^2 = \mathbb{E}\left\{f_t(\sigma_t Z)^2\right\}, \tag{5.157}$$

where the expectation is taken with respect to $Z \sim \mathcal{N}(0, 1)$. The initialization $\theta^{(0)} = u$ implies $\sigma_1^2 = f_0(1)^2$.

2. For $i \in S$, we have $\mathbf{A}_{ij} = \kappa$ if $j \in S$ as well, and $\mathbf{A}_{ij} = \mathbf{Z}_{ij}/\sqrt{n}$ having zero mean and variance $1/n$ otherwise. Hence

$$\theta_i^{(t+1)} = \kappa \sum_{j \in S} f_t(\theta_j^{(t)}) + \frac{1}{\sqrt{n}} \sum_{j \in [n] \setminus S} \mathbf{Z}_{ij} f_t(\theta_j^{(t)}). \tag{5.158}$$

By the same argument as above, the second part gives rise to a zero-mean Gaussian contribution, with variance $\sigma_t^2$, and the first has nonzero mean and negligible variance. We conclude that $\theta_i^{(t)}$ is approximately $\mathcal{N}\left(\mu_t, \sigma_t^2\right)$ with $\sigma_t$ given recursively by (5.157). Applying the law of large numbers to the nonzero mean contribution, we get the recursion

$$\mu_{t+1} = \kappa \, \mathbb{E}\big\{ f_t(\mu_t + \sigma_t \, Z) \big\}, \tag{5.159}$$

where the expectation is taken with respect to $Z \sim \mathcal{N}(0,1)$, and the initialization $\theta^{(0)} = u$ implies $\mu_1 = \kappa \, f_0(1)$ (recall that $\kappa$ is defined as the limit of $k/\sqrt{n}$).

A few important remarks are in order here:

- *The above derivation is of course incorrect!* The problem is that the central limit theorem cannot be applied to the right-hand side of (5.156) because the summands are not independent. Indeed, each term $f_t(\theta_j^{(t)})$ depends on all the entries of the matrix $\mathbf{A}$.
- *The conclusion that we reached is incorrect.* It is not true that, asymptotically, $\theta_i^{(t)}$ is approximately Gaussian, with the above mean and variance.
- *Surprisingly, the conclusion is correct for a slightly modified algorithm,* namely, a message-passing algorithm that will be introduced in the next section. This is a highly nontrivial phenomenon

## 5.6.2  A message-passing algorithm

We modify the nonlinear power iteration (5.156) by transforming it into a message-passing algorithm, whose underlying graph is the complete graph with $n$ vertices. The iteration variables are "messages" $\theta_{i \to j}^{(t)}$ for each $i \neq j$ (with $\theta_{i \to j}^{(t)} \neq \theta_{j \to i}^{(t)}$). These are updated using the rule

$$\theta_{i \to j}^{(t)} = \sum_{k \in [n] \setminus j} \mathbf{A}_{ik} f_t(\theta_{k \to i}^{(t)}). \tag{5.160}$$

The only difference with respect to the iteration (5.156) is that we exclude the term $k = j$ from the sum. Despite this seemingly negligible change (one out of $n$ terms is dropped), the statistical properties of this algorithm are significantly different from those of the nonlinear power iteration (5.156), even in the limit $n \to \infty$. In particular, the Gaussian limit derived heuristically in the previous section holds for the message-passing algorithm. Informally, we have, as $n \to \infty$,

$$\theta_{i \to j}^{(t)} \sim \begin{cases} \mathcal{N}\left(\mu_t, \sigma_t^2\right) & \text{if } i \in S, \\ \mathcal{N}\left(0, \sigma_t^2\right) & \text{if } i \notin S, \end{cases} \tag{5.161}$$

where $\mu_t$ and $\sigma_t$ are determined by the state evolution equations (5.157) and (5.157).

Let us stress that we have not yet chosen the functions $f_t(\cdot)$: we defer this choice, as well as an analysis of state evolution, to the next section. Before this, we note that—as in the case of sparse regression—an approximate message passing (AMP) version of this algorithm can be derived by writing $\theta_{i\to j}^{(t)} = \theta_i^{(t)} + \delta\theta_{i\to j}^{(t)}$ and linearizing in the latter correction. This calculation leads to the simple AMP iteration

$$\theta^{(t+1)} = \mathbf{A}f_t(\theta^{(t)}) - \mathsf{b}_t f_{t-1}(\theta^{(t-1)}), \tag{5.162}$$

where the "Onsager term" $\mathsf{b}_t$ is given in this case by

$$\mathsf{b}_t = \frac{1}{n}\sum_{i=1}^n f_t'(\theta_i^{(t)}). \tag{5.163}$$

### 5.6.3   Analysis and optimal choice of $f_t(\cdot)$

We now consider the implications of state evolution for the performance of the above message-passing algorithms. For the sake of simplicity, we will refer to the AMP form (5.162), but analogous statements hold for the message-passing version (5.160). Informally, state evolution implies that

$$\theta^{(t)} \approx \mu_t\, u_S + \sigma_t\, z\,, \tag{5.164}$$

where $z \sim \mathcal{N}(0, \mathbf{I}_n)$, and this statement holds asymptotically in the sense of finite-dimensional marginals.

In other words, we can interpret $\theta^{(t)}$ as a noisy observation of the unknown vector $u_S$, corrupted by Gaussian noise. This suggests that we choose $f_t(\cdot)$ as the posterior expectation denoiser. Namely, for $y \in \mathbb{R}$,

$$f_t^{\mathrm{opt}}(y) = \mathbb{E}\{U \,|\, \mu_t\, U + \sigma_t\, Z = y\}, \tag{5.165}$$

where $U \sim \mathrm{Bernoulli}(p)$ for $p = k/n = \kappa/\sqrt{n}$, and $Z \sim \mathcal{N}(0,1)$ independently of $U$. A simple calculation yields the explicit expression

$$f_t^{\mathrm{opt}}(y) = \frac{\delta}{\delta + (1-\delta)\exp\left(-\dfrac{\mu_t}{\sigma_t^2}y + \dfrac{\mu_t^2}{\sigma_t^2}\right)}\,. \tag{5.166}$$

This is indeed empirically the best choice for the nonlinearity $f_t^{\mathrm{opt}}(\cdot)$. We shall next rederive it from a different point of view, which also allow us to characterize its behavior.

Reconsider the Gaussian limit (5.164). It is clear that the quality of the information contained in $\theta^{(t)}$ depends on the signal-to-noise ratio $\mu_t/\sigma_t$. Note that $u_S$ is very sparse; hence the vector $\theta^{(t)}$ is indistinguishable from a zero-mean Gaussian vector unless $\mu_t/\sigma_t \to \infty$. Indeed, unless this happens, the entries $\theta_i^{(t)}$, $i \in S$, are hidden in the tail of the zero-mean entries $\theta_i^{(t)}$, $i \in [n] \setminus S$ (see Fig. 5.26). It turns out that by optimally choosing $f_t(\cdot)$, this happens if and only if $\kappa > 1/\sqrt{e}$. In other words, the

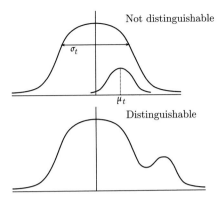

**Fig. 5.26** For small signal-to-noise ratios, all information about assignments is lost. As $\mu_t/\sigma_t \to \infty$, meaningful assignments are disambiguated from noise.

message-passing algorithm succeeds with high probability as long as $k$ is larger than $(1 + \varepsilon)\sqrt{n/e}$, for any $\varepsilon > 0$.

To determine whether $\widetilde{\mu}_t \equiv (\mu_t/\sigma_t) \to \infty$, note that—without loss of generality—we can rescale the functions $f_t(\cdot)$ so that $\sigma_t = 1$ for all $t$ (simply replacing $f_t(z)$ by $f_t(z)/\mathbb{E}\{f_t(Z)^2\}^{1/2}$ in (5.160) or in (5.162)). After this normalization, (5.159) yields

$$\widetilde{\mu}_{t+1} = \kappa \frac{\mathbb{E}\{f_t(\widetilde{\mu}_t + Z)\}}{\mathbb{E}\{f_t(Z)^2\}^{1/2}} \ . \tag{5.167}$$

Note that

$$\mathbb{E}\{f_t(\widetilde{\mu}_t + Z)\} = \int f_t(x)\, \frac{1}{\sqrt{2\pi}}\, e^{-(x - \mu_t)^2/2}\, dx$$

$$= e^{-\widetilde{\mu}_t^2/2} \mathbb{E}\{f_t(Z)\, e^{\widetilde{\mu}_t x}\} \le e^{\widetilde{\mu}_t^2} \mathbb{E}\{f_t(Z)^2\}^{1/2}, \tag{5.168}$$

where the last inequality follows from the Cauchy–Schwartz inequality. The inequality is saturated by taking $f_t(x) = e^{\widetilde{\mu}_t x - (\widetilde{\mu}_t^2/2)}$, which yields the state evolution recursion

$$\widetilde{\mu}_{t+1} = \kappa\, e^{\widetilde{\mu}_t^2/2} \ . \tag{5.169}$$

It is immediate to study this recursion, and conclude that $\widetilde{\mu}_t \to \infty$ if and only if $\kappa > 1/\sqrt{e}$.

The above analysis indeed yields the correct threshold for a message-passing algorithm, as proved in Deshpande and Montanari (2013). (For proving the theorem below, a "cleaning" step is added to the message-passing algorithm.)

**Theorem 5.9 (Deshpande and Montanari (2013))** *There exists an algorithm with time complexity $O(n^2 \log n)$ that outputs an estimate $\widehat{S}$ such that if*

$$k > (1 + \epsilon)\sqrt{n/e},$$

*then $\widehat{S} = S$ with probability converging to 1 as $n \to \infty$.*

In other words, the message-passing algorithm is able to find cliques smaller by a factor $1/\sqrt{e}$ with respect to spectral methods, with no increase in complexity. A natural research question is the following:

- *Is it possible to find planted cliques of size* $(1 - \varepsilon)\sqrt{n/e}$ *in time* $O(n^2 \log n)$?

Deshpande and Montanari (2013) provide a partially positive answer to this question by showing that no "local algorithm" (a special class of a linear-time algorithm) can beat message-passing algorithms for a sparse-graph version of the planted clique problem.

Let us conclude by showing how the last derivation agrees in fact with the guess (5.166) for the optimal nonlinearity. Note that $\delta = \kappa/\sqrt{n} \to 0$ as $n \to \infty$. In this limit,

$$f_t^{\mathrm{opt}}(y) \approx C_t \, \exp\left(\frac{\mu_t}{\sigma_t^2} y\right). \tag{5.170}$$

This coincides with the choice optimizing the state evolution threshold, once we set $\sigma_t = 1$ (which entails no loss of generality).

# References

Alon, N., Krivelevich, M., and Sudakov, B. (1998). Finding a large hidden clique in a random graph. In *Proceedings of the Ninth Annual ACM–SIAM Symposium on Discrete Algorithms*, pp. 594–598. Society for Industrial and Applied Mathematics.

Amelunxen, D., Lotz, M., McCoy, M. B., and Tropp, J. A. (2013). Living on the edge: a geometric theory of phase transitions in convex optimization. arXiv:1303.6672.

Barbier, J. and Krzakala, F. (2014). Replica analysis and approximate message passing decoder for superposition codes. arXiv:1403.8024.

Bayati, M., Lelarge, M., and Montanari, A. (2015). Universality in polytope phase transitions and message passing algorithms. *Annals of Applied Probability*, **25**, 753–822.

Bayati, M. and Montanari, A. (2011). The dynamics of message passing on dense graphs, with applications to compressed sensing. *IEEE Transactions on Information Theory*, **57**(2), 764–785.

Bayati, M. and Montanari, A. (2012). The LASSO risk for Gaussian matrices. *IEEE Transactions on Information Theory*, **58**(4), 1997–2017.

Beck, A. and Teboulle, M. (2009). A fast iterative shrinkage-thresholding algorithm for linear inverse problems. *SIAM Journal on Imaging Sciences*, **2**(1), 183–202.

Berthet, Q. and Rigollet, P. (2013). Complexity theoretic lower bounds for sparse principal component detection. In *Conference on Learning Theory*, pp. 1046–1066.

Bickel, P. J., Ritov, Y., and Tsybakov, A. B. (2009). Simultaneous analysis of Lasso and Dantzig selector. *American Journal of Mathematics*, **37**, 1705–1732.

Bolthausen, E. (2014). An iterative construction of solutions of the TAP equations for the Sherrington–Kirkpatrick model. *Communications in Mathematical Physics*, **325**(1), 333–366.

Bühlmann, Peter and Van De Geer, Sara (2011). *Statistics for High-Dimensional Data: Methods, Theory and Applications*. Springer.

Candés, E. and Recht, B. (2013). Simple bounds for recovering low-complexity models. *Mathematical Programming*, **141**(1–2), 577–589.

Candés, E. J. and Tao, T. (2005). Decoding by linear programming. *IEEE Transactions on Information Theory*, **51**, 4203–4215.

Candés, E. J. and Tao, T. (2007). The Dantzig selector: Statistical estimation when $p$ is much larger than $n$. *Annals of Statistics*, 2313–2351.

Capitaine, M., Donati-Martin, C., and Féral, D. (2009). The largest eigenvalues of finite rank deformation of large Wigner matrices: convergence and nonuniversality of the fluctuations. *Annals of Probability*, **37**(1), 1–47.

Capitaine, M., Donati-Martin, C., and Féral, D. (2012). Central limit theorems for eigenvalues of deformations of Wigner matrices. *Annales de l'Institut Henri Poincaré, Probabilités et Statistiques*, **48**(1), 107–133.

Chandrasekaran, V., Recht, B., Parrilo, P. A., and Willsky, A. S. (2012). The convex geometry of linear inverse problems. *Foundations of Computational Mathematics*, **12**(6), 805–849.

Chen, S. S. and Donoho, D. L. (1995). Examples of basis pursuit. In *Proceedings of Wavelet Applications in Signal and Image Processing III*, San Diego, CA.

Chen, S. S., Donoho, D. L., and Saunders, M. A. (1998). Atomic decomposition by basis pursuit. *SIAM Journal on Scientific Computing*, **20**(1), 33–61.

Daubechies, I., Defrise, M., and De Mol, C. (2004). An iterative thresholding algorithm for linear inverse problems with a sparsity constraint. *Communications on Pure and Applied Mathematics*, **57**(11), 1413–1457.

Derrida, B. (1981). Random-energy model: an exactly solvable model of disordered systems. *Physical Review B*, **24**(5), 2613.

Deshpande, Y. and Montanari, A. (2013). Finding hidden cliques of size $sqrtN/e$ in nearly linear time. arXiv:1304.7047.

Donoho, D.L., Javanmard, A., and Montanari, A. (2013a). Information-theoretically optimal compressed sensing via spatial coupling and approximate message passing. *IEEE Transactions on Information Theory*, **59**(11), 7434–7464.

Donoho, D.L., Johnstone, I., and Montanari, A. (2013b, June). Accurate prediction of phase transitions in compressed sensing via a connection to minimax denoising. *IEEE Transactions on Information Theory*, **59**(6), 3396–3433.

Donoho, D. and Montanari, A. (2013). High dimensional robust $m$-estimation: Asymptotic variance via approximate message passing. arXiv:1310.7320.

Donoho, D. L. (2006). High-dimensional centrally symmetric polytopes with neighborliness proportional to dimension. *Discrete & Computational Geometry*, **35**(4), 617–652.

Donoho, D. L. and Johnstone, I. M. (1994a). Minimax risk over $l_p$ balls. *Probability Theory and Related Fields*, **99**, 277–303.

Donoho, D. L. and Johnstone, I. M. (1994b). Neo-classical minimax problems, thresholding, and adaptation. *Bernoulli*, 39–62.

Donoho, D. L. and Johnstone, I. M. (1995). Adapting to unknown smoothness via wavelet shrinkage. *Journal of the American Statistical Association*, **90**, 1200–1224.

Donoho, D. L. and Johnstone, I. M. (1998). Minimax estimation via wavelet shrinkage. *Annals of Statistics*, **26**, 879–921.

Donoho, D. L and Johnstone, J. M (1994*c*). Ideal spatial adaptation by wavelet shrinkage. *Biometrika*, **81**(3), 425–455.

Donoho, D. L., Maleki, A., and Montanari, A. (2009). Message-passing algorithms for compressed sensing. *Proceedings of the National Academy of Sciences*, **106**(45), 18914–18919.

Donoho, D. L., Maleki, A., and Montanari, A. (2010). Message passing algorithms for compressed sensing: I. motivation and construction. In *Information Theory Workshop (ITW), 2010 IEEE*, pp. 1–5. IEEE.

Donoho, D. L., Maleki, A., and Montanari, A. (2011). The noise-sensitivity phase transition in compressed sensing. *IEEE Transactions on Information Theory*, **57**(10), 6920–6941.

Donoho, D. L. and Tanner, J. (2005*a*). Neighborliness of randomly projected simplices in high dimensions. *Proceedings of the National Academy of Sciences*, **102**(27), 9452–9457.

Donoho, D. L. and Tanner, J. (2005*b*). Sparse nonnegative solution of underdetermined linear equations by linear programming. *Proceedings of the National Academy of Sciences*, **102**(27), 9446–9451.

Féral, D. and Péché, S. (2007). The largest eigenvalue of rank one deformation of large Wigner matrices. *Communications in Mathematical Physics*, **272**(1), 185–228.

Füredi, Z. and Komlós, J. (1981). The eigenvalues of random symmetric matrices. *Combinatorica*, **1**(3), 233–241.

Gauss, C. F. (1823). *Theoria combinationis observationum erroribus minimis obnoxiae*. H. Dieterich.

Guo, D., Baron, D., and Shamai, S. (2009). A single-letter characterization of optimal noisy compressed sensing. In *Communication, Control, and Computing, 2009. Allerton 2009. 47th Annual Allerton Conference on*, pp. 52–59. IEEE.

Herrmann, F. J., Friedlander, M. P., and Yilmaz, Özgür (2012). Fighting the curse of dimensionality: Compressive sensing in exploration seismology. *IEEE Signal Processing Magazine*, **29**(3), 88–100.

James, W. and Stein, C. (1961). Estimation with quadratic loss. In *Proceedings of the Fourth Berkeley Symposium on Mathematical Statistics and Probability*, Volume 1, pp. 361–379.

Javanmard, A. and Montanari, A. (2013). State evolution for general approximate message passing algorithms, with applications to spatial coupling. *Information and Inference*, iat004.

Jerrum, M. (1992). Large cliques elude the Metropolis process. *Random Structures & Algorithms*, **3**(4), 347–359.

Johnstone, I. M. (2011). *Gaussian Estimation: Sequence and Wavelet Models*. Draft version, December 27, 2011.

Juditsky, A. and Nemirovski, A. (2011). First-order methods for nonsmooth convex large-scale optimization, I: general purpose methods. In *Optimization for Machine Learning* (ed. S. Sra, S. Nowozin, and S. J. Wright), pp. 121–148. MIT Press.

Kabashima, Y., Wadayama, T., and Tanaka, T. (2009). A typical reconstruction limit for compressed sensing based on $l_p$-norm minimization. *Journal of Statistical Mechanics: Theory and Experiment*, **2009**(09), L09003.

Krzakala, F., Mézard, M., Sausset, F., Sun, Y. F., and Zdeborová, L. (2012). Statistical-physics-based reconstruction in compressed sensing. *Physical Review X*, **2**(2), 021005.

Metzler, C. A., Maleki, A., and Baraniuk, R. G. (2014). From denoising to compressed sensing. arXiv:1406.4175.

Mezard, M. and Montanari, A. (2009). *Information, Physics, and Computation.* Oxford University Press.

Oymak, S.and Thrampoulidis, C. and Hassibi, B. (2013). The squared-error of generalized lasso: A precise analysis. arXiv:1311.0830.

Rangan, S. (2011). Generalized approximate message passing for estimation with random linear mixing. In *IEEE International Symposium on Information Theory*, St. Petersburg, pp. 2168–2172.

Rangan, S., Goyal, V., and Fletcher, A. K. (2009). Asymptotic analysis of map estimation via the replica method and compressed sensing. In *Advances in Neural Information Processing Systems*, pp. 1545–1553.

Richardson, T. and Urbanke, R. (2008). *Modern Coding Theory.* Cambridge University Press.

Rudelson, M. and Vershynin, R. (2008). On sparse reconstruction from Fourier and Gaussian measurements. *Communications on Pure and Applied Mathematics*, **61**(8), 1025–1045.

Som, S. and Schniter, P. (2012). Compressive imaging using approximate message passing and a markov-tree prior. *IEEE Transactions on Signal Processing*, **60**(7), 3439–3448.

Stojnic, M. (2013). A framework to characterize performance of LASSO algorithms. arXiv:1303.7291.

Tan, J., Ma, Y., and Baron, D. (2014). Compressive imaging via approximate message passing with image denoising. arXiv:1405.4429.

Tibshirani, R. (1996). Regression shrinkage and selection with the Lasso. *Journal of the Royal Statistical Society B*, **58**, 267–288.

Tropp, J. A., Laska, J. N., Duarte, M. F., Romberg, J. K., and Baraniuk, R. G. (2010). Beyond Nyquist: efficient sampling of sparse bandlimited signals. *IEEE Transactions on Information Theory*, **56**(1), 520–544.

Tsybakov, A. B. and Zaiats, V. (2009). *Introduction to Nonparametric Estimation.* Volume 11. Springer.

Wasserman, L. (2004). *All of Statistics: A Concise Course in Statistical Inference.* Springer.

Wasserman, L. (2006). *All of Nonparametric Statistics.* Springer.

# 6
# Error correcting codes and spatial coupling

Rüdiger URBANKE

EPFL IC ISC LTHC,
INR 116, Bâtiment INR,
Station 14,
CH-1015 Lausanne, Switzerland

*Notes written in collaboration with*
Rafah El-Khatib, EPFL, Switzerland
Jean Barbier, École Normale Supérieure, France
Ayaka Sakata, RIKEN, Japan

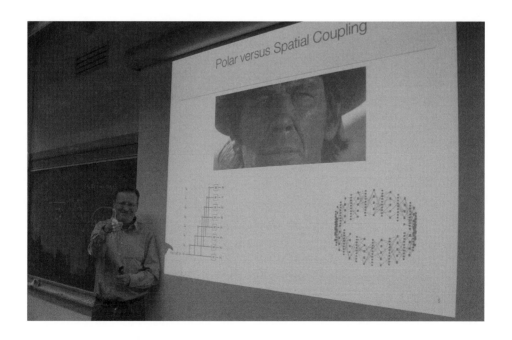

# Chapter Contents

## 6.1  Polar codes

### 6.1.1  Motivation

Consider the transmission scheme depicted in Fig. 6.1, where one bit $x \in \{0, 1\}$ is sent over a channel that either erases $x$ with probability (w.p.) $\epsilon$ or passes the bit unchanged w.p. $1 - \epsilon$. This channel is called the binary erasure channel with parameter $\epsilon$ and we will denote it by $\mathrm{BEC}(\epsilon)$. The receiver thus receives the symbol $y \in \{0, 1, ?\}$, where

$$y = \begin{cases} x & \text{w.p. } 1 - \epsilon, \\ ? & \text{w.p. } \epsilon. \end{cases} \tag{6.1}$$

We want to recover the transmitted bit at the receiver, and for this purpose the receiver forms an "estimate" of $x$ given the received symbol $y$; we denote this estimate by $\hat{x}(y)$. Our goal is to minimize the quantity $\Pr(\{\hat{x}(y) \neq x\})$; i.e., we want to minimize the *probability of error*.

If we are only sending a single bit, then we cannot hope to estimate the transmitted bit reliably in the case that it was erased. There is simply not enough information available. The picture changes if we are sending a *block* of bits.

Consider therefore the slightly more general setting shown in Fig. 6.2, where a vector $\mathbf{x} \in \{0, 1\}^n$ is sent on the same channel, the $\mathrm{BEC}(\epsilon)$. At the receiver, the vector $\mathbf{y} \in \{0, 1\}^n$ is received such that each component of this vector follows the rules in (6.1); i.e., each component is erased independently from all other components with probability $\epsilon$. It is easy to determine the expected number of erased and non-erased bits for this scenario, namely,

$$\mathbb{E}[|\{y_i = ?\}|] = n\epsilon,$$
$$\mathbb{E}[|\{y_i \neq ?\}|] = n(1 - \epsilon).$$

The standard deviation associated with these values is $\sigma = \sqrt{n\epsilon(1 - \epsilon)}$. Similar to before, we are interested in determining the transmitted vector $\mathbf{x}$ given the received vector $\mathbf{y}$, and for this purpose we form the estimate $\hat{\mathbf{x}}(\mathbf{y})$, and, as before,

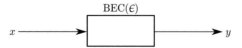

**Fig. 6.1** A transmission scheme of scalar quantities over the binary erasure channel with parameter $\epsilon$.

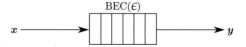

**Fig. 6.2** A transmission scheme of vector quantities over the binary erasure channel with parameter $\epsilon$.

$\Pr(\{\hat{\mathbf{x}}(\mathbf{y}) \neq \mathbf{x}\})$ is the probability of error. If this probability of error is "small," then we say that we achieve a reliable transmission.

It is now natural to ask the following question: How many bits can we reliably transmit over such a channel, measured as a function of the vector length $n$, when $n \to +\infty$? We derive first an upper bound and then a matching lower bound.

### 6.1.2   Upper bound

Assume that we are aided by a genie that tells the transmitter the positions that will be erased ahead of time. More formally, let $S \subseteq [n]$ be the set of erasures and assume that we know this set before sending the vector $\mathbf{x}$. Some thought then shows that the optimal strategy consists of sending our information in the positions $[n] \setminus S$ and filling the positions $S$ with dummy bits. The receiver then simply reads off the positions in $[n] \setminus S$ to recover the transmitted information; in this way, the estimate at the receiver is perfect and we never make an error. This is also clearly the maximum amount of information that we can transmit reliably in this scenario.

Since $\mathbb{E}[|[n] \setminus S|] = n(1 - \epsilon)$, it follows that the fraction of channel uses on which information can be sent reliably is equal to $1 - \epsilon$. This fraction is called the "transmission rate" and the highest possible rate for which reliable transmission is possible is called the "capacity." Since we have just shown that the capacity of the genie-aided transmission is $1 - \epsilon$, it follows that the real capacity (i.e., the capacity without the genie) is *at most* $1 - \epsilon$.

### 6.1.3   Lower bound

To prove that the capacity is equal to $1 - \epsilon$, we now derive a matching lower bound by describing a scheme that allows reliable transmission all the way up to a rate of $1 - \epsilon$. The typical way to prove this lower bound is by using a so-called "random coding" argument. This argument proceeds by showing that "randomly" chosen codes from a suitably defined "ensemble" of codes perform well with high probability.[1] This argument has the advantage that it is relatively simple and short. However, on the downside, the argument is nonconstructive and in addition does not take the complexity of the scheme into account.

Instead, we will describe an explicit scheme that in addition is also of low complexity. It is called the "polar coding" scheme. This scheme is fairly recent but the basic idea has already proven to be fundamental in a variety of areas Arikan (2009).

Let us first make an observation. For the BEC($\epsilon = 0$), the capacity is $1 - \epsilon = 1$. That is, we can fill in the entire vector $\mathbf{x}$ with information bits and recover them reliably. If, on the other hand, the erasure probability is $\epsilon = 1$, then the capacity is 0 and there is no use sending any information in $\mathbf{x}$. Both of these cases are thus *easy* to deal with in the sense that we know what to do. This observation extends to cases where $\epsilon \sim 0$ and $\epsilon \sim 1$. More precisely, we lose very little in the case when $\epsilon \sim 1$ by

---

[1] A *code* is a subset of the set of all binary $n$-tuples, and typically this subset is chosen in such a way that the individual codewords are well separated. This ensures that even with some of the components being erased, the receiver can still figure out which of the codewords was sent.

not using the channel, and if $\epsilon$ is very small (compared with $n$), then we can still use all components of the block and most of the time the whole block will arrive without erasures. Only once in a while will we not be able to recover the block, and this simply results in a small probability of error.

Let us now introduce the basic idea of *polarization*. Consider the transmission scheme in Fig. 6.3. Two bits, call them $U_1$ and $U_2$, are chosen uniformly at random from $\{0, 1\}$ and are encoded into two other bits, denoted by $X_1$ and $X_2 \in \{0, 1\}$, as follows (all operations are over the binary field):

$$X_1 = U_1 + U_2, \tag{6.2}$$
$$X_2 = U_2. \tag{6.3}$$

Equivalently, we can describe this relationship in matrix form:

$$[U_1\ U_2] \begin{bmatrix} 1 & 0 \\ 1 & 1 \end{bmatrix} = [X_1\ X_2].$$

Assume that we receive $\mathbf{Y} = (Y_1\ Y_2)$ and that we want to estimate $U_1$ given $\mathbf{Y} = (Y_1\ Y_2)$, where $U_2$ is unknown (and has a uniform prior). We denote by $\hat{U}_1(\mathbf{Y})$ this estimate. Note that from (6.2) we know that $U_1 = X_1 + U_2$ (there are no signs in the binary field). If we combine this with (6.3), we see that $U_1 = X_1 + X_2$. Note further that $Y_1$ and $Y_2$ are the result of transmitting $X_1$ and $X_2$, respectively, through independent erasure channels.

We therefore see that we can reconstruct $U_1$ if and only if neither $Y_1$ nor $Y_2$ is an erasure, so that $X_1 = Y_1$ and $X_2 = Y_2$. Hence, we have

$$\hat{U}_1(\mathbf{Y}) = \begin{cases} Y_1 + Y_2, & \text{if } Y_1 \neq ? \wedge Y_2 \neq ?, \\ ?, & \text{otherwise.} \end{cases}$$

Note that the bits $X_1$ and $X_2$ are sent over independent erasure channels and that

$$\Pr(\{Y_1 = ?\}) = \epsilon, \tag{6.4}$$
$$\Pr(\{Y_2 = ?\}) = \epsilon. \tag{6.5}$$

Therefore,

$$\Pr(\{\hat{U}_1(\mathbf{Y}) = U_1\}) = \Pr(\{Y_1 \neq ? \wedge Y_2 \neq ?\}) = (1 - \epsilon)^2,$$
$$\Pr(\{\hat{U}_1(\mathbf{Y}) = ?\}) = 1 - (1 - \epsilon)^2 = \epsilon(2 - \epsilon) > \epsilon.$$

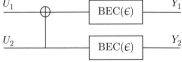

**Fig. 6.3** A one-step polarization transformation for the BEC($\epsilon$)s.

As we can see, the probability that $U_1$ is erased is strictly larger than $\epsilon$ (unless $\epsilon = 1$). So this does not seem to be a very good scheme. Why then would we use this transform, which is called the *polar* transform? As we will see shortly, estimating the bit $U_2$ is in fact easier than the original problem, and estimating the bit $U_1$ is more difficult, as we just discussed. The key point is that both of these tasks are closer to the two trivial scenarios ($\epsilon = 0$ and $\epsilon = 1$), and by recursing this transform we will be able to approach these trivial cases more and more closely. Once we are sufficiently close, no extra coding will be necessary, since we know how to deal with these two cases.

Let us now look at the problem of estimating $U_2$. For this task, we will assume that a genie gives us the true value of $U_1$. We will soon see that in fact we have this information at the receiver as long as we decode the various bits in the appropriate order. Therefore, this assumption is in fact realistic. Let us summarize: we want to estimate $U_2$ given $U_1$ and $\mathbf{Y}$. Reconsider our two basic equations. First, rewrite (6.2) as $U_2 = X_1 + U_1$ and note that, by assumption, $U_1$ is known. Further, write (6.3) as $U_2 = X_2$. We therefore see that we have two estimates of $U_2$ available at the receiver and that these two estimates are conditionally independent since $X_1$ and $X_2$ are transmitted over two independent channels (and $U_1$ is a known constant). We conclude that we will be able to recover $U_2$ as long as at least one of $Y_1$ and $Y_2$ is not erased. In summary, we have

$$\hat{U}_2(U_1, \mathbf{Y}) = \begin{cases} Y_2, & \text{if } Y_2 \neq ?, \\ Y_1 + U_1, & \text{if } Y_2 =? \wedge Y_1 \neq ?, \\ ?, & \text{otherwise}, \end{cases}$$

and

$$\Pr(\{\hat{U}_2(U_1, \mathbf{Y}) =?\}) = \epsilon^2 < \epsilon. \tag{6.6}$$

Assume now that we estimate $U_1$ and $U_2$ successively using the following estimators:

$$\hat{U}_1 = \hat{U}_1(\mathbf{Y}),$$

$$\hat{U}_2 = \begin{cases} \hat{U}_2(\hat{U}_1, \mathbf{Y}) & \text{if } U_1 \neq ?, \\ ? & \text{otherwise}. \end{cases}$$

Then,

$$\Pr(\{\hat{U}_1(\mathbf{Y}) =? \vee \hat{U}_2(\hat{U}_1(\mathbf{Y}), \mathbf{Y}) =?\}) = \Pr(\{\hat{U}_1(\mathbf{Y}) =? \vee \hat{U}_2(U_1, \mathbf{Y}) =?\}) \tag{6.7}$$

$$\leq \Pr(\{\hat{U}_1(\mathbf{Y}) =?\}) + \Pr(\{\hat{U}_2(U_1, \mathbf{Y}) =?\})$$

$$= \epsilon(2 - \epsilon) + \epsilon^2 = 2\epsilon. \tag{6.8}$$

This has the following interpretation. In terms of this union bound, the successive decoder is as good as the scenario shown in Fig. 6.4, where we have two independent BECs with different parameters.

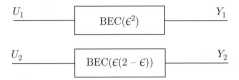

**Fig. 6.4**  A transmission scheme equivalent to the one-polarization step of two $BEC(\epsilon)$s.

**Remark 6.1**   *The scheme in Fig. 6.3 is obtained from that in Fig. 6.1 using the following relationship. Given two BECs with parameter $\epsilon$ each, we obtain a BEC with parameter $\epsilon(2 - \epsilon)$, which is called the "$-$ channel" and a BEC with parameter $\epsilon^2$, which is called the "$+$ channel.". Note further that the sum of the capacities of these two channels is $1 - \epsilon(2 - \epsilon) + 1 - \epsilon^2 = 2(1 - \epsilon)$. In other words, the average capacity of these two channels is equal to the original capacity. So we have lost nothing in terms of capacity by using this transform with the particular successive decoding algorithm. If we look at the transform itself, this is not too surprising. After all, this transform is invertible, and hence lossless.*

**Remark 6.2 (Example)**   *Consider the scheme in Fig. 6.3 with $\epsilon = 0.5$. Then the equivalent scheme in Fig. 6.4 consists of two cascaded channels $BEC(0.75)$ and $BEC(0.25)$. Notice that the average erasure probability over the two channels is $(0.75 + 0.25)/2 = \epsilon$.*

This procedure of starting with two independent channels, combining them, and then separating them again into two channels constitutes one "polarization step."

Rather than performing only a single step, we can now recurse. Let us look explicitly at one further step, as shown in Fig. 6.5. Note that in the second step we combine "like" channels and that we decode successively in a very particular order, namely $U_1$, $U_2$, $U_3$, $U_4$. The erasure probabilities that we get for the four resulting "synthetic" channels are as follows:

- $U_1$ "sees" the BEC w.p. $\delta(2 - \delta)$ where $\delta = \epsilon(2 - \epsilon)$.
- $U_2$ "sees" the BEC w.p. $\delta^2$ where $\delta = \epsilon(2 - \epsilon)$.

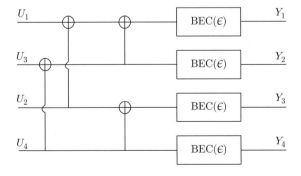

**Fig. 6.5**  A two-polarization step of four $BEC(\epsilon)$s.

- $U_3$ "sees" the BEC w.p. $\delta(2 - \delta)$ where $\delta = \epsilon^2$.
- $U_4$ "sees" the BEC w.p. $\delta^2$ where $\delta = \epsilon^2$.

Clearly, we can recurse this procedure $n \in \mathbb{N}$ times to create from $2^n = N$ independent channels with parameter $\epsilon$, $N$ "new" (sometimes called synthetic) channels with parameters $\epsilon_i$, $i \in 0 \dots N - 1$. The parameters evolve at each polarization step according to the rules

$$z \to z(2 - z), \tag{6.9}$$

$$z \to z^2.$$

The evolution of the erasure probabilities upon this recursion can be seen as an expansion of the tree diagram in Fig. 6.6. Notice that the mean of each column, with respect to the uniform distribution, is constant, namely, equal to $\epsilon$. This is true since

$$\frac{z^2 + z(2 - z)}{2} = z.$$

This implies that the overall capacity stays preserved.

Recall now the motivation for using this transform. We know how to deal with trivial and perfect channels and we hope that by applying a sufficient number of these transforms, the resulting synthetic channels will all become either trivial or perfect. If this is indeed the case, and since we know that the overall capacity is preserved, it must be true that the proportion of perfect channels is equal to the capacity of the original channel. Therefore, if we send our bits over the perfect channels and fix the trivial channels to some known value, we will be able to transmit reliably arbitrarily close to capacity. It remains to show that this "polarization" of the channels toward these extreme points is indeed the case.

Toward this goal, let us look at the second moment associated with this transformation $\rho_n^n$:

$$\rho_n^2 = \frac{z^4 + z^2(2 - z)^2}{2} = z^4 + 2z^2(1 - z).$$

Consider $f(z) = z^4 + 2z^2(1 - z) - z^2 = z^2(z^2 - 2z + 1) = z^2(z - 1)^2$, which represents the difference of the second moment after the transform and before the transform.

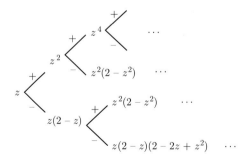

**Fig. 6.6** A tree diagram that tracks the erasure probabilities obtained by polarization.

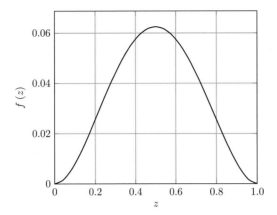

**Fig. 6.7** The plot of $f(z)$.

Figure 6.7 shows the plot of $f(z)$. Note that $f(z) > 0$, $z \in (0,1)$, and that $f(0) = f(1) = 0$. This means the following: Consider the $n$th column and let $\mu_n$ and $\rho_n^2$ denote the mean and second moment, respectively. We have seen that for $n \in \{0, 1, \dots\} = \mathbb{N}$,

$$\mu_n = \epsilon,$$
$$\rho_n^2 \text{ is increasing and } \rho_n^2 \leq 1.$$

Thus, $\lim_{n \to +\infty} \rho_n^2 = \rho_\infty^2$ exists. Note further that as long as a nonzero probability mass lies strictly bounded away from 0 and 1, then the increase in the second moment is strict. It is therefore clear that the limiting distribution must be the one where all the mass is located either at 0 or at 1. This is made precise in the following statement:

**Lemma 6.3**   *Let $z_{nj}$, $j \in \{1, 2, \dots, 2^n\}$, denote the $2^n$ numbers in the nth column. For $\delta \in [0, \frac{1}{2}]$, define*

$$S_n(\delta) = \{j : \delta \leq z_{nj} \leq 1 - \delta\}.$$

*Then for any $\delta > 0$, $\lim_{n \to +\infty} |S_n(\delta)| 2^{-n} = 0$. In words, all but a sublinear fraction of channels are either "good" or "bad."*

**Remark 6.4**   *For a fixed $\delta > 0$, we call a channel "good" if it belongs to the set $G_n(\delta) = \{j : |z_{nj}| < \delta\}$, and "bad" if it belongs to the set $B_n(\delta) = \{j : |1 - z_{nj}| < \delta\}$.*

**Proof**   Fix $\delta \in [0, \frac{1}{2}]$. Since $\lim_{n \to +\infty} \rho_n^2 = \rho_\infty^2$ exists, then for all $\Delta > 0$, there exists $n_0 \in \mathbb{N}$ such that

$$\rho_n^2 > \rho_\infty^2 - \Delta \min\{f(\delta), f(1 - \delta)\}$$

for all $n \geq n_0$. We claim that for all $n \geq n_0$, $|S_n(\delta)| 2^{-n} \leq \Delta$. Since this is true for all $\Delta > 0$, the claim will follow.

Assume that $|S_n(\delta)|2^{-n} > \Delta$. This means that there are at least $\Delta 2^n$ numbers $z_{nj}$ in the range $[\delta, 1 - \delta]$. It follows that the second moment must go up in the next iteration by at least $\Delta \min\{f(\delta), f(1 - \delta)\}$. But this would imply that $\rho_{n+1}^2 > \rho_\infty^2$, which is a contradiction. $\qquad\square$

**Remark 6.5** *A more careful analysis shows that, for $0 \le \beta < \frac{1}{2}$, with $N = 2^n$, we have*

$$\lim_{n \to +\infty} |S_n(2^{-2^{\beta n}})|2^{-n} = 0,$$

$$\lim_{n \to +\infty} |G_n(2^{-2^{\beta n}})|2^{-n} = 1 - \epsilon,$$

$$\lim_{n \to +\infty} |B_n(2^{-2^{\beta n}})|2^{-n} = \epsilon.$$

This gives rise to the scheme shown in Fig. 6.8. Consider a polar code that is polarized $n$ times, where $n$ is chosen to be "sufficiently" large. The code thus has $2^n$ channels, input bits $U_1, \ldots, U_{2^n}$, and output bits $Y_1, \ldots, Y_{2^n}$. "Freeze" the channels $j \in B_n$ and put the information bits in the channels $j \in G_n$. Here "freezing" means that we put a fixed value in these positions and this value is known both to the transmitter as well as the receiver. In fact, we are free to choose the value, and generically we will choose it to be 0. Decode the bits $U_1, \ldots, U_{2^n}$ successively from 1 to $N$. Of course, if a bit $U_k$ is frozen, then we already know its value and no actual decoding has to be done. Only if $U_k$ belongs to the good set $G_n$ will we need to decode. But in this case the error probability will by definition be very small. The associated computational complexity is $\mathcal{O}(n2^n) = \mathcal{O}(N \log_2 N)$.[2] Since good channels are very good, the union

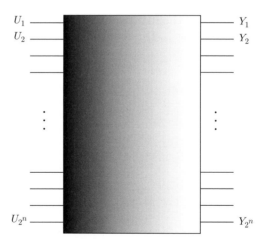

**Fig. 6.8**   A polarization scheme of $n$ steps.

[2] That this is indeed the case takes some thinking. A closer look shows that the "graphical model" that describes the relationship between the input and the output, i.e., the graphical model corresponding to the binary matrix that describes this relationship, has $\mathcal{O}(N \log_2 N)$ nodes and edges, and decoding can be accomplished by computing one message for each edge in this model.

bound on the error probability decays, and it decays like $N2^{-\sqrt{N}}$. Finally, note that $\lim_{n\to+\infty}|G_n|2^{-n} = 1 - \epsilon$. This means that the fraction of channels that we can use for information transmission is $1 - \epsilon$. This fraction is equal to the upper bound on the capacity we previously derived. Hence, we have matching upper and lower bounds on the capacity and have therefore determined the capacity exactly. In addition, we have found a low-complexity capacity-achieving scheme!

### 6.1.4 Summary

For the BEC($\epsilon$), the capacity is $1 - \epsilon$. For polar codes, the encoding and decoding complexity is $\mathcal{O}(N \log_2 N)$, where $N = 2^n$ is the blocklength.

We mention a final point. Assume that we want to transmit information at a rate $R = C(1 - \delta)$. Here $\delta$ represents the so-called *gap* (of the rate) *to the capacity*. Assume further that we want to achieve a certain fixed block probability of error. Then what blocklength is required to achieve this goal? More precisely, if we let $\delta$ tend to 0, then how does the blocklength have to scale with $\delta$? This question was addressed by Strassen (1962) and by Polyanskiy *et al.* (2010). The result is that for any code, the blocklength must grow at least as the square of the reciprocal of the gap to capacity, i.e., $N = \Theta(1/\delta^2)$, and there exist coding schemes that achieve this lower bound.

How do polar codes stack up in terms of their finite-length scaling? It was shown in Hassani *et al.* (2014) that for general channels we have

$$\frac{1}{\delta^{3.56}} \le N \le \frac{1}{\delta^7}.$$

For transmission over the BEC, we need $N = 1/\delta^{3.67}$ and for the binary symmetric channel (BSC), we need $N = 1/\delta^{4.2}$. So this means that polar codes require roughly the square of the blocklength compared with optimal codes.

**Remark 6.6** *A BSC($\epsilon$) is a channel that takes a bit $x \in \{0, 1\}$. It flips the bit to $1 - x$ w.p. $\epsilon$ and leaves it unchanged w.p. $1 - \epsilon$.*

So far, we have only talked about polar codes for the binary erasure channel. But everything we mentioned can be extended to more general channels, such as the BSC or the so-called additive white Gaussian-noise channel (AWGNC). In the same manner as for the BEC, we can construct low-complexity capacity-achieving polar codes or such channels.

Another caveat concerning polar codes concerns the question of "universality." Consider a polar code $\mathcal{C}_1$ for the BEC($\epsilon$) and a polar code $\mathcal{C}_2$ for the BSC($p$). Assume that the parameters $\epsilon$ and $p$ are chosen in such a way that the capacities of the two channels are equal. We know from our previous discussion that for both scenarios we can construct capacity-achieving polar codes. This means that if we pick $N = 2^n$ sufficiently large, then the fraction of good indices in both cases is close to capacity.

Denote by $B_{n,1}$ and $B_{n,2}$ the sets of good channels corresponding to $\mathcal{C}_1$ and $\mathcal{C}_2$, respectively. Similarly, denote by $G_{n,1}$ and $G_{n,2}$ the sets of bad channels. It is now natural to ask if

$$B_{n,1} = B_{n,2}?$$
$$G_{n,1} = G_{n,2}?$$

In words, we are asking if the *same* synthetic channels are good for the two scenarios. If this is the case, then the code is *universal*, meaning that one and the same code is good for both scenarios.

Unfortunately, the answer has been found to be negative, and so polar codes are not universal.[3]

## 6.2 Applications

Before we continue and describe codes based on sparse graphs, it might be interesting to consider some standard application scenarios. This will make it clearer what range of parameters is typically of interest.

Consider transmission over the AGWNC. That is,

$$y_i = h_i x_i + z_i,$$

where $x_i \in \{-1, +1\}$ is the bit that we want to transmit, $y_i$ is the received value, $z_i \sim \mathcal{N}(0, \sigma^2)$ (Gaussian noise of zero mean and unit variance), and $h_i$ is the so-called fading coefficient, describing the path loss of signal strength caused by the transmission medium.

In *wireless* transmission settings, we are using electromagnetic waves emitted and captured by *antennas*, as the transmission medium. In practice, the following values are typical:

- blocklength: $N \sim 10^3 - 10^4$ bits,
- rate: $R \sim 0.5$,
- block error probability: $P_B \sim 10^{-2}$,
- throughput: $10^4 - 10^6$ bits/s,
- processing power consumption: $10\,\mathrm{mW}$.

Thus, we have at our disposal about $10^{-7}\,\mathrm{J}$ to process one bit.

Another transmission scheme is that over the BSC. In that case,

$$y_i = x_i \oplus z_i,$$

where $x_i \in \{0, 1\}$, $z_i \in \{0, 1\}$, and $\mathrm{P}(z_i = 1) = p$.

This is a first-order approximation to model optical transmission. In such settings, the following values are typical:

- blocklength: $N \sim 10^4 - 10^6$ bits,
- rate: $R \sim \frac{239}{255}$ for historical reasons,

---

[3] In fact, recent results show that they can be made universal at the price of increasing the blocklength.

- bit error probability: $P_B \sim 10^{-15}$, basically "one error per day,"
- throughput: $100\,\text{Gbits/s}$,
- processing power consumption: $100\,\text{W}$,
- interchip data rate: $5 \times 10^{12}\,\text{bits/s}$ (within chip, for message passing).

Thus, we have at out disposal about $10^{-9}\,\text{J}$ to process one bit, quite a limited quantity.

### 6.2.1 Metrics

How can one "measure" codes so that we can compare various competing schemes in a meaningful manner? The following are some useful metrics:

- **Construction complexity**: How difficult is it to find a code? For polar codes, this can be done quite efficiently.
- **Encoding and decoding complexity**: How many operations do we need to encode and decode one information bit? As we have seen, for polar codes both encoding and decoding can be done in $\mathcal{O}(N \log_2 N)$ (real) operations, where $N$ is the blocklength. This is also efficient.
- **Finite-length performance**: What blocklengths do we need in order to get "close" to capacity. This is one of the few weaknesses of polar codes. We have seen that the required blocklength is roughly the square of what is optimally achievable.
- **Throughput**: How many bits can we decode per clock cycle. For some high-speed applications such as optical transmission on the backbone network, this is important. The standard decoder for polar codes is inherently sequential and so does not have a very high throughput. But it can made more parallel if we are willing to pay a higher processing cost.
- **Universality**: Is one and the same code good for many channels? Standard polar codes are not universal, but they can be made universal if we are willing to consider longer codes.
- **Proofs**: How simple is it to explain the scheme? Polar codes are by far the simplest of all known capacity-achieving schemes. In addition, they have an explicit construction rather than only probabilistic guarantees.

## 6.3 Low-density parity-check codes

### 6.3.1 Linear codes

Linear codes are codes such that the (weighted) sum of any two codewords is again a codeword. As a consequence, such codes have a compact algebraic description, either as the image of a linear map or as the kernel of a linear map. In the first case, we typically consider the so-called *generator* matrix $G$ and we represent the code as the space spanned by the rows of $G$. More precisely, let $G$ be an $k \times n$ matrix over a field $\mathbb{F}$. Here, $n$ is the *blocklength*[4] and $0 \le k \le n$ is the *dimension* of the code. Although more

---

[4] In the previous section, the blocklength was denoted by $N$ and $N = 2^n$, where $n$ denoted the number of polar steps. For the most part we will revert now to the more standard notation where $n$

general cases are possible and indeed can be useful, we will restrict our discussion to the case where $\mathbb{F}$ is the binary field. The code *generated* by $G$ is then

$$C(G) = \{x \in \mathbb{F}^n : x = uG, u \in \mathbb{F}^k\} = \{x \in \mathbb{F}^n : Hx^T = 0\}. \qquad (6.10)$$

Here the second representation of the code is in terms of the kernel of the so-called *parity-check* matrix $H$. Note that if $G$ has rank $k$ (and thus $|C(G)| = 2^k$), then by the rank-nullity theorem, $H$ has rank $n - k$.

**Example 6.7 (Polar codes)** As an important example, the *polar* codes that we discussed in the prequel are linear codes. We can find the generator matrix corresponding to polar codes by starting with the binary matrix

$$G_1 = \begin{bmatrix} 1 & 0 \\ 1 & 1 \end{bmatrix}.$$

Let $G_n$, $n \in \mathbb{N}$, be the $n$th Kronecker product of $G_1$. The generator matrix corresponding to a polar code of length $N = 2^n$ is then the matrix that corresponds to picking those rows of $G_n$ that correspond to the "good" channels.

Almost all codes used in practice are linear. There are two reasons for this. First, it can be shown that for most scenarios linear codes suffice if we want to get close to capacity. Second, linear codes are typically much easier to deal with in terms of complexity.

### 6.3.2 MAP decoding

In order to decode the output of the noisy channel, an appropriate choice is the maximum a posteriori (MAP) estimator:

$$\hat{x}^{\mathrm{MAP}} = \underset{c \in C(G)}{\operatorname{argmax}} P_{X|Y}(x|y) = \underset{c \in C(G)}{\operatorname{argmax}} P_{Y|X}(y|x) \frac{P_X(x)}{P_Y(y)} = \underset{c \in C(G)}{\operatorname{argmax}} P_{Y|X}(y|x) P_X(x),$$

$$(6.11)$$

where $P_X(x)$ is the prior over X and $P_{Y|X}(y|x)$ is the likelihood of the output $Y$ of the noisy channel given $X$. The MAP decoder outputs the mode of the posterior distribution and thus minimizes the block-error probability. This is why we would like to implement it. (In order to achieve capacity, it is in fact not necessary to do MAP decoding.)

### 6.3.3 Low-density parity-check codes

Low-density parity-check codes (LDPC) are linear codes defined by a parity-check matrix $H$ that has few nonzero entries, more precisely, the number of nonzero entries grows only linearly with the dimension $n$ of the matrix.

denotes the blocklength except when we talk about polar codes. We hope that the resulting confusion will stay bounded.

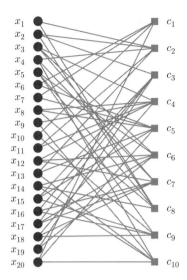

**Fig. 6.9** Factor graph of an LDPC with $N = 20$ variables and 10 factors.

A particularly useful description is in term of a factor graph (see Fig. 6.9). In this factor graph, there are $n$ *variable* nodes representing the components of the codeword and there are $n - k$ factor nodes, each representing one of the $n - k$ linear constraints implied by the parity-check matrix. There is an *edge* between a factor node and a variable node if that particular node participates in the constraints represented by the factor node. Since the parity matrix $H$ is *sparse*, the number of edges in this factor graph grows only linearly with the length of the code.

The code is then the set of all binary $n$-tuples that satisfy each of the $n - k$ constraints. As an example, looking at Fig. 6.9, the first constraint is $C1 = \mathbb{I}(x_6 \oplus x_7 \oplus x_{10} \oplus x_{20} = 0)$ where $\mathbb{I}(\bullet)$ is the indicator function.

The code rate, i.e., the fraction of information bits contained in the $n$ transmitted bits, is equal to $R \geq R_{\text{design}} = (\#\text{variables} - \#\text{factors})/\#\text{variables}$. In general, $R_{\text{design}}$ is only a lower bound on the actual rate, since some of the constraints can be linearly dependent. But it was shown by Miller and Cohen (2003) that, with high probability, the rate of a randomly chosen regular code is very close to this lower bound (see also Richardson and Urbanke, 2009). A regular code is one where all variables have constant degree $d_l$ and all check nodes have degree $d_r$.

### 6.3.4 Configuration model

One possible way of generating a $(d_l, d_r)$-regular code is to define an *ensemble* of such codes and to create a specific instance by sampling uniformly at random from this ensemble. A canonical way of achieving this is via the so-called *configuration model*. In this model, we associate $d_l$ "sockets" with each variable node and $d_r$ sockets with each check node. Note that there are in total $nd_l$ variable-node sockets and an equal number, namely, $(n - k)d_r$, check-node sockets. We get from this model a graph by picking uniformly at random a permutation on $nd_l$ elements and by matching up sockets according to this permutation. This ensemble is convenient for practical implementations, since it is easy to sample from, and it is also well suited for theoretical analysis.

For applications, it is also important to be able to have nodes of various degrees. The corresponding ensembles of codes are called *irregular* ensembles. To specify such an ensemble, we need to specify how many nodes there are of what degree, or, equivalently, how many edges there are that connect to nodes of various degrees.

One useful representation of the statistics of the degrees is in terms of a polynomial representation. For example, the polynomials corresponding to Fig. 6.10 are

$$\Lambda(x) = \frac{1}{4}x^2 + \frac{2}{4}x^4 + \frac{1}{4}x^5, \qquad \lambda(x) = C\frac{\mathrm{d}\Lambda(x)}{\mathrm{d}x} = \frac{2}{15}x + \frac{8}{15}x^3 + \frac{5}{15}x^4,$$

$$P(x) = \frac{1}{2}x^5 + \frac{1}{2}x^7, \qquad \rho(x) = C\frac{\mathrm{d}P(x)}{\mathrm{d}x} = \frac{5}{12}x^4 + \frac{7}{12}x^6. \qquad (6.12)$$

Here, $\Lambda$ ($P$) is the normalized distribution from the node perspective (where $\Lambda$ specifies the variable-node degrees and $P$ specifies the check-node degrees). The coefficient in front of $x^i$ is the fraction of nodes of degree $i$. The normalized derivatives of these quantities, namely $\lambda$ ($\rho$), represent the same quantities, but this time from the perspective of the edges; i.e., they represent the probabilities that a randomly chosen edge is connected to a node of a particular degree.

### 6.3.5 From bit-MAP to belief propagation decoding

For the BEC, bit-MAP decoding can be done by solving a system of linear equations, i.e., in complexity $O(n^3)$: one must solve $Hx = 0$, i.e., $H_\epsilon x_\epsilon \oplus H_{\bar{\epsilon}} x_{\bar{\epsilon}} = 0$, where $H_\epsilon$ is submatrix of the parity-check matrix spanned by the columns corresponding to the

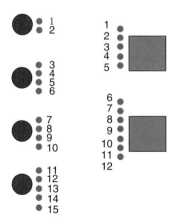

**Fig. 6.10** Instance of a (3,6)-random factor graph, where 3 (respectively, 6) is the average degree of nodes (respectively, factors).

erased components of $x$, i.e., $x_\epsilon$, and $H_{\bar{\epsilon}}$ is the complement. Thus, $H_\epsilon x_\epsilon = H_{\bar{\epsilon}} x_{\bar{\epsilon}} = s$ has to be solved to find back the missing part $x_\epsilon$. But we are interested in an algorithm that is applicable for general binary-input memoryless output-symmetric (BMS) channels, where MAP decoding is typically intractable. We therefore consider a message-passing algorithm that is applicable also in the general case. More precisely, we consider the sum–product (also called the belief propagation, BP) algorithm. This algorithm performs bit-MAP decoding on codes whose factor graph is a tree, and performs well on locally tree like graphs such as the random ones (see Fig. 6.11) From (6.11), we get

$$\hat{x}_i^{\text{MAP}} = \operatorname*{argmax}_{x_i \in \pm 1} \sum_{\{x_j : j \neq i\}} \left( \prod_j p(y_j|x_j) \right) \mathbb{1} \left( x \in \mathcal{C} \right) \tag{6.13}$$

BP is a message-passing algorithm that finds a fixed point to the following set of equations (here for the case of a parity check code):

$$m_{i \to \mu}(x_i) = \frac{p(y_i|x_i)}{z_{i \to \mu}} \prod_{\nu \in \partial_i \setminus \mu} \hat{m}_{\nu \to i}(x_i), \tag{6.14}$$

$$\hat{m}_{\mu \to i}(x_i) = \frac{1}{\hat{z}_{\mu \to i}} \sum_{\{x_j : j \in \partial_\mu \setminus i\}} \mathbb{1} \left( \left[ \bigoplus_{\{x_j : j \in \partial_\mu \setminus i\}} x_j \right] \oplus x_i = 0 \right) \prod_{\{x_j : j \in \partial_\mu \setminus i\}} m_{j \to \mu}(x_j),$$

where $\partial_\mu \setminus i$ stands for the ensemble of variable indices of the variables that are neighbors of factor $\mu$ except $i$, the messages $\{m_{i \to \mu}, \hat{m}_{\mu \to i}\}$ are the so-called cavity messages (which are probability distributions, $\{z_{i \to \mu}, \hat{z}_{\mu \to i}\}$ are the normalization constants) from which we can infer their most probable state by maximization of the marginals $\{m(x_i)\}$ allowing bit-MAP decoding:

$$m(x_i) = \frac{1}{z_i} \prod_{\nu \in \partial_i} \hat{m}_{\nu \to i}(x_i). \tag{6.15}$$

What is the performances of the BP algorithm on the BEC? Here is the experiment we consider. Fix the ensemble. In the above example, it is the $(3, 6)$-regular ensemble.

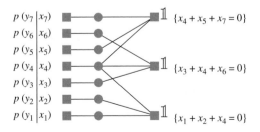

**Fig. 6.11** Instance of a random factor graph. The BP decoder allows us to estimate the marginal or mode of each input component.

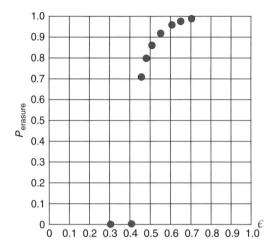

**Fig. 6.12**   Performance of BP decoding over a BEC channel.

Now pick very long instances of this ensemble. Pick a random codeword and transmit over a BEC with erasure probability $\epsilon$. Run the BP decoder until convergence. Record the error probability and average over many instances. Plot the average bit-error probability versus $\epsilon$. Naturally, as $\epsilon$ decreases, the error probability decreases. What is most interesting is that at some specific point we see a jump in the error probability from a nonzero value down to zero. This is the *BP threshold* (see Fig. 6.12).

### 6.3.6   Asymptotic analysis: density evolution

Density evolution (DE) is a general method that allows us to analyze decoding in the limit where the numbers of nodes and factors both become large but their ratio remains constant. We do this by looking at how the erasure probability behaves at each of the two types of the nodes. Consider a $(d_l, d_r)$-regular code, i.e., one in which every variable node has degree $d_l$ and and every check node has degree $d_r$. We focus on the BEC. At the variable node, if there is an incoming message that is not an erasure, then the variable node is exactly determined. This is because we are transmitting over the BEC and either we have perfect information or we have absolutely useless information. On the check-node side, even if only one incoming message is an erasure, the check-node output has no way of knowing whether it is 0 or 1. Denoting by $y$ (respectively, $x$) the probabilities that a factor (respectively, node) is undetermined, we obtain Fig. 6.13, giving the probabilities for a node (factor) to output no information after one iteration. So, if we perform $l$ iterations, we get a sequence of erasure probabilities as shown in Fig. 6.14, which is how LDPC codes were analyzed in Gallager (1962). A somewhat different procedure was used in Luby *et al.* (2001). In their analysis, they looked at the so-called peeling decoder. This decoder is entirely equivalent to the BP decoder (when transmitting over the BEC). In this decoder, as long as there is a degree-one check node, we use this check node to determine one more

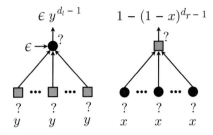

Fig. 6.13  DE iteration over a factor and a node.

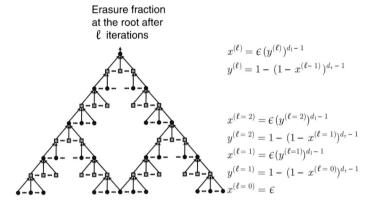

Fig. 6.14  Representation of the DE dynamics over an infinite tree, allowing the computation of the asymptotic probability of decoding at the root.

bit and then remove the used check node as well as the determined variable. We then follow the evolution of the graph. This can be done by writing down a system of differential equations. This method is called the Wormald method (Wormald, 1995). Note that in the density evolution approach, we assume that we first fix the number of iterations and let the length of the code tend to infinity (so that there are no loops in the graph up to the desired size). We then let the number of iterations tend to infinity. In the Wormald approach, on the other hand, we take exactly the opposite limit. Luckily, both approaches give exactly the same threshold: DE corresponds to the limit $\lim_{l \to \infty} \lim_{n \to \infty}$, but in fact we can take the limit in any order, or jointly, and we will always get the same threshold: the approach is robust. The density evolution is decreasing and bounded from below and will thus converge. For large codes, the behavior of almost all of them in the ensemble is accurately predicted by DE: it is the concentration property. DE can be applied to the BEC to predict the fraction of bits that cannot be recovered by BP decoding as a function of the erasure probability (see Fig. 6.15). It is predicted that there exists a critical threshold ($\epsilon \approx 0.429$ for the BEC) under which BP will recover the full codeword and above which it becomes impossible

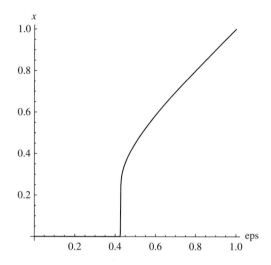

**Fig. 6.15** DE prediction for the fraction of lost bits after BP decoding for the BEC as a function of the erasure probability.

to recover everything. It perfectly matches the experimental threshold (Fig. 6.12), but the curves are different. We will understand why in the next section.

### 6.3.7   EXIT curves

Instead of plotting the $x$ value on the vertical axis, it is often more convenient to plot the EXIT value (see Fig. 6.16). The EXIT value has a simple interpretation. It is the error probability of the best estimate we can obtain using all the internal messages at a node but without the channel observation at this bit. This is why we have $y$ to the power $d_l$ and not $d_l - 1$, but we do not have the factor $\epsilon$ corresponding to the channel erasure fraction. We will see soon why the EXIT value is the right quantity to plot. Rather than running the recursion, we can immediately find the value to which the recursion converges. This is because this final value must be a solution to the fixed-point (FP) equation $x = f(\epsilon, x)$, where $f(\cdot)$ denotes a recursive

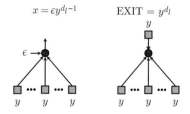

**Fig. 6.16** DE forward (stable) fixed points.

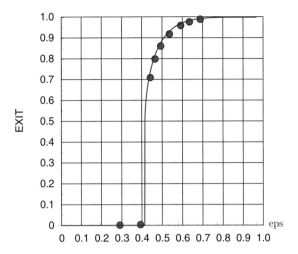

**Fig. 6.17** All DE fixed points.

DE equation. The forward fixed points of DE (see Fig. 6.17), which follows the true decoding dynamics, with initial condition $x^{(l=0)} = \epsilon$, are

$$y^{(l)} = 1 - (1 - x^{(l-1)})^{d_r - 1},$$ (6.16)

$$x^{(l)} = \epsilon(y^{(l)})^{d_l - 1},$$ (6.17)

$$x^{(l)} = \epsilon(1 - (1 - x^{(l-1)})^{d_r - 1})^{d_l - 1}.$$ (6.18)

Then, the fixed points of DE (see Fig. 6.18) are obtained by removing the time-step index:

$$x = \epsilon(1 - (1 - x)^{d_r - 1})^{d_l - 1},$$

$$\epsilon = \frac{x}{(1 - (1 - x)^{d_r - 1})^{d_l - 1}}.$$ (6.19)

Note that there are in general several values of $x$ that satisfy the FP equation for a given $\epsilon$, but there is always just a single value of $\epsilon$ for a given $x$, which is easily seen by solving for $\epsilon$ from the FP equation above. This makes it easy to plot this curve. But note also that in this picture we have additional fixed points. These fixed points are unstable and we cannot get them by running DE. The previous DE equations can be easily extended to the irregular graph case:

$$x^{(l=0)} = \epsilon,$$

$$y^{(l)} = 1 - \rho(1 - x^{(l-1)}),$$

$$x^{(l)} = \epsilon\lambda(y^{(l)}),$$

$$x^{(l)} = \epsilon\lambda(1 - \rho(1 - x^{(l-1)})).$$

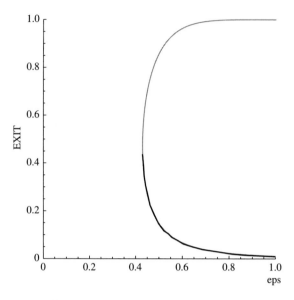

**Fig. 6.18** The experimental error probability curve improves and gets closer to the BP threshold as N increases. The improvement rate as N increases can be optimized numerically.

These distributions $\rho(\cdot)$ and $\lambda(\cdot)$ can be optimized over by finite-size scaling techniques, in order to reach capacity of the channel in the large-blocklength limit. For example, we can take a family of the form

$$\lambda_\alpha(x) = 1 - (1 - x)^\alpha, \ \rho_\alpha(x) = x^{1/\alpha}$$

and try to find the best parameter $\alpha$ such that the error probability decreases as rapidly as possible to zero below the channel capacity as $N$ increases, see Fig. 6.19. In addition, the distributions must verify the matching condition

$$\epsilon\lambda(1 - \rho(1 - x)) - x \leq 0.$$

Capacity-achieving degree distributions should satisfy the strict conditions $\epsilon\lambda(1 - \rho(1 - x)) - x = 0$ and have an average degree $\to \infty$. For instance, we can write $\lambda(x) = \sum_i w_i x^{i-1}$, with $\sum_i w_i = 1$, $w_i \geq 0$. In this case, $\lambda(\cdot)$ can be inverted, and the matching condition becomes

$$\int_0^\epsilon (1 - \rho(1 - x)) \, dx \leq \int_0^\epsilon \lambda^{-1}\left(\frac{x}{\epsilon}\right) dx,$$

$$\epsilon - \frac{1}{O_R} + \int_0^{1-\epsilon} \rho(x) \, dx \leq \epsilon\left(1 - \frac{1}{O_L}\right),$$

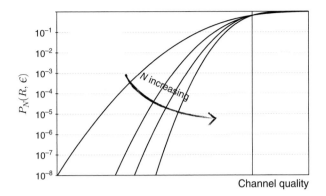

**Fig. 6.19** The $x$ value versus the EXIT value.

where $O_L = \int_0^\epsilon \lambda(x)\,\mathrm{d}x$ is the average node degree and $O_R = \int_0^\epsilon \rho(x)\,\mathrm{d}x$ is the average check degree. This implies

$$\epsilon \leq \frac{O_L}{O_R}\left(1 - \frac{\int_0^{1-\epsilon}\rho(x)\,\mathrm{d}x}{\int_0^1 \rho(x)\,\mathrm{d}x}\right). \tag{6.20}$$

Again, the capacity is reached only in the case of strict equality. In the case where there are $n$ nodes and $m$ checks in the graph, the condition $O_L n = O_R m$ must be true, and then the rate

$$R = \frac{n-m}{m} = 1 - \frac{m}{n} = 1 - \frac{O_L}{O_R} = 1 - \epsilon_{\mathrm{Sh}},$$

where $\epsilon_{\mathrm{Sh}}$ is the Shannon threshold satisfying $\epsilon_{\mathrm{Sh}} = O_L/O_R$. This implies for the matching condition

$$\epsilon \leq \epsilon_{\mathrm{Sh}}\left(1 - P(1 - \epsilon_{\mathrm{Sh}})\right) \tag{6.21}$$

where $P$ is a polynomial that approaches 0 as $O_R \to \infty$.

### 6.3.8 Some basic facts

What we have seen here works not only for the BEC but for a large class of practically relevant channels. However, only for the BEC do we have a proof that these codes achieve capacity. For the general case, we need to optimize numerically. So far, we have looked at ensembles and excluded many practical concerns. To find a particular code for a standard, much care and work is needed. These are the codes that are these days included in standards. Codes are not universal, but need to be constructed with a particular channel in mind.

## 6.4   Spatially coupled codes

So far, we have discussed the simplest form of LDPC ensembles, namely, ensembles that are defined by degree distributions but are otherwise completely unstructured. Such ensembles can have good performance (e.g., we have seen that for the BEC they can achieve capacity), but "real" codes typically have additional structure that allows optimization of various performance metrics. We will now discuss one such structure, which is called *spatial coupling*. As we will see, this structure will allow us to construct capacity-achieving ensembles for a much broader class of channels, and it is nicely grounded in basic facts from statistical physics.

### 6.4.1   Protographs

There are many ways of describing LDPC ensembles and many flavors of such ensembles. One particularly useful way of describing an ensemble is in terms of so-called *protographs*. This language will be useful when describing the more complex case of spatially-coupled ensembles.

Protographs were introduced by Thorpe (2003). They give a convenient and compact way of specifying ensembles, and the additional structure they impose is useful in practice. The creation of a "real" graph from protographs is illustrated in Fig. 6.20. For simplicity, $M = 5$ copies are introduced in this figure, but $M$ is typically of the order of hundreds or thousands. The edges denoted by dashed lines in Fig. 6.20 are "edge bundles." An edge bundle is a set of "like" edges that connect the same variable node and the same check node in each protograph. In a protograph, we connect the $M$ copies by permuting the edges in each edge bundle by means of a permutation chosen uniformly at random as shown in Fig. 6.20(b). Strictly speaking, the ensemble generated in this way is different from the ensemble generated by the configuration model,

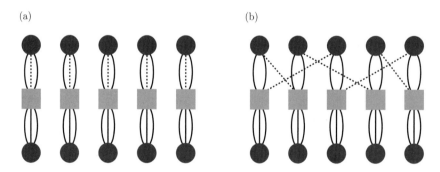

(a)                                              (b)

**Fig. 6.20**   In the protograph construction, we start with a single "protograph"; see, e.g., the left-most graph in (a). We then "lift" this protograph to a larger graph by taking $M$ copies (in our specific case, $M = 5$). Finally, we connect the various copies by taking "like" edges (which we call an *edge bundle*) and by permuting the edges in the edge bundle via a permutation picked uniformly at random. One particular edge bundle is shown in (a) by dotted lines, and the result of the permutation is shown in (b).

but these models are asymptotically equivalent in the sense that density evolution as discussed before gives the correct asymptotic predictions in both cases.

### 6.4.2 Construction of spatially coupled codes

Let us now introduce spatially coupled ensembles. There are as many flavors and variations of spatially coupled codes as there are for uncoupled codes. The exact version we consider here is not so important, since they all behave more or less the same. Hence, let us consider two variants that are easy to describe and are typical. The first is a protograph-based construction, whereas the second is purely random.

*Protograph construction*

In protograph construction, we start by taking a certain number of like protographs and placing them next to each other on a line as shown in Fig. 6.21(a). We then "connect" *neighboring* copies in a regular fashion as shown in Fig. 6.21(b). This gives us a protograph that has a spatial structure, explaining the origin of the name "spatially coupled." Note that toward the middle of the chain, the degree structure of the graph is exactly the same as that of the protograph with which we started. Only toward the boundary, as a result of boundary effects, do we have a different degree structure. Note that a variable node in the picture is connected to 3 different positions. We therefore say that the "connection width" is 3 and we write $w = 3$. At the boundaries, the code has more available information in the sense that the number of edges is less than in the middle part as shown in Fig. 6.21. As we will see, this boundary condition plays a crucial role.

Note that Fig. 6.21(b) is not yet the graph (code) itself but just a protograph representing the code. As mentioned in Section 6.4.1, to generate the real code from a given protograph, we need to "lift" the graph $M$ times and then randomly permute edges in the same edge bundle.

Note also that coupled codes constructed in this way from protographs exhibit excellent performance and are ideally suited for implementation by virtue of the additional structure. But they are more difficult to analyze than the random construction that we discuss below.

(a)  (b)

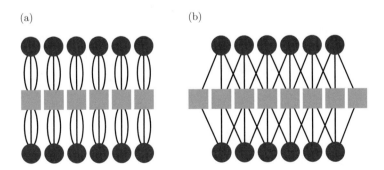

**Fig. 6.21** Spatially coupled codes (b) constructed from a set of $(3, 6)$ protographs (a).

### Random construction

In the random construction, we have the same spatial structure for the nodes, but edges connecting neighbors are placed in a more random fashion. More precisely, we randomly connect check nodes and variable nodes within a window of size $w$ as shown in Fig. 6.22. Again, we ensure that the degree distribution away from the boundary is equal to that of the original code. Note that in fact, when $w = L$, we impose no spatial constraints on the connectivity, and we recover the standard uncoupled LDPC ensemble. This randomly constructed coupled ensemble performs slightly worse in terms of its finite-length performance, but it is easier to analyze since it has fewer parameters.

## 6.4.3   Why spatial coupling?

Before we proceed with a theoretical analysis of spatially coupled ensembles, let us quickly show that such ensembles behave quite differently from uncoupled ensembles when we let the degrees tend to infinity. Since the local degree distribution is the same, this will show that the spatial structure indeed leads to some interesting behavior.

### Degree dependence of uncoupled ensembles

Figure 6.23 shows the fixed points of density evolution for the uncoupled case for (a) the $(3, 6)$ LDPC ensemble and (b) the $(100, 200)$ LDPC ensemble. Note that both have a rate of one-half. The solid and dashed lines represent stable and unstable fixed points, respectively, and the vertical lines represent the BP threshold; for (a), we have $\epsilon_{\mathrm{BP}} = 0.42944$ and for (b) 0.0372964. It can be seen from Fig. 6.23 (and shown analytically) that as we increase the degree, the BP threshold decreases, and it reaches 0 when the degree tends to infinity. Is this decrease of the threshold due to the fact that the associated code gets worse as the degrees become larger, or is it the fault of the (suboptimal) BP decoder? A closer look reveals that in fact the code itself gets better as the degree increases. But the decoder becomes more and more suboptimal.

### Spatial coupling might help

Let us now repeat the above experiment with spatially coupled ensembles. We will see that they behave very differently.

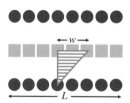

**Fig. 6.22**   Random construction of spatially coupled codes. Edges are defined randomly in the shaded area.

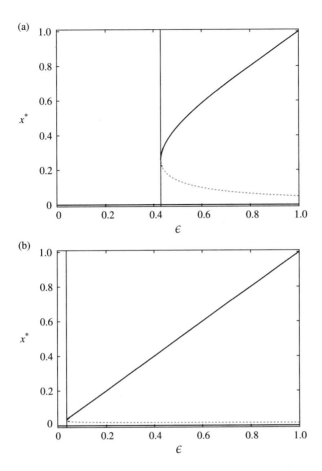

**Fig. 6.23** Fixed points of uncoupled $(3, 6)$ code (a) and $(100, 200)$ code (b). The vertical lines represent threshold $\epsilon_{\mathrm{BP}} \simeq 0.42944$ (a) and $0.0372964$ (b).

Consider a coupled ensemble constructed via the protograph approach. To make the argument particularly simple, assume that all the edges between factor nodes and variables nodes are in fact double edges, as shown in Fig. 6.24. For example, the protograph shown in this figure therefore represents a $(4, 8)$-regular ensemble.

Consider now the decoder procedure. We want to show that the BP threshold does not tend to zero for such an ensemble even if we increase the degrees and let them tend to infinity.

To show this, note that we can get a lower bound on the decoding threshold by "weakening" the decoder. We weaken the decoder in the following way. Instead of allowing the decoder to use all available information, assume that when we decode the bits in the first position, we are not allowed to use the information we received in any of the positions to the right.

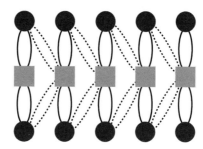

**Fig. 6.24** A coupled $(2, 4)$ ensembles with double edges.

This means that for the given example we concentrate only on the double edges connected to a factor node (denoted by solid lines in Fig. 6.24) and ignore other edges (denoted by dashed lines). Note that if we concentrate on the bits in the left-most position, this means that we are decoding a $(2, 4)$-regular code, which is also known as "cycle code." The BP threshold of such a code is known and, for example, for the BEC it is equal to $\epsilon_{\text{BP}} = \frac{1}{3}$.

Therefore, we know that we can decode the left-most bits using the BP decoder if we transmit over a BEC with erasure probability not exceeding $\frac{1}{3}$. Now assume that these positions are known. We can then remove (the effect of) these bits from the graph. But if we do so, what is left looks again exactly like the original situation except that now the chain is shorter by one. We can therefore recurse our argument. In summary, we have just argue that the BP threshold of this chain is at least one-third.

The punchline is now the following. Exactly the same argument holds if we increase the degrees and look at the spatially coupled $(2k, 4k)$-regular ensemble, regardless of the value of $k$. Therefore, the BP threshold does not tend to zero for coupled ensembles even if we let the degrees tend to infinity. This argument only shows that the threshold is lower-bounded by a constant and it does not permit us to determine the actual threshold. In fact, we will shortly see that the actual threshold improves as the degree gets larger.

## 6.5   Density evolution for coupled codes

Let us now get to the analysis of coupled ensembles using the same method, namely, density evolution, which we used in the uncoupled case.

In the uncoupled case, the variable $x$, which represents the erasure fraction along an outgoing edge from the variable node, is a scalar, and density evolution tracks the evolution of this scalar as a function of the iteration number.

For the coupled case, the *state* is a vector, since variables at different positions will not experience the same "environment." Recall that at the boundary we have a slightly different degree distribution and the decoder problem is easier there. As we will see, the decoder will be able to decode at the boundary first and this progress will then propagate toward the interior of the code along a "decoding wave."

Because of this lack of "homogeneity" along the spatial dimension, we need a vector $\boldsymbol{x}$ to describe the state, where $x_i$ describes the erasure probability at position $i \, (= 1, \ldots, L)$. Recall that we know the values at the boundary, and hence the erasure probability at the boundary is 0.

In the randomly constructed code, each edge can be connected to positions in a certain range. More precisely, consider Fig. 6.25: variable nodes assigned $\{x_i\}$ are always connected to positions "to the right" and check nodes assigned $\{y_i\}$ are always connected to variable nodes "on the left." We therefore need to average over the incoming messages from this range, and the density evolution equations for the coupled ensemble are given by

$$
x_i = \epsilon \left( \frac{1}{w} \sum_{j=0}^{w-1} y_{i+j} \right)^{d_l - 1}, \tag{6.22}
$$

$$
y_i = 1 - \left( 1 - \frac{1}{w} \sum_{k=0}^{w-1} x_{i-k} \right)^{d_r - 1}, \tag{6.23}
$$

where $i = 1, \ldots, L$. Note that there is an $x_i$ and an $y_i$ value for each position of the chain, and the equations for these values are coupled through the averaging operations.

Combining (6.22) and (6.23) and adding an index for the iteration number, we get

$$
x_i^{(l)} = \epsilon \left( 1 - \frac{1}{w} \sum_{j=0}^{w-1} \left( 1 - \frac{1}{w} \sum_{k=0}^{w-1} x_{i+j-k}^{(l-1)} \right)^{d_r - 1} \right)^{d_l - 1}. \tag{6.24}
$$

To simplify our notation, and also to abstract from the specific case we are considering, let us define the functions $\{f_i(\cdot)\}$ and $g(\cdot)$:

$$
f_i = \left( 1 - \frac{1}{w} \sum_{k=0}^{w-1} x_{i-k} \right)^{d_r - 1}, \qquad g(\{x_{i \in I(i)}\}) = \left( 1 - \frac{1}{w} \sum_{j=0}^{w-1} f_{i+j} \right)^{d_l - 1}, \tag{6.25}
$$

where $I(i)$ denotes the set of indices connected to $i$. In this way, we get simple expressions, like

$$
x_i = \epsilon g(\{x_{i \in I(i)}\}). \tag{6.26}
$$

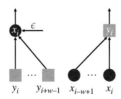

**Fig. 6.25** Density evolution in spatially coupling code with width $w$.

We call a vector $x = \{x_i\}(i = 1, \ldots, L)$ whose components are the erasure fractions at the various indices a *constellation*. At all the indices outside the constellation, $i < 1$ and $i > L$, we assume that the corresponding $x_i$ values are 0, i.e., we have perfect knowledge. A constellation $x$ that when inserted into the DE equation results in $x$ is called a *fixed point* of the DE equation.

In Fig. 6.26, the $\epsilon$ dependence of the time evolution of the DE equation for the coupled ensemble, according to (6.24), is shown. The plot (a) corresponds to $\epsilon = 0.3$, (b) to $\epsilon = 0.48$, and (c) to $\epsilon = 0.6$. Note that for $\epsilon < \epsilon_{BP} \simeq 0.4294$, DE proceeds in essentially exactly the same way as for the uncoupled case if we look at the $x_i$ values in the center of the chain. At the boundary, we see somewhat better values owing to the boundary condition. And, as expected, the DE is able to drive the erasure fraction in each section to zero and BP is successful.

At $\epsilon = 0.48$, which is considerably larger than the BP threshold $\epsilon_{BP} = 0.4294$ of the uncoupled ensemble (and close to the optimal threshold of 0.5 of the best code and decoding algorithm), a small "wave front" is formed at both boundaries after a few iterations owing to the fact that at the boundaries more knowledge is available; see Fig. 6.26(b). These wave fronts move toward the center of the coupled code at a constant speed and by doing so decrease the value of $x_i$ for $i$ located in the central part

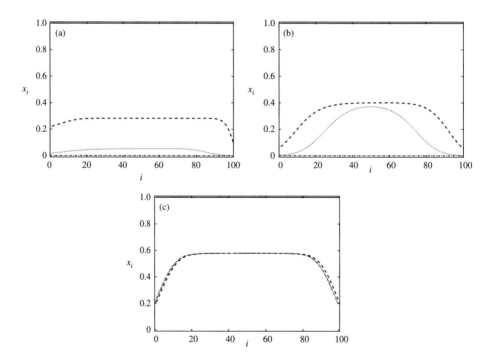

**Fig. 6.26** Time evolution of $\{x_i\}$ by the DE equation for the $(3, 6)$ coupled code at (a) $\epsilon = 0.3$, (b) $\epsilon = 0.48$, and (c) $\epsilon = 0.6$. The length $L = 100$ and the width $w = 20$. The constellations evolve in the order of solid line → dashed line → dotted line → dashed–dotted line.

of the coupled code until the whole constellation is decoded. This is the interesting new phenomenon that happens owing to the spatial structure. In other words, owing to the spatial structure, the wave front can smoothly connect the desired fixed point of $x_i = 0$ to the undesired fixed point that is found by the BP decoder of the uncoupled system, and the undesired fixed point is guided at a constant speed toward the desired one until decoding is accomplished. As we increase the parameter $\epsilon$ up to a critical threshold, call it $\epsilon_{\text{Area}}$, the speed of the wave is linearly decreased and it reaches the value zero at $\epsilon_{\text{Area}}$.

At $\epsilon$ above $\epsilon_{\text{Area}}$, we get a nontrivial fixed point of DE and decoding is no longer successful. In the middle of the chain, the $x_i$ values are exactly as large as they would be for the same $\epsilon$ value in the uncoupled case. Only at the boundary do we get somewhat better values because of the boundary condition.

### 6.5.1   Summary

In the following sections, we will see that spatially coupled ensembles can be decoded up to $\epsilon_{\text{Area}}$ and this value is essentially equal to $\epsilon_{\text{MAP}}$ of the underlying ensemble. This phenomenon is called *threshold saturation*. In order to exactly achieve $\epsilon_{\text{MAP}}$, we have to let the chain length $L$ tend to infinity (this makes decoding harder) and the interaction width $w$ tend to infinity as well (with $w \ll L$). But, in practice, even for moderate values of $L$ of perhaps 10 or 20 and very small values of $w$, perhaps 2 or 3, the decoding thresholds are already very close to $\epsilon_{\text{MAP}}$; for instance, for the BEC, the difference between $\epsilon_{\text{MAP}}$ and $\epsilon_{\text{Area}}$ with $w = 3$ is only about $10^{-5}$ for the $(3, 6)$-regular ensemble.

Note finally that one can show that the MAP threshold $\epsilon_{\text{MAP}}$ is an increasing function of the degrees and converges to the Shannon threshold exponentially fast with increasing degrees. This is contrary to the BP threshold $\epsilon_{\text{BP}}$ for uncoupled codes, which typically decreases with increasing degrees.

## 6.6   Threshold saturation

Let us now discuss why threshold saturation happens and how we can prove the above assertions.

We will limit our discussion to the simplest case, namely transmission over the BEC. Currently, there are proofs of the threshold saturation phenomenon for the following cases: sparse graph codes and transmission over any BMS channel, any system whose state (for the uncoupled system) is a scalar or a vector, and compressive sensing.

For the BEC, there are currently three known proof strategies: via the Maxwell construction, via EXIT charts, and via potential functions. These proofs share important features, but each also has its own advantages.

Historically speaking, the proof of threshold saturation via the Maxwell construction was the first proof of spatially coupled codes achieving capacity under BP decoding when transmitting over the BEC. Later, the same approach led to the proof of spatially coupled codes universally achieving capacity under BP decoding over the whole class of BMS channels. The details of the proof for the BEC can be found in Kudekar *et al.* (2011), whereas the general case is described in Kudekar *et al.* (2010).

Recall that we are interested in finding the largest channel parameter $\epsilon$ such that the DE recursion of the coupled system, when started with the all-ones vector inside the range $[1, L]$, converges to the all-zeros vector. We denote this parameter by $\epsilon_{\text{Area}}$, and called it the area threshold.

### 6.6.1   Proof by Maxwell construction

The proof by the Maxwell construction consists of three parts: show the existence of a special fixed point of the coupled DE equations explained in Section 6.6.1, prove that any such FP must have a channel parameter that is very close to the area threshold $\epsilon_{\text{Area}}$, and, finally, show that for any channel parameter $\epsilon$ below $\epsilon_{\text{Area}}$, the DE equations converge to the all-zero constellation $\boldsymbol{x} = \boldsymbol{0}$.

#### *Definition of area threshold $\epsilon_{Area}$*

Consider Fig. 6.27, which shows the so-called EXIT curve for the $(3, 6)$-regular uncoupled ensemble. Recall that this EXIT curve is the curve that we get if we project the fixed points of density evolution. The branch plotted as a solid line corresponds to the stable fixed points, whereas the branch plotted as dots corresponds to the unstable fixed points. Recall that (6.19) gives an explicit description of these fixed points; i.e., it expresses the channel parameter $\epsilon$ as an explicit function of the erasure probability $x$ emitted by variable nodes, call this function $\epsilon(x)$. Explicitly, the EXIT curve is the curve given in parametric form as $\{x^{d_l}, \epsilon(x))\}_{x=0}^{1}$.

In terms of this EXIT curve, the area threshold is defined as follows. Integrate the area enclosed under the top (stable) branch of the EXIT curve starting from the right $(\epsilon = 1)$ until the channel parameter such that this area is equal to the rate of the code. For the example shown in Fig. 6.27, the rate is equal to $\frac{1}{2}$ and $\epsilon_{\text{Area}} \simeq 0.48818$.

Recall that for this example the BP threshold $\epsilon_{\text{BP}} \simeq 0.4299$, so that the area threshold is (considerably) larger than the BP threshold. This is always the case. It

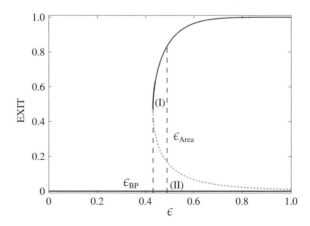

**Fig. 6.27**   Definition of the area threshold $\epsilon_{\text{Area}}$.

is also easy to see that the area threshold is always lower than the Shannon threshold since the EXIT curve is upper-bounded by 1; $\text{EXIT}(\epsilon) \leq 1$ for any $\epsilon \in [0,1]$, and so the area threshold is upper-bounded by $\epsilon_{\text{Area}} \leq 1 - \text{rate}$.

By simple explicit calculation, it can be further shown that the area contained "inside" the C-shaped EXIT curve is also equal to the rate. This implies that the area (I) and (II) shown in Fig. 6.27 are equal to each other. Therefore, an equivalent definition of the area threshold is to say that it is that point where a vertical line makes the two areas to be of equal size.

Because of the similarity between the definition of area threshold and the Maxwell construction in thermodynamics, this line is called the Maxwell construction of the BP EXIT curve Méasson *et al.* (2005). Indeed, whereas in the original Maxwell construction the areas represent work, in the coding context the areas represent information that, on the one hand, a genie has to provide to the BP decoder in order to convert it into a MAP decoder (area (I)) and, on the other, the number of "confirmations" that the BP decoder receives during the decoding process that proves that the information provided by the genie is indeed correct. When these two areas are in balance, the BP decoder can with high probability decode (just like the MAP decoder could do) and can at the end certify that all the information provided by the genie is indeed correct.

### Existence of a special fixed point

The special fixed point of $\boldsymbol{x}$ that we need is illustrated in Fig. 6.28(a). What we need is a fixed point that is unimodal, where $x_i$ is close to 0 close to the boundary, and close to $x^*$ in the middle, respectively; here $x^*$ is the fixed point of DE for the uncoupled system under the same channel parameter. Further, the number of positions $i$ whose $x_i$ value is in the range $[\delta, x^* - \delta]$, $\delta > 0$, must be $O(w)$. Note that the two stable fixed points of DE for the uncoupled system, namely, 0 and $x^*$, are essentially the lower and upper bounds on $\boldsymbol{x}$ and that $\boldsymbol{x}$ should smoothly interpolate between them.

For simplicity, we consider DE for one-sided constellations $(x_{-L}, \ldots, x_0) \in [0,1]^{L+1}$, as shown in Fig. 6.28(b), where $L$ is the length of the chain and $x_{-L} < \cdots < x_0$. The DE equation for one-sided constellations is obtained from the usual

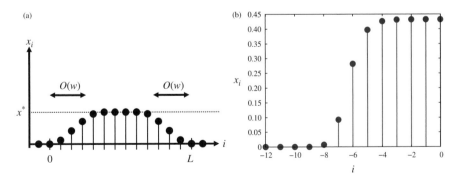

**Fig. 6.28** (a) Schematic picture of the special fixed point. (b) One-sided constellation of the special fixed point for $(3,6)$-coupled code with $L = 6$ and $w = 3$.

coupled DE equation (6.24) by setting $x_i = 0$ for $i < -L$ and $x_i = x_0$ for $i > 0$. We define the average value (entropy) of the one-sided constellation as

$$\bar{x} = \frac{1}{L+1} \sum_{i=-L}^{0} x_i. \tag{6.27}$$

We can establish the existence of the fixed point with the desired properties by the use of Schauder's fixed-point theorem, which states that any continuous mapping $f$ from a convex compact subset $S$ of a Euclidean space to $S$ itself has a fixed point. In fact, when applying the fixed-point theorem, we do not fix the parameter $\epsilon$, but this parameter is part of the fixed point itself. Therefore, as a consequence of Schauder's theorem, after some proper definition of the fixed-point equation, we are guaranteed the existence of a constellation $\boldsymbol{x}^*$ with the desired properties that is a fixed point for some channel parameter $\epsilon^*$. Although one can establish a priori bounds on the range of $\epsilon^*$, its exact value is not known. This is the point of the next step in the proof.

### Saturation

Next, we show that when we have the special fixed point, its channel parameter must be very close to $\epsilon_{\text{Area}}$. The basic idea is very simple. Recall our discussion of the EXIT curve for the uncoupled system. In this case, we mentioned that the area enclosed "within" this EXIT curve is equal to the rate of the code. For the uncoupled case, this was the result of a simple explicit computation, since the EXIT curve was known in parametric form and the integration can be carried out without problems. But there also exists a more conceptual proof that does not rely on explicit calculations and that shows that any time one has a smooth EXIT curve, the area it encloses must be equal to the rate of the code. Why is this true? It turns out that the EXIT curve can be interpreted as the derivative of an entropy term with respect to the channel parameter, and so, when we integrate, by the fundamental theorem of calculus, the area is just the difference between the values of this entropy term at the two endpoints. This difference can be determined explicitly and it happens to be equal to the rate of the code. More is true: assume that instead of having a real EXIT curve, where we recall that each point corresponds to a fixed point of DE, we have a smooth curve where every point corresponds to an "approximate" fixed point of density evolution. Here, "approximate" means that the difference between this point and the point we get after one iteration is small in the appropriate metric. In this case, the same conceptual argument tells us that the area enclosed by this curve is "close" to the rate of the code, where the measure of "closeness" is related to how close the points are to being fixed points.

The idea is hence the following. Given the special fixed $(\epsilon^*, \boldsymbol{x}^*)$, we will construct from it a whole family of approximate fixed points such that this family gives rise to an approximate EXIT curve. The shape of theis curve is shown as a solid curve in Fig. 6.29(b). In particular, the sharp vertical drop happens exactly at the parameter $\epsilon^*$ and the whole EXIT curve will look just like the curve we get from the Maxwell construction. Then using the fact that the integral must be equal to the rate of the code will tell us that the sharp vertical drop must happen exactly at the area threshold.

But how can we construct from this single special fixed point a whole family? Rather than considering the whole construction, let us discuss only the most interesting

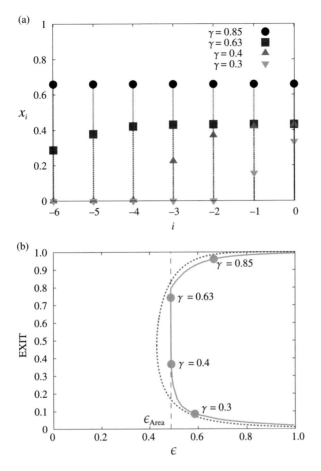

**Fig. 6.29** (a) Examples of an interpolated family for the $(3, 6)$ coupled code with $L = 6$ and $w = 2$. (b) The corresponding EXIT curve. The dashed line is the EXIT curve of the uncoupled case.

part, namely, the part corresponding to the sharp vertical drop. Recall that one of the conditions on the special fixed point was that in the "middle," the fixed point was essentially flat and had a value essentially equal to what the uncoupled ensemble would have for this channel parameter. Furthermore, toward the boundary, the values had to be essentially equal to 0. This means that we can insert any number of further sections in the middle with the appropriate value or any number of further sections at the boundary with the value 0 and still have an appropriate fixed point. All of them will be appropriate fixed points corresponding to the same channel value, but their average value will depend on how wide we make the middle part. By changing this width, we get points on the vertical line. Since the width can only be changed in discrete steps but we need a continuous curve, we also need to interpolate the discrete steps. In addition, to get the points on the top horizontal portion of the EXIT curve, we also need to interpolate. This is shown in Fig. 6.29(a).

## Convergence

We now get to the last part of the argument. By now, we have established that such a special fixed point can only exist if its channel parameter is very close to the area threshold. We will now argue that if we start DE with a channel value below this area threshold, then it must converge to the all-zero constellation.

To see this, we consider the following experiment. We apply DE at $\epsilon_{\mathrm{BP}} < \epsilon < \epsilon_{\mathrm{Area}}$ to a constellation of length $L$ whose initial condition is the all-one vector inside the constellation and 0 outside. DE produces a sequence of monotonically decreasing (pointwise) constellations that are bounded by 0 (again pointwise) from below. We denote the fixed point by $\boldsymbol{\xi}^*$ and assume that $\boldsymbol{\xi}^*$ is nontrivial, i.e., is not all-zero, as shown in Fig. 6.30. It is clear that at each point in the constellation, the value of the fixed point is no larger than the fixed point we would get for the uncoupled case at $\epsilon$, since at the boundary the decoder has access to additional information.

Now let us compare this fixed point with our special fixed point, where we pick the length for the latter sufficiently large that the special fixed point dominates $\boldsymbol{\xi}^*$ everywhere pointwise. Note that $\boldsymbol{\xi}^*$ is a fixed point for the parameter $\epsilon$, but the special fixed point is for the parameter $\epsilon_{\mathrm{Area}}$ and $\epsilon < \epsilon_{\mathrm{Area}}$. So, if we now apply DE to the special fixed point but with the parameter $\epsilon$, then the special fixed point must be strictly decreasing pointwise and it must in fact converge to the all-zero constellation, since otherwise we would get another nontrivial fixed point that would again satisfy all the requirements of a special fixed point (some further arguments are required to prove this), and we know that the only channel parameter for which such a special fixed point exists is very close to $\epsilon_{\mathrm{Area}}$, a contradiction. But since our putative fixed point $\boldsymbol{\xi}^*$ is dominated by our special fixed point and the special fixed point collapses to the all-zero constellation, it must in fact be true that $\boldsymbol{\xi}^*$ is also the all-zero constellation.

### 6.6.2   Proof by EXIT charts

#### EXIT charts

EXIT charts were introduced by ten Brink (1999) as a convenient way of visualizing DE. For transmission over the BEC, the EXIT chart method is equivalent to DE. EXIT charts are quite different from the EXIT curves that we have already introduced, despite their similar names. The reason both objects are described as "EXIT" is that

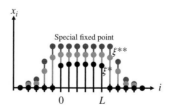

**Fig. 6.30** Assumed fixed point $\boldsymbol{\xi}^*$ at $\epsilon_{\mathrm{BP}} < \epsilon < \epsilon_{\mathrm{Area}}$, the special fixed point, and a fixed point $\boldsymbol{\xi}^{**}$ obtained by applying DE for the special fixed point at $\epsilon$.

in both cases we measure the same thing (namely, whether the "other" bits in a code are able to determine the bit we are considering via the code constraints), but for EXIT charts we make local measurements, whereas for EXIT curves we measure the performance of the whole code.

An EXIT chart consists of two curves. One curve corresponds to the message-passing rules at the variable nodes and the other to the message-passing rules at the check nodes. In addition, it is customary, and convenient, that we plot one curve with its input on the horizontal axis and its output on the vertical axis, while the other curve is plotted with its output on the horizontal axis and its input on the vertical axis. Figure 6.31 shows the EXIT charts of uncoupled $(d_l, d_r)$-LDPC for $d_l = 3$ and $d_r = 6$ at three channel parameters, where the two curves are given by

$$x = \epsilon y^{d_l - 1}, \tag{6.28}$$
$$y = 1 - (1 - x)^{d_r - 1}. \tag{6.29}$$

On this EXIT chart, the DE trajectory can be regarded as a staircase pattern bounded by these two curves. The DE points converge to zero if and only if the two curves do not cross (Fig. 6.31(a)). The threshold for the uncoupled case is given by the channel

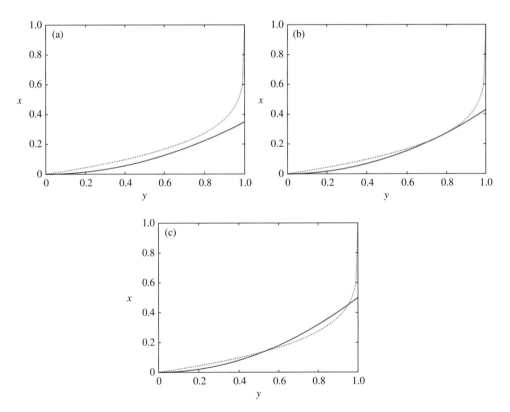

**Fig. 6.31** EXIT charts of $(3, 6)$ uncoupled LDPC for (a) $\epsilon = 0.35$, (b) $\epsilon = \epsilon_{\mathrm{BP}} \simeq 0.4294$, and (c) $\epsilon = 0.5$.

parameter such that the two EXIT curves just touch but do not cross (Fig. 6.31(b)). At $\epsilon > \epsilon_{\mathrm{BP}}$, the two curves touch at three points (Fig. 6.31(c)).

### *Proof by EXIT charts for coupled code*

In the coupled systems, the criterion for the threshold, namely, that the two curves may touch but are not allowed to cross, is relaxed. To determine the threshold for the coupled system, the two EXIT curves are now allowed to cross, but not by too much, and indeed the threshold relates to a balance of areas enclosed by the two curves. One can show that the condition for the threshold is exactly the same as the matching of areas condition that we have seen in the Maxwell construction.

The first step of the proof consists in considering an appropriately chosen continuous version of the constellation. For the random coupled ensemble, we have introduced the window $w$, within which the random connections are generated. This discrete system is difficult to analyze. Instead, one can consider the limit when $w$ goes to infinity; of course, the length of the chain has to go to infinity as well. If we increase the length $w$ and scale the length of the code by the same proportion, then in the limit we can treat the constellation as a continuous curve rather than a set of spikes. The DE equation for this continuous constellation is given by integrating over a window instead of taking discrete sums.

The next step is to analyze this continuous system. It is convenient to think of systems of infinite length, i.e., with the horizontal axis extending from $-\infty$ to $-\infty$. Further, instead of considering a two-sided constellation, we consider a one-sided constellation, i.e., we focus only on the "left" part of the constellation from $-\infty$ to 0. For a continuous system, it has been proved that there are three different scenarios depending on the balance of the areas in the EXIT chart picture of the uncoupled system. We omit here the trivial case where the curves do not overlap at all, since in this case it is easy to show that we will decode.

Consider first the scenario where the channel parameter is below $\epsilon_{\mathrm{Area}}$, but above the $\epsilon_{\mathrm{BP}}$ of the uncoupled ensemble (Fig.6.32(a)). In this case, the curves do overlap but only slightly, and the area on the left (dark gray) is larger than the area on the right (light gray). One can show that there does not exist a fixed point of DE but there does exist a one-sided constellation $\boldsymbol{x}$. If we apply DE to this constellation, we get the same $\boldsymbol{x}$, but the position is shifted to the right. Given that our one-sided constellation represents the left part of an actual constellation, saying that the wave is propagating to the right means that the decoder is working and in each step decodes a further part of the constellation. The shift that we see in each iteration corresponds to the decoding speed and so tells us how many iterations we will need. To summarize, below the area threshold, we get a decoding wave that moves to the right, which means that the decoder is working.

Assume next that the areas are exactly in balance, which means that we are transmitting exactly at $\epsilon_{\mathrm{Area}}$ (Fig. 6.32(b)). In this case, one can prove that the continuous version of DE has a fixed point. This fixed point can be regarded as a stationary wave or a wave with zero speed.

Finally, consider the case where we are transmitting above $\epsilon_{\mathrm{Area}}$ (Fig. 6.32(c)). The curves overlap so much that the area on the left is smaller than the area on the

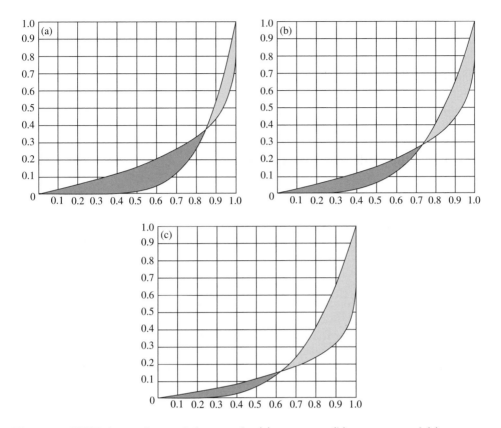

**Fig. 6.32** EXIT charts of a coupled system for (a) $\epsilon < \epsilon_{\text{Area}}$, (b) $\epsilon = \epsilon_{\text{Area}}$, and (c) $\epsilon > \epsilon_{\text{Area}}$. These figures are quoted from online lecture materials from the EFPL (EPFL-IPG, 2013).

right. In this case, one can then show that there does not exist a nontrivial fixed point but only a continuous constellation $\boldsymbol{x}$, so that after one round of DE, we get the same constellation back, but shifted to the left. This means that the decoder does not work.

In a final step, one needs to reconnect the continuous system to the actual discrete system and show that if $w$ is not too small, then the behavior of the discrete system is predicted well by the behavior of the continuous system (Kudekar *et al.*, 2012).

### 6.6.3 Proof by potential functions

*Potential functions*

The potential function for an uncoupled LDPC code is defined as

$$U(x, \epsilon) = \int_0^x (z - f(g(z); \epsilon))g'(z)\,\mathrm{d}z$$

$$= g(x) - G(x) - F(g(x); \epsilon), \tag{6.30}$$

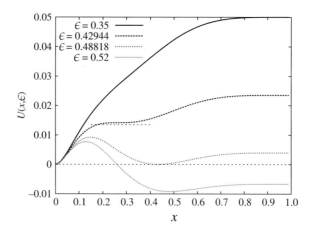

**Fig. 6.33** Potential function of $(3, 6)$ uncoupled LDPC.

where $f(g(z); \epsilon) = \epsilon\lambda(1 - \rho(1 - z))$, $g(z) = 1 - \rho(1 - z)$, $F(x; \epsilon) = \int_0^g (z; \epsilon)\,\mathrm{d}z$, and $G(x) = \int_0^x g(z)\,\mathrm{d}z$. The functions $\lambda(\cdot)$ and $\rho(\cdot)$ are node-perspective degree distributions. The potential function corresponds to the Bethe free energy. Figure 6.33 shows the $x$ dependence of the potential function for $(3, 6)$ uncoupled LDPC, where $\lambda(x) = x^2$ and $\rho(x) = x^5$. When $\epsilon$ is smaller than $\epsilon_{\mathrm{BP}} \simeq 0.42944$, the potential function is an increasing function for all $x \geq 0$. At $\epsilon = \epsilon_{\mathrm{BP}}$, the potential function has zero gradient at a certain $x$, and at $\epsilon > \epsilon_{\mathrm{BP}}$, a local minimum appears and its value touches to the line 0 at $\epsilon_{\mathrm{Area}} \simeq 0.48818$. This means that the state $x = 0$ always exists as the unique solution when $\epsilon < \epsilon_{\mathrm{BP}}$. At $\epsilon > \epsilon_{\mathrm{BP}}$, another locally stable solution appears at $x > 0$, although the solution $x = 0$ is the globally stable state. At $\epsilon > \epsilon_{\mathrm{Area}}$, the solution $x = 0$ is no longer a globally stable solution. In physical terminology, the area threshold and BP threshold correspond to the first transition point and the spinodal point, respectively.

### *Potential functions for a coupled system*

The potential function of the coupled system, which is defined for a vector constellation $\boldsymbol{x}$, is introduced in analogy to that for the uncoupled system. We consider the potential function for a one-sided constellation, with $x_i = 0$ when $i$ is not in $[-L, i_0]$ and with the value of $x_i$ increasing with $i$ up to $i_0 = \lfloor (w - 1)/2 \rfloor$, given by Yedla *et al.* (2012) as

$$U(\boldsymbol{x}; \epsilon) = \int_C \boldsymbol{g}'(z)(z - \boldsymbol{A}^{\mathsf{T}}\boldsymbol{f}(\boldsymbol{A}\boldsymbol{g}(z; \epsilon))) \cdot \mathrm{d}z$$

$$= \boldsymbol{g}(\boldsymbol{x})^{\mathsf{T}}\boldsymbol{x} - G(\boldsymbol{x}) - F(\boldsymbol{A}\boldsymbol{g}(\boldsymbol{x}); \epsilon), \qquad (6.31)$$

where $\boldsymbol{g}'(\boldsymbol{x}) = \mathrm{diag}([g'(x_i)])$, $[\boldsymbol{f}(\boldsymbol{x}; \epsilon)]_i = f(x_i; \epsilon)$, $[\boldsymbol{g}(\boldsymbol{x})]_i = g(x_i)$,

$$G(\boldsymbol{x}) = \int_C \boldsymbol{g}(z) \cdot \mathrm{d}z = \sum_i G(x_i),$$

and

$$F(\boldsymbol{x}; \epsilon) = \int_C \boldsymbol{f}(\boldsymbol{z}; \epsilon) \cdot \mathrm{d}\boldsymbol{z} = \sum_i F(x_i; \epsilon).$$

The matrix $\boldsymbol{A}$ is an $(L + 3w + i_0 + 1) \times (L + 3w + i_0 + 1)$ matrix given by

$$\boldsymbol{A} = \frac{1}{w} \begin{bmatrix} 1 & 1 & \cdots & 1 & 0 & \cdots & 0 \\ 0 & 1 & 1 & \cdots & 1 & \ddots & \vdots \\ \vdots & \ddots & \ddots & \ddots & \ddots & \ddots & 0 \\ 0 & \cdots & 0 & 1 & 1 & \cdots & 1 \\ 0 & 0 & \cdots & 0 & 1 & \ddots & 1 \\ 0 & 0 & \cdots & 0 & 0 & 1 & \vdots \\ 0 & 0 & \cdots & 0 & 0 & 0 & 1 \end{bmatrix}, \tag{6.32}$$

which gives the matrix notation for the DE equation (6.24).

We set $K_{f,g} = \|g'\|_\infty + \|g'\|_\infty^2 \|f'\|_\infty + \|g''\|_\infty$, where $\|h\|_\infty = \sup_{x \in [0,1]} |h(x)|$ for functions $h : [0,1] \to \mathbb{R}$ (Yedla *et al.*, 2012). Using the potential function, it can be shown that when $\epsilon < \epsilon_{\mathrm{Area}}$ and $w > K_{f,g}/\Delta E(\epsilon)$, the coupled potential of a non-zero vector decreases as a result of the shifting. This implies that the value of the constellation must be reduced by recursion, and the fixed point is the zero vector. The outline is as follows. Let us assume that $\boldsymbol{x} \neq \boldsymbol{0}$ is the unique fixed point of the one-sided DE equation. We introduce a right-shifted constellation from $\boldsymbol{x}$ as $\boldsymbol{Sx}$, where $[\boldsymbol{Sx}]_1 = 0$ and $[\boldsymbol{Sx}]_i = x_{i-1}$ for $i \geq 2$. The difference in the potential functions between $\boldsymbol{Sx}$ and $\boldsymbol{x}$ is given by

$$U(\boldsymbol{Sx}; \epsilon) - U(\boldsymbol{x}; \epsilon) = -U(x_{i_0}; \epsilon). \tag{6.33}$$

Meanwhile, Taylor expansion of $U(\boldsymbol{Sx}; \epsilon)$ around $\boldsymbol{x}$ gives

$$U'(\boldsymbol{x}; \epsilon) \cdot (\boldsymbol{Sx} - \boldsymbol{x}) \leq (U(\boldsymbol{Sx}; \epsilon) - U(\boldsymbol{x}; \epsilon)) + \frac{K_{f,g}}{w}$$

$$< -U(x_{i_0}; \epsilon) + \Delta E(\epsilon) \leq 0, \tag{6.34}$$

where we exploit $w > K_{f,g}/\Delta E(\epsilon)$.

When all components of $\boldsymbol{Sx} - \boldsymbol{x}$ are non-positive, at least one component of $U'(\boldsymbol{x}; \epsilon)$ should be positive to satisfy eq.(6.34). The derivative is given by $[U'(\boldsymbol{x}; \epsilon)]_i = g'(x_i)[\boldsymbol{x} - \boldsymbol{A}^\mathsf{T} f(\boldsymbol{A}g(\boldsymbol{x}; \epsilon))]_i$ and $g'(x) \geq 0$, and hence $[\boldsymbol{A}^\mathsf{T} f(\boldsymbol{A}g(\boldsymbol{x}; \epsilon))] < x_i$ should hold. This means that one more iteration reduces the value of the $i$th component. This leads to a contradiction and means that the fixed point of the one-sided constellation is only $\boldsymbol{x} = \boldsymbol{0}$ at $w > K_{f,g}/\Delta E(\epsilon)$.

### 6.6.4 Summary

We have shown three proofs of the threshold saturation phenomena. These three different criteria each have their own advantage. The first proof not only determines the threshold of the coupled system but also makes it clear that this is equal to the MAP threshold. The EXIT chart approach is convenient for people who are already familiar with EXIT charts for uncoupled systems and this criterion is very easy to apply. Finally, the potential function approach leads to the currently simplest proof of the threshold saturation phenomenon.

In more detail, let us summarize some of the main points of these three proofs. In the proof by the Maxwell construction, at the area threshold, a special fixed point exists, which has long tails, a quick transition, and a large flat part. This special fixed point cannot exist below the area threshold (nor can it exist at larger channel parameters). This proof has a connection with problems in statistical physics and the picture is exactly the same if we consider transmission over general BMS channels, although the proofs are more complicated. Further, it strongly suggests that the area threshold is also the MAP threshold of the underlying ensemble. Indeed, this has recently been shown to be true (Giurgiu *et al.*, 2013).

In the EXIT chart approach, at the area threshold, a stationary wave exists. Below the area threshold, a propagating wave, traveling at a constant speed, shows that decoding will be successful. EXIT charts and the matching condition are frequently used to analyze systems that have a one-dimensional state, and they are often use to approximately model more general systems (e.g., in Gaussian approximation). For any such system, if we replace the matching condition with the area balance condition, then we get the equivalent criterion for coupled systems. If the original state is one-dimensional, then this criterion is exact; otherwise, it is an approximation in the same way that the matching condition for EXIT charts is an approximation for uncoupled systems.

Finally, in the potential function approach, at the area threshold, the potential function has zero gradient. Below the area threshold, the potential energy is strictly decreasing, implying convergence to perfect decoding. This leads to the currently simplest known proof for one-dimensional systems. It can be extended to systems whose state is no longer a scalar but rather a vector, and even to infinite-dimensional systems, such as general BMS channels.

## References

Arikan, E. (2009). Channel polarization: a method for constructing capacity-achieving codes for symmetric binary-input memoryless channels. *IEEE Transactions on Information Theory*, **55**(7), 3051–3073.

EPFL-IPG (2013). http://ipg.epfl.ch/doku.php?id=en:publications:scc_tutorial.

Gallager, R. G. (1962). Low-density parity-check codes. *IRE Transactions on Information Theory*, **8**(1), 21–28.

Giurgiu, A., Macris, N., and Urbanke, R. (2013). Spatial coupling as a proof technique. arXiv:1301.5676.

Hassani, S. H., Alishahi, K., and Urbanke, R. L. (2014). Finite-length scaling of polar codes. *IEEE Transactions on Information Theory*, **60**(10), 5875–5898.

Kudekar, S., Méasson, C., Richardson, T., and Urbanke, R. (2010). Threshold saturation on BMS channels via spatial coupling. In *Proceedings of 6th International Symposium on Turbo Codes and Iterative Information Processing (ISTC)*, pp. 309–313. IEEE.

Kudekar, S., Richardson, T., and Urbanke, R. (2012). Wave-like solutions of general one-dimensional spatially coupled systems. arXiv:1208.5273.

Kudekar, S., Richardsony, T., and Urbanke, R. (2011). Threshold saturation via spatial coupling: why convolutional LDPC ensembles perform so well over the BEC. *IEEE Transactions on Information Theory*, **57**(2), 803–834.

Luby, M. G., Mitzenmacher, M., Shokrollahi, M. A., and Spielman, D. A. (2001). Efficient erasure correcting codes. *IEEE Transactions on Information Theory*, **47**, 569–584.

Méasson, C., Montanari, A., and Urbanke, R. (2005). Maxwell construction: the hidden bridge between iterative and maximum a posteriori decoding. arXiv:cs/0506083.

Miller, G. and Cohen, G. (2003). The rate of regular ldpc codes. In *Proceedings of IEEE International Symposium on Information Theory*, p. 89.

Polyanskiy, Y., Poor, H. V., and Verdu, S. (2010). Channel coding rate in the finite blocklength regime. *IEEE Transactions on Information Theory*, **56**(5), 2307–2359.

Richardson, T. and Urbanke, R. (2009). *Modern Coding Theory*. Cambridge University Press.

Strassen, V. (1962). Asymptotische Abschaetzungen in Shannon Informationstheorien. *Transactions of 3rd Prague Conference on Information Theory*, 689–732.

ten Brink, S. (1999). Convergence of iterative decoding. *Electronics Letters*, **35**(10), 806–808.

Thorpe, J. (2003). Low-density parity-check (LDPC) codes constructed from protographs. *IPN Progress Report*, **42–154**, 1–7.

Wormald, N. (1995). Differential equations for random processes and random graphs. *Annals of Applied Probability*, **5**, 1218–1235.

Yedla, A., Jian, Y., Nguyen, P. S., and Pfister, H. D. (2012). A simple proof of threshold saturation for coupled scalar recursions. In *Proceedings 7th International Symposium on Turbo Codes and Iterative Information Processing (ISTC)*, pp. 51–55. IEEE.

# 7

# Constraint satisfaction: random regular $k$-SAT

Amin COJA-OGHLAN

University of Frankfurt
Mathematics Institute
10 Robert Mayer Strasse, Frankfurt 60325, Germany

*Lecture notes taken by*
Seyed Hamed Hassani, ETH Zürich, Switzerland
Felicia Rassmann, University of Frankfurt, Germany

*Statistical Physics, Optimization, Inference, and Message-Passing Algorithms.* First Edition.
F. Krzakala et al. © Oxford University Press 2016. Published in 2016 by Oxford University Press.

# Chapter Contents

## 7.1   A first approach

To what extent can we make the physics predictions about thresholds in constraint satisfaction problems mathematically rigorous? As an example, we investigate **random regular $k$-SAT**. In contrast to the well-known (uniformly) **random $k$-SAT** problem, the regular case has not been studied much (for references, see Section 7.3). But the regular case is much simpler both from the physics perspective and mathematically, and very precise rigorous results can be obtained in this case.

The random regular $k$-SAT problem can be stated as follows. There are $n$ Boolean variables $x_1, \ldots, x_n$ and $m$ clauses built from these variables. We call $m/n$ the **density**. All clauses have length $k \geq 3$, and we let $d$ denote the literal degree. That is, we are interested in random $k$-SAT formulas where each literal appears exactly $d$ times. This means that every variable appears $d$ times positively and $d$ times negatively. Thus, we let $\Phi$ be a formula chosen uniformly at random from all formulas with this property (note that we do not make an assumption about the distribution of variables in each clause). By construction, $km = 2dn$.

We are interested in the following questions:

1. For what values of $d$ and $k$ is $\Phi$ satisfiable *with high probability* (w.h.p.)?
2. How many satisfying assignments exist?
3. What does the geometry of the solution space look like? (Can we detect the clustering threshold etc. rigorously in this model?)

If we follow the physics predictions, what matters about the random formula is its local structure. How does the neighborhood of a variable look like? In the model of random regular formulas, we are in the lucky situation that in the factor graph all neighborhoods look exactly the same, i.e., we always see the exact same tree locally. From belief propagation, which is utterly simple for this model, we thus get the same marginal distribution for every variable. Also, survey propagation provides the same result for all variables. To summarize:

- All BP marginals are equal to $\frac{1}{2}$, the SP marginals are $\frac{1}{2}(1 - \delta)$ (corresponding to "true"), $\frac{1}{2}(1 - \delta)$ (corresponding to "false") and $\delta$ (corresponding to "star") as we will see in the second part of the chapter.

The cavity method yields the following predictions for the geometry of the solution space (error terms that tend to 0 as $k$ tends to infinity are omitted):

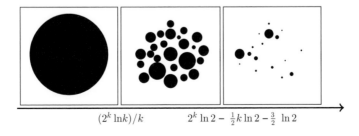

$(2^k \ln k)/k \qquad\qquad 2^k \ln 2 - \frac{1}{2}k \ln 2 - \frac{3}{2}\ln 2$

To what extent can we verify these predictions rigorously? In this chapter, we establish rigorous upper and lower bounds on the satisfiability transition that is predicted to occur at $2d/k \sim 2^k \ln 2 - \frac{1}{2}k \ln 2 - \frac{1}{2}(1 + \ln 2)$.

**Theorem 7.1** *There is a sequence $\delta_k > 0$ with $\lim_{k \to \infty} \delta_k = 0$ such that for any $\varepsilon > 0$ the following is true:*

1. *Assume that*

$$\frac{2d}{k} < 2^k \ln 2 - \frac{k}{2} \ln 2 - 1 - \frac{\ln 2}{2} - \delta_k.$$

   *Then w.h.p. $\Phi$ admits an assignment $\sigma$ that satisfies all but $\varepsilon m$ clauses.*
2. *If, on the other hand,*

$$\frac{2d}{k} > 2^k \ln 2 - \frac{k}{2} \ln 2 + \delta_k,$$

   *then w.h.p. $\Phi$ does not admit an assignment $\sigma$ that satisfies all but $\varepsilon m$ clauses.*

To prove Theorem 7.1, we start by computing $\mathbb{E}[Z]$ where $Z$ is the number of satisfying assignments. To this end, we introduce the so-called **configuration model**:

- Create $d$ "clones" of each literal:

$$x_1 \rightsquigarrow (x_1, 1), \ldots, (x_1, d),$$

$$\bar{x}_1 \rightsquigarrow (\bar{x}_1, 1), \ldots, (\bar{x}_1, d),$$

  and the same for all the other variables.
- Choose a random bijection $\pi : [m] \times [k] \to L \times [d]$, where $L = \{x_1, \bar{x}_1, \ldots\}$.
- Set $\Phi_{ij} = \pi(i, j)$, where $\Phi_{ij}$ is the $j$th literal of the $i$th clause.

This model actually generates random regular formulas. We can think of the clones as a deck of cards. To create the formula, we just shuffle the deck randomly and put the cards down in the resulting order to "fill in" the $k$-clauses one by one.

We now want to calculate $\mathbb{E}[Z] = \sum_\sigma \mathbb{P}(\sigma \models \Phi)$, where $\sigma \models \Phi$ means that $\sigma$ satisfies $\Phi$. A nice property of the model is that $\mathbb{P}(\sigma \models \Phi)$ is independent of the actual assignment $\sigma$, because every literal appears exactly $d$ times and so each assignment satisfies exactly fifty percent of the literals. Hence,

$$\mathbb{E}[Z] = 2^n \mathbb{P}(\mathbf{1} \models \Phi),$$

where $\mathbf{1}$ is the assignment that assigns "true" to every variable.

To calculate $\mathbb{P}(\mathbf{1} \models \Phi)$, we introduce a "pebble experiment" where we only discriminate between "true" and "false" literals (but forget, for example, about the variable that underlies the literal) and visualize them as pebbles in the shape of circles and squares.

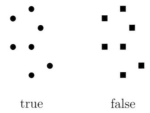

true                false

We toss the pebbles into the clauses randomly and note that $\Phi$ is satisfied if and only if each clause receives at least one circle.

As exactly half of all pebbles are circles (owing to regularity), it would be tempting to claim that $\mathbb{P}(\mathbf{1} \models \Phi) = (1-2^{-k})^m$. This formula does, in fact, hold in the "uniform" random $k$-SAT model (i.e., without regularity). But in the random regular formula, we have $\mathbb{P}(\mathbf{1} \models \Phi) < (1-2^{-k})^m$. This is because regularity introduces a subtle dependence between the clauses, since the total number of circles and squares has to be the same.

**Proposition 7.2**  *Let $q \in (0,1)$ be the solution to*

$$\frac{1}{2} = \frac{q}{1 - (1-q)^k}. \tag{7.1}$$

*Then*

$$\frac{1}{n} \ln \mathbb{E}[Z] \sim \ln 2 + \frac{2d}{k} \ln \left(1 - (1-q)^k\right) + 2d\, \mathrm{D}\left(\tfrac{1}{2}\|q\right), \tag{7.2}$$

*where*

$$\mathrm{D}(p\|q) = p\ln\frac{p}{q} + (1-p)\ln\frac{1-p}{1-q}$$

*is the Kullback–Leibler divergence and $f \sim g$ means $f/g \xrightarrow{n\to\infty} 1$.*

Observe that the right-hand side of (7.2) is a linear function of $d$. This function turns out to coincide with the "replica-symmetric solution."

**Proof** The pebble experiment has the disadvantage that there are dependences amongst the clauses. To circumvent this problem, we introduce yet another experiment with new random variables on a different probability space. The idea is to take independent variables and to choose the appropriate probabilities such that the constellation that comes out likely is the one we wanted to see in the pebble experiment. In other words, we introduce a bias that adjusts the number of circles so that we can work with independent events.

Thus, let $(\phi_{ij})_{i\in[m],j\in[k]}$ be independent random variables representing literal number $j$ in clause number $i$, such that $\phi_{ij} \in \{\bullet,\blacksquare\}$ and $\mathbb{P}(\phi_{ij} = \bullet) = q$, with $q$ as in (7.1). We consider the two events

$$S = \{\forall\, i \in [m] \,\exists\, j \in [k] : \phi_{ij} = \bullet\}$$

and

$$B = \{\#\{(i,j) : \phi_{ij} = \bullet\} = dn\}.$$

What we are interested in is the conditional probability $\mathbb{P}(S|B) = \mathbb{P}(\mathbf{1} \models \Phi)$, since, given that $B$ occurs, the distribution of the random variables $\phi_{ij}$ is identical with the distribution of the literal shapes in the pebble experiment.

To calculate $\mathbb{P}(S|B)$, we compute $\mathbb{P}(S)$, $\mathbb{P}(B)$, and $\mathbb{P}(B|S)$. This yields $\mathbb{P}(S|B)$ because

$$\mathbb{P}(S|B) = \frac{\mathbb{P}(S \cap B)}{\mathbb{P}(B)} = \frac{\mathbb{P}(S)}{\mathbb{P}(B)} \cdot \mathbb{P}(B|S). \tag{7.3}$$

Clearly,

$$\mathbb{P}(S) = \left(1 - (1-q)^k\right)^m, \tag{7.4}$$

owing to independence. Furthermore, $\mathbb{P}(B) = \mathbb{P}\left(\mathrm{Bin}(km, q) = \frac{1}{2}km\right)$. As $2dn = km$, we obtain from Stirling's formula that

$$\frac{1}{n} \ln \mathbb{P}(B) \sim -2d\,\mathrm{D}\left(\frac{1}{2}\|q\right). \tag{7.5}$$

Finally, to calculate $\mathbb{P}(B|S)$, consider the random variable $X = \#\{(i,j) : \phi_{ij} = \bullet\}$ that counts the number of circles. It is binomially distributed. Moreover, $X$ is a sum of independent random variables if we condition on $S$ and we have chosen $q$ such that under this conditioning the event $B$ occurs if $X$ equals its expected value: because of (7.1),

$$\mathbb{E}[X|S] = km \cdot \frac{q}{1 - (1-q)^k} = \frac{km}{2} = dn.$$

With the local limit theorem for the sum of independent random variables that tells us the asymptotic probability for the sum of discrete random variables to take a certain value (Flajolet and Sedgewick, 2009), we can calculate the probability of this event:

$$\mathbb{P}(B|S) = \mathbb{P}(X = dn|S) \sim \frac{c}{\sqrt{n}}$$

for a certain number $c = c(d,k) > 0$. Hence,

$$\frac{1}{n} \ln \mathbb{P}(B|S) \xrightarrow{n\to\infty} 0. \tag{7.6}$$

Plugging (7.4)–(7.6) into (7.3) completes the proof. $\qquad\square$

The first moment $\mathbb{E}[Z]$ and Markov's inequality give an upper bound on the satisfiability threshold. In fact, together with a simple calculation, Proposition 7.2 implies the second part of Theorem 7.1. To prove the first part, we calculate the second moment $\mathbb{E}\left[Z^2\right]$. If the second moment is approximately the square of the first moment, say,

$$\mathbb{E}\left[Z^2\right] \leq C \cdot \mathbb{E}[Z]^2 \tag{7.7}$$

for a number $C = C(d,k) > 0$ that may depend on $d,k$ but not on $n$, then the *Paley–Zygmund inequality* yields

$$\mathbb{P}\left[\Phi \text{ is satisfiable}\right] \geq \mathbb{P}\left[Z > 0\right] \geq \frac{\mathbb{E}[Z]^2}{\mathbb{E}[Z^2]} \geq \frac{1}{C} > 0. \tag{7.8}$$

Thus, (7.7) implies that the probability that $\Phi$ is satisfiable is bounded away from 0. However, (7.7) does not imply (at least not directly) that this probability tends to 1 as $n \to \infty$. Nonetheless, standard concentration inequalities (e.g., the Azuma–Hoeffding inequality) yield the following lemma:

**Lemma 7.3** *For any $\varepsilon > 0$, the following is true. Assume that*

$$\frac{1}{n} \ln \mathbb{E}[Z^2] \sim \frac{2}{n} \ln \mathbb{E}[Z]. \tag{7.9}$$

*Then, with probability tending to 1 as $n \to \infty$, the random formula $\Phi$ admits an assignment $\sigma$ that satisfies at least $(1-\varepsilon)m$ clauses.*

Thus, the goal in the rest of this section is to establish (7.9). We can cast the second moment as

$$\mathbb{E}\left[Z^2\right] = \sum_{\sigma,\tau} \mathbb{P}(\sigma \models \Phi, \tau \models \Phi) = \sum_{a=0}^{n} \mathbb{E}[Z_a], \tag{7.10}$$

with $Z_a$ the number of pairs $(\sigma,\tau)$ of satisfying assignments such that $\mathrm{dist}(\sigma,\tau) = a$. By a similar argument as above in the calculation of the first moment, $\mathbb{P}(\sigma \models \Phi, \tau \models \Phi)$ is the same for all assignments $\sigma$ and $\tau$ with the same distance. Thus, for a given $a$, we pick $\sigma = \mathbf{1}$ and $\tau$ the assignment that has 0s on the first $a$ positions and 1s on the last $n - a$ positions.

To model satisfied literals in pairs of assignments, pebbles are not quite sufficient. Instead, we introduce four types of "dominos" where the pebbles in the first row represent the values of the variables under the assignment $\sigma$ and those in the second row represent the values under $\tau$. Because of regularity, $a = \alpha n$ determines the fraction of dominos of each type (see Fig. 7.1).

We are now performing a similar change of experiments as before to get rid of the dependences among the dominos arising from prescribing the total number of dominos of each type. Thus, let $(\phi_{ij})_{i \in [m], j \in [k]}$ be independent random variables representing

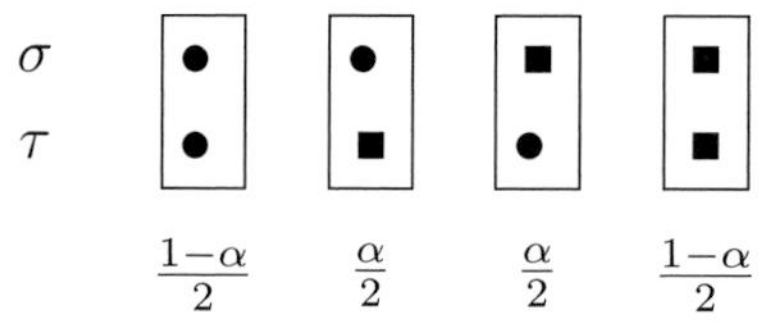

$$\begin{array}{ccccc} \sigma & \bullet & \bullet & \blacksquare & \blacksquare \\ \tau & \bullet & \blacksquare & \bullet & \blacksquare \\ & \frac{1-\alpha}{2} & \frac{\alpha}{2} & \frac{\alpha}{2} & \frac{1-\alpha}{2} \end{array}$$

**Fig. 7.1** Marginal probabilities of the four types of dominos.

the domino for the $j$th literal of the $i$th clause. With $q, \hat{q} \in (0,1)$ to be determined in due course, we let

$$\phi_{ij} = \begin{cases} (\bullet, \bullet), & \text{with probability } q, \\ (\bullet, \blacksquare), & \text{with probability } \hat{q} - q, \\ (\blacksquare, \bullet), & \text{with probability } \hat{q} - q, \\ (\blacksquare, \blacksquare), & \text{with probability } 1 - 2\hat{q} + q, \end{cases}$$

independently for all $i, j$.

As before, we define an event $S$ that corresponds to the formula being satisfied under both assignments. More precisely, $S$ is the event that for each clause $i$ there exist indices $j, j'$ such that the first component of $\phi_{ij}$ is a circle and such that the second component of $\phi_{ij'}$ is a circle. (Of course, it is allowed that $j = j'$.) Because the $\phi_{ij}$ are mutually independent, we obtain from inclusion/exclusion

$$\mathbb{P}(S) = \left(1 - 2(1 - \hat{q})^k + (1 - 2\hat{q} + q)^k\right)^m. \tag{7.11}$$

In addition, let $B$ be the event that each type of domino occurs with the frequency indicated in Fig. 7.1. By Stirling's formula,

$$\frac{1}{n} \ln \mathbb{P}(B) \sim -2d \, \mathrm{D}\left(\left(\frac{1-\alpha}{2}, \frac{\alpha}{2}, \frac{\alpha}{2}, \frac{1-\alpha}{2}\right) \middle\| (q, \hat{q} - q, \hat{q} - q, 1 - 2\hat{q} + q)\right) \tag{7.12}$$

Finally, as in the calculation of the first moment, we need to determine $\mathbb{P}[B|S]$. To facilitate this computation, we choose $q$ and $\hat{q}$ such that

$$\frac{1}{km} \mathbb{E}[\#\{(i,j) : \phi_{ij} = (\bullet, \bullet)\}] = \frac{1-\alpha}{2} \quad \text{and} \quad \frac{1}{km} \mathbb{E}[\#\{(i,j) : \phi_{ij} = (\bullet, \blacksquare)\}] = \frac{\alpha}{2}. \tag{7.13}$$

It is fairly straightforward to calculate the two expectations in (7.13) explicitly. The result is that (7.13) holds if $q, \hat{q} \in (0,1)$ are the solutions to the equations

$$\frac{q}{1 - 2(1 - \hat{q})^k + (1 - 2\hat{q} + q)^k} = \frac{1-\alpha}{2} \tag{7.14}$$

and

$$\frac{(\hat{q} - q)(1 - (1 - \hat{q})^{k-1})}{1 - 2(1 - \hat{q})^k + (1 - 2\hat{q} + q)^k} = \frac{\alpha}{2}. \tag{7.15}$$

With this choice of $q, \hat{q}$, the local limit theorem shows that

$$\lim_{n \to \infty} \frac{1}{n} \ln \mathbb{P}[B|S] = 0.$$

Thus, (7.11)–(7.15) yield a closed-form expression for

$$f(\alpha) = \lim_{n \to \infty} \frac{1}{n} \ln \mathbb{E}[Z_{\alpha n}].$$

Indeed, with $q, \hat{q}$ the solutions to (7.14) and (7.15), we have

$$f(\alpha) = h(\alpha) + p(\alpha),$$

where

$$h(\alpha) = \ln 2 - \alpha \ln \alpha - (1 - \alpha) \ln(1 - \alpha),$$
$$p(\alpha) = \frac{2d}{k} \ln(1 - 2(1 - \hat{q})^k + (1 - 2\hat{q} + q)^k)$$
$$+ 2d\, \mathrm{D}\left(\left(\frac{1 - \alpha}{2}, \frac{\alpha}{2}, \frac{\alpha}{2}, \frac{1 - \alpha}{2}\right) \middle\| (q, \hat{q} - q, \hat{q} - q, 1 - 2\hat{q} + q)\right).$$

Hence, (7.10) yields

$$\frac{1}{n} \ln \mathbb{E}[Z^2] \sim \max_{\alpha \in (0,1)} f(\alpha), \tag{7.16}$$

Furthermore, by direct inspection, we find that

$$\frac{2}{n} \ln \mathbb{E}[Z] \sim f(1/2). \tag{7.17}$$

Plotting $f(x)$ for different values of $m/n$, we get the graphs shown in Fig. 7.2. The top curve corresponds to $m/n < (2^k \ln k)/k$, and the lower curves to $m/n > (2^k \ln k)/k$. As long as the peak at $x = \frac{1}{2}$ exceeds the other peaks, the second-moment method works, because $f(\frac{1}{2}) = (2/n) \ln \mathbb{E}[Z]$. The small peak corresponds to the expected cluster size. When the line drops below the axis, it means that w.h.p. there exists no pair of assignments with this distance.

To establish (7.9) we need to show that the maximum in (7.16) is attained at $\alpha = \frac{1}{2}$. A necessary condition for this is that $\frac{1}{2}$ is a stationary point of $f$. The following lemma shows that this is indeed the case:

**Lemma 7.4**  *We have $f'(\frac{1}{2}) = 0$.*

**Proof**  Clearly, the entropy function $h(\alpha)$ satisfies $h'(\frac{1}{2}) = 0$. Thus, we are left to verify that $p'(\frac{1}{2}) = 0$. The proof of this is based on a combinatorial argument. Let $\mathcal{M}$

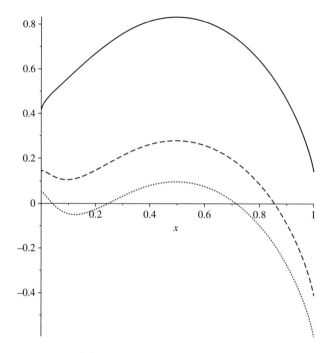

**Fig. 7.2** $f(x)$ versus $x$ for three different values of $m/n$.

be the set of all maps $\phi : [m] \times [k] \to \{\bullet, \blacksquare\}$ such that $|\phi^{-1}(\bullet)| = |\phi^{-1}(\blacksquare)| = \frac{1}{2}km$. Let $\vec{\xi}_1, \vec{\xi}_2 \in \mathcal{M}$ be two elements chosen uniformly and independently. In addition, let

$$Y^{c_1 c_2} = \frac{1}{km} \# \left\{ (i,j) \in [m] \times [k] : \vec{\xi}_1(i,j) = c_1, \vec{\xi}_2(i,j) = c_2 \right\} \qquad (c_1, c_2 \in \{\bullet, \blacksquare\}).$$

By construction, the following relations hold between these random variables:

$$Y^{\bullet\bullet} + Y^{\bullet\blacksquare} = Y^{\bullet\bullet} + Y^{\blacksquare\bullet} = \frac{1}{2},$$

$$Y^{\blacksquare\blacksquare} = 1 - Y^{\bullet\bullet} - Y^{\bullet\blacksquare} - Y^{\blacksquare\bullet} = Y^{\bullet\bullet}.$$

Hence, $Y^{\bullet\bullet}$ determines the three others. Thus, let $\mathcal{X}(y)$ be the event that $Y^{\bullet\bullet} = y$.

Clearly, $\mathbb{E}[Y^{\bullet\bullet}] = \frac{1}{4}$. Furthermore, a standard inequality (Azuma–Hoeffding) implies that there is a number $C_1 = C_1(k) > 0$ such that

$$\frac{1}{m} \ln \mathbb{P}[\mathcal{X}(y)] \le -C_1 (y - \tfrac{1}{4})^2. \tag{7.18}$$

By a similar token, we have $\mathbb{E}[Y^{\bullet\bullet}|S] = \frac{1}{4}$ and there is $C_2 = C_2(k) > 0$ such that

$$\frac{1}{m} \ln \mathbb{P}[\mathcal{X}(y)|S] \le -C_2 (y - \tfrac{1}{4})^2. \tag{7.19}$$

As $\alpha = Y^{\bullet\bullet} + Y^{\bullet\bullet} = 1 - 2Y^{\bullet\bullet}$, with $y = y(\alpha) = \frac{1}{2}(1-\alpha)$, we find

$$p(\alpha) \sim \frac{1}{m} \ln \frac{\#\mathcal{X}(y) \cap S}{\#\mathcal{X}(y)} = \frac{1}{m} \ln \#\mathcal{X}(y) \cap S - \frac{1}{m} \ln \#\mathcal{X}(y). \tag{7.20}$$

Finally, (7.18) and (7.19) show that both summands in (7.20) attain their maximum at $\alpha \sim \frac{1}{2}$. Thus, $p'(\frac{1}{2}) = 0$. $\square$

In addition, to show that $\alpha = \frac{1}{2}$ is a local maximum, we calculate $f''(\frac{1}{2})$. Clearly, $h''(\frac{1}{2}) = -4$. Moreover, with respect to the second differential of $p$, we have the following estimate.

**Lemma 7.5** *There is a constant $k_0 > 0$ such that $\sup_{|\alpha - \frac{1}{2}| < 2^{-k/3}} p''(\alpha) < 1$ for all $k > k_0$.*

**Proof** As a first step, we calculate the differentials of the implicitly defined $q = q(\alpha)$ and $\hat{q} = \hat{q}(\alpha)$. To accomplish this, we use the inverse function theorem. Consider the function

$$\mu(q, \hat{q}, \alpha) =$$
$$\left( \frac{q}{1 - 2(1-\hat{q})^k + (1 - 2\hat{q} + q)^k} - \frac{1-\alpha}{2}, \frac{(\hat{q} - q)(1 - (1-\hat{q})^{k-1})}{1 - 2(1-\hat{q})^k + (1 - 2\hat{q} + q)^k} - \frac{\alpha}{2}, \alpha \right). \tag{7.21}$$

Then $(q(\alpha), \hat{q}(\alpha), \alpha) = \mu^{-1}(0, 0, \alpha)$. Thus, we need to calculate the first two derivatives of the function $\alpha \mapsto \mu^{-1}(0, 0, \alpha)$. By the inverse function theorem, the Jacobian of $\mu^{-1}$ satisfies $D\mu^{-1} = (D\mu)^{-1}$. In the sequel, let $\varepsilon$ denote a term such that $|\varepsilon| \le k^C 2^{-k}$ (not necessarily the same one for each occurrence of $\varepsilon$). If $|\alpha - \frac{1}{2}|, |q - \frac{1}{4}|, |\hat{q} - \frac{1}{2}| < k^{-2}$, then

$$D\mu = \begin{pmatrix} 1 + \varepsilon & \varepsilon & \frac{1}{2} \\ -1 + \varepsilon & 1 + \varepsilon & -\frac{1}{2} \\ 0 & 0 & 1 \end{pmatrix} = \begin{pmatrix} \Delta & & \frac{1}{2} \\ & & -\frac{1}{2} \\ 0 & & 1 \end{pmatrix}.$$

Hence,

$$D\mu^{-1} = (D\mu)^{-1} = \begin{pmatrix} \Delta^{-1} & -\Delta^{-1}\begin{pmatrix} \frac{1}{2} \\ -\frac{1}{2} \end{pmatrix} \\ 0 & 1 \end{pmatrix}. \tag{7.22}$$

Of course, the inverse of the $2 \times 2$ matrix $\Delta$ is just

$$\Delta^{-1} = \frac{1}{\det \Delta} \begin{pmatrix} \Delta_{22} & -\Delta_{12} \\ -\Delta_{21} & \Delta_{11} \end{pmatrix} = \begin{pmatrix} 1+\varepsilon & -1+\varepsilon \\ \varepsilon & 1+\varepsilon \end{pmatrix}.$$

Thus, we find that for $|\alpha - \frac{1}{2}| \le 2^{-k/3}$,

$$q'(\alpha) = -\frac{1}{2} + \varepsilon, \qquad \hat{q}'(\alpha) = \varepsilon. \tag{7.23}$$

Moreover, applying the chain rule to (7.22), we obtain

$$q''(\alpha) = \varepsilon, \qquad \hat{q}''(\alpha) = \varepsilon. \tag{7.24}$$

Hence, the second derivative of the function

$$s(\alpha) = 1 - 2(1 - \hat{q}(\alpha))^k + (1 - 2\hat{q}(\alpha) + q(\alpha))^k$$

satisfies

$$(\ln s(\alpha))'' = \frac{s(\alpha)s''(\alpha) - (s'(\alpha))^2}{s(\alpha)^2} \le \varepsilon^2. \tag{7.25}$$

To calculate the second derivative of

$$\psi(\alpha) =$$
$$D\left(\left(\frac{1-\alpha}{2}, \frac{\alpha}{2}, \frac{\alpha}{2}, \frac{1-\alpha}{2}\right) \middle\| (q(\alpha), \hat{q}(\alpha) - q(\alpha), \hat{q}(\alpha) - q(\alpha), 1 - 2\hat{q}(\alpha) + q(\alpha))\right), \tag{7.26}$$

we observe that

$$|q(\alpha) - \tfrac{1}{4}| \le \varepsilon, \qquad |\hat{q}(\alpha) - \tfrac{1}{2}| \le \varepsilon \qquad \text{if} \quad |\alpha - \tfrac{1}{2}| \le 2^{-k/3}; \tag{7.27}$$

this follows directly from the equations (7.14) and (7.15) that define $q(\alpha)$ and $\hat{q}(\alpha)$. Moreover, to estimate the differentials of the Kullback–Leibler divergence, we observe that

$$\frac{\partial}{\partial x} x \ln \frac{x}{y} = 1 + \ln \frac{x}{y}, \qquad \frac{\partial}{\partial y} x \ln \frac{x}{y} = -\frac{x}{y}, \tag{7.28}$$

$$\frac{\partial^2}{\partial x^2} x \ln \frac{x}{y} = \frac{1}{x}, \qquad \frac{\partial^2}{\partial y^2} x \ln \frac{x}{y} = \frac{x}{y^2}, \qquad \frac{\partial^2}{\partial x \partial y} x \ln \frac{x}{y} = -\frac{1}{y}. \tag{7.29}$$

Let

$$\Psi(y_1, \ldots, y_8) = D\left((y_1, y_2, y_3, y_4) \| (y_5, y_6, y_7, y_8)\right).$$

In addition, let

$$y_1(\alpha) = y_4(\alpha) = \frac{1-\alpha}{2}, \quad y_2(\alpha) = y_3(\alpha) = \frac{\alpha}{2},$$

$$y_5(\alpha) = q(\alpha), \qquad y_6(\alpha) = y_7(\alpha) = \hat{q}(\alpha) - q(\alpha), \quad y_8(\alpha) = 1 - 2\hat{q}(\alpha) + q(\alpha).$$

Then, by the chain rule,

$$\psi''(\alpha) = \sum_{i=1}^{8} \frac{\partial \Psi}{\partial y_i} \cdot y_i''(\alpha) + \sum_{i,j=1}^{8} \frac{\partial^2 \Psi}{\partial y_i \partial y_j} \cdot y_i'(\alpha)y_j'(\alpha). \tag{7.30}$$

From (7.24), (7.27), and (7.28), we immediately obtain

$$\sum_{i=1}^{8} \frac{\partial \Psi}{\partial y_i} \cdot y_i''(\alpha) \le \varepsilon \sum_{i=5}^{8} \frac{\partial \Psi}{\partial y_i} \le \varepsilon^2. \tag{7.31}$$

Similarly, (7.23) and (7.29) yield

$$\sum_{i,j=1}^{8} \frac{\partial^2 \Psi}{\partial y_i \partial y_j} \cdot y_i'(\alpha)y_j'(\alpha) \le \varepsilon^2. \tag{7.32}$$

Finally, the assertion follows from (7.25) and (7.30)–(7.32).    □

Lemma 7.5 shows that $\frac{1}{2}$ is a local maximum of the function $f$. To complete the second-moment argument, we need to verify that $\frac{1}{2}$ is indeed the global maximum of $f$.

**Lemma 7.6**    *Assume that there is a sequence $\varepsilon_k > 0$ with $\lim_{k \to \infty} \varepsilon_k = 0$ such that*

$$\frac{m}{n} < 2^k \ln 2 - \frac{k}{2} \ln 2 - 1 - \frac{\ln 2}{2} - \varepsilon_k.$$

*Then, for $\alpha \in [0,1] \setminus [\frac{1}{2} - 2^{-k/3}, \frac{1}{2} + 2^{-k/3}]$, we have $f(\alpha) < f(\frac{1}{2})$.*

**Proof**    We consider the auxiliary function

$$g(\alpha) = h(\alpha) + \frac{2d}{k} \ln \left( 1 - 2^{1-k} + ((1-\alpha)/2)^k \right).$$

We claim that

$$f(\alpha) \le g(\alpha).$$

Indeed, if we plug $q = (1-\alpha)/2$ and $\hat{q} = \frac{1}{2}$ into (7.11) and (7.12), then we obtain

$$f(\alpha) - h(\alpha) \sim \frac{1}{n} \ln \mathbb{P}[S|B] = \frac{1}{n} \ln \left[ \frac{\mathbb{P}[S]}{\mathbb{P}[B]} \cdot \mathbb{P}[B|S] \right]$$

$$= \frac{1}{n} \ln \mathbb{P}[S] - \frac{1}{n} \ln \mathbb{P}[B] + \frac{1}{n} \ln \mathbb{P}[B|S]$$

$$\sim \frac{1}{n} \ln \mathbb{P}[S] + \frac{1}{n} \ln \mathbb{P}[B|S] \le \frac{1}{n} \ln \mathbb{P}[S] \sim g(\alpha) - h(\alpha).$$

Thus, to complete the proof, we just need to show that

$$\sup_{\alpha \notin [\frac{1}{2}-2^{-k/3}, \frac{1}{2}+2^{-k/3}]} g(\alpha) \le f(\tfrac{1}{2}),$$

which is a mere exercise in basic calculus. □

Lemmas 7.4–7.6 imply that

$$\max_{\alpha \in (0,1)} f(\alpha) = f(\tfrac{1}{2}).$$

Thus, (7.16), (7.17), and Lemma 7.3 yield the first part of Theorem 7.1.

The second-moment argument presented here can be combined with the "planting trick" from Achlioptas and Coja-Oghlan (2008) to obtain rigorous results about the geometry of the set of (nearly) satisfying assignments. For instance, this leads to a rigorous verification of the clustering picture, and of the existence of "frozen variables" (details omitted).

## 7.2   Beyond the condensation threshold

When the formulas are sufficiently sparse, i.e.,

$$\frac{m}{n} < 2^k \ln 2 - \frac{k}{2} \ln 2 - \frac{\ln 2}{2} - 1 - \varepsilon_k, \tag{7.33}$$

we know that $\ln Z \sim \ln \mathbb{E}[Z]$. However, this is not true for all densities. More precisely, the physics calculations predict that there is a *condensation threshold* beyond which the actual number $Z$ is smaller than the expected number $\mathbb{E}[Z]$ by an exponential factor w.h.p. In effect, the argument from Section 7.1 cannot extend beyond the condensation threshold.

The condensation threshold can be interpreted in terms of the geometry of the set of satisfying assignments. For $d$ below the condensation threshold, we expect that this set decomposes into well-separated clusters, each of which only contains an exponentially small fraction of all satisfying assignments. By contrast, for $d$ beyond the condensation threshold, a bounded number of clusters dominate. Now, as we saw in Section 7.1, the second-moment method applied to $Z$ works only if two random solutions "look completely uncorrelated." When we are below the condensation threshold, if we sample twice from the space of solutions, then it is highly likely that the two solutions are picked from two different clusters. Thus, at least to this extent, they look uncorrelated. However, in the condensation regime, where there are finitely many clusters that almost cover the whole solution space, it is likely that the two sampled solutions are from the same cluster. In this case, they are heavily correlated.

The objective of this section is to overcome the problem caused by such correlations by using a physics-inspired workaround: we forget the big clusters. More specifically, the idea is to switch from counting solutions to counting clusters. We represent each cluster by a map $\sigma : V \rightarrow \{0, 1, \star\}$ in a way such that each variable either takes a value in $\{0, 1\}$ or is given the value $\star$. The intention is that within each cluster, the frozen variables are assigned their corresponding Boolean value (either 0 or 1), and the unfrozen variables are set to $\star$.

Formally, we define a map $\sigma : V \rightarrow \{0, 1, \star\}$ to be a *cover* of a $k$-CNF $\Phi = \Phi_1 \wedge \ldots \wedge \Phi_m$ if the following two conditions are satisfied: extend $\sigma$ to a map from the set of literals $(L)$ to $\{0, 1, \star\}$ by letting $\sigma(\bar{x}) = \overline{\sigma(x)}$, with the convention that $\bar{0} = 1$, $\bar{1} = 0$, and $\bar{\star} = \star$. Then

**CV1:** Every clause contains either
* a literal $l$ with $\sigma(l) = 1$, or
* two literals $l_1, l_2$ with $\sigma(l_1) = \sigma(l_2) = \star$.
**CV2:** For a literal $l$, if $\sigma(l) = 1$ then there is a clause $\Phi_i$ that blocks $l$; i.e., the clause $\Phi_i$ contains $l$, and all the other literals in $\Phi_i$ are set to 0.

**Observation:** If $\sigma(l) = \star$ for all the literals in $L$, then it is a cover.

Let us now see how we can map the set of satisfying assignments to the set of covers via a *whitening process*, explained as follows. Let $\tau$ be a SAT assignment. Construct $\hat{\tau} : V \rightarrow \{0, 1, \star\}$ as follows:

- Initially $\hat{\tau} = \tau$.
- While there is a $l$ with $\tau(l) = 1$ that violates CV2, set $\hat{\tau}(l) = \star$.

**Result:** $\hat{\tau}$ is a cover!

We bear in mind that through the whitening process, variables that are not frozen should in principle be able to take either of values $0, 1$. Indeed, the statistical mechanics intuition is that the whitening process maps all satisfying assignments in one cluster to the same cover. Thus, the whitening process diminishes the internal entropy of the clusters. In effect, a big cluster is as good as any other. Now, we are going to apply the first- and second-moment method to the number of covers.

One thing we have to take care of is that we do not want to count the all-$\star$ cover.

We say that a cover $\sigma$ is a $(\nu, \mu)$-cover if the cardinality of the set $\{x \in V : \sigma(x) = \star\}$ is $\nu n$, and the total number of critical clauses (with one literal assigned 1 and all others assigned 0) is $\mu m$. Define $\Sigma_{\nu, \mu} = \#(\nu, \mu)$-covers of $\Phi$.

One technical point is the presence of short cycles. The issue is that the factor graph of the random formula, whose vertices are the clauses and variables of the formula and in which a variable is connected with all the clauses in which it appears, is liable to have a (small) number of short cycles. These may lead the whitening process to yield different covers for two satisfying assignments in the same cluster, i.e., the relationship between covers and clusters is not quite one-to-one. However, w.h.p. one cluster only gives rise to a sub-exponential number of covers.

As mentioned above, every satisfying assignment $\tau$ induces a cover $\hat{\tau}$ (using the whitening process). However, it is not necessarily true that every cover $\sigma$ comes from an actual satisfying assignment. Thus, we call a cover *valid*, if $\sigma = \hat{\tau}$ for a SAT assignment

$\tau$, i.e., if it comes from a SAT assignment. Denote the number of valid $(\nu, \mu)$-covers by $\Sigma'_{\nu,\mu}$. We also define

$$\Sigma := \max_{\substack{\nu \leq 2^{-k/2} \\ \mu \leq k^2/2^k}} \Sigma_{\nu,\mu}.$$

It is easy to see from the definition of $\Sigma$ that the all-$\star$ cover is not included in the set where we are maximizing inside.

**Theorem 7.7** *Assume $k \geq k_0$ for a fixed constant $k_0$. We have the following:*

- *If $d, k$ are such that $\lim_{n \to \infty} (1/n) \ln \mathbb{E}[\Sigma] \geq 0$, then $\Phi$ has an assignment that satisfies $(1 - o(1))m$ of the clauses.*
- *If $\lim_{n \to \infty} (1/n) \ln \mathbb{E}[\Sigma] < 0$, then there is an $\epsilon > 0$ such that any assignment violates at least $\epsilon m$ clauses with high probability.*

A consequence of this theorem is that it yields for any $k \geq k_0$ a threshold $d_k$ such that the following holds w.h.p.: For a randomly chosen formula with $d \leq d_k$, there is an assignment satisfying $1 - o(1)$ fraction of all its clauses, and for a random formula with $d > d_k$, every assignment violates at least a nonzero fraction $\epsilon$ of its clauses. It is possible to expand the threshold degree $d_k$ asymptotically in the limit of large $k$. The result is that

$$\frac{2d_k}{k} = 2^k \ln 2 - \frac{k}{2} \ln 2 - \frac{1 + \ln 2}{2} + o_k(1).$$

Recalling that $2d_k/k = m/n$, we can verify that this threshold is consistent with predictions based on the cavity method from statistical mechanics that we mentioned above.

**Proof**  Let us now explain the idea behind the proof. The main objective is as follows. First, we compute $(1/n) \ln \mathbb{E}[\Sigma]$. We then apply the second-moment method to $\Sigma$ by using the notion of covers.

So this is the agenda:

- Compute $\mathbb{E}[\Sigma]$  ✓
- Compute $\mathbb{E}[\Sigma^2]$  ✗
- Get from covers to solutions  ✓
- Observation out of this: "planting covers."

**Computing $\mathbb{E}[\Sigma_{\nu,\mu}]$:** This is done by using the so called "configuration model." Because each of the $2n$ literals appears precisely $d$ times in a formula $\Phi$, we can generate a random $\Phi$ as follows:

- Create a deck of cards with $2dn$ cards, $d$ copies of each literal.
- Create the formula by shuffling the deck of cards and putting them down one by one.

It is easy to see that this experiment indeed yields a formula that is chosen uniformly at random from all the possible regular formulas. As we know by now, each literal

$l$ in a formula appears in precisely $d$ clauses, which we denote by pairs $(l, j)$ where $j \in \{1, \ldots, d\}$. Let us refer to these pairs $(l, j)$ as the clones of the literal $l$. In order to get a handle on various events (such as a particular map $\sigma$ being a cover), we introduce a kind of "shape code" to represent the role that a particular literal clone plays. The basic shapes that we work with are diamond, signified throughout by ♦, triangle ▲, circle ●, and square ■. The idea is that ♦, ● represent clones $(l, j)$ of literals $l$ that take the value true, while ■ stands for a clone of a literal set to false, and ▲ means that the literal has got the joker value ⋆. Furthermore, the distinction between ♦ and ● is going to be that ♦ represents a clone $(l, j)$ of a true literal $l$ that occurs in a critical clause, i.e., all the other clones in that clause come from literals set to false. By contrast, ● means that the clone $(l, j)$ occurs in a clause that contains at least one more clone of shape either ● or ▲. So, for instance, it is forbidden to have all squares, and thus if $k - 1$ are square, the $k$th has to be diamond ($k = 3 : (\blacklozenge, \blacksquare, \blacksquare)$).

With this shape code in mind, we call a map $\zeta : L \times [d] \to \{\blacklozenge, \blacksquare, \blacktriangle, \bullet\}$ a *shade* if the following is true:

**SD1.** For any variable $x$, one of the following three statements is true:
- $\zeta(x, j) = \zeta(\bar{x}, j) = \blacktriangle$ for all $j \in [d]$.
- $\zeta(x, j) \in \{\blacklozenge, \bullet\}$ and $\zeta(\bar{x}, j) = \blacksquare$ for all $j \in [d]$.
- $\zeta(\bar{x}, j) \in \{\blacklozenge, \bullet\}$ and $\zeta(x, j) = \blacksquare$ for all $j \in [d]$.

**SD2.** There is no literal $l$ such that $\zeta(l, j) = \bullet$ for all $j \in [d]$.

In addition, we say that $\zeta$ is a $(\mu, \nu)$-shade if $|\zeta^{-1}(\blacklozenge)| = \mu m|$ and $|\zeta^{-1}(\blacktriangle)| = 2\nu dn$.

There is a close connection between shades and covers in the way that from a shade we can construct a cover and vice versa. So one way to compute $\Sigma_{\mu,\nu}$ would be to compute the number of $(\mu, \nu)$-shades, which we denote by $N_{\mu,\nu}$, and relate this to $\Sigma_{\mu,\nu}$. (Indeed, the expectation of $\Sigma_{\mu,\nu}$ is equal to the product of $N_{\mu,\nu}$ and some other computable term.) This task is nontrivial (but still doable). It resembles in some way an assignment problem (see Fig. 7.3), and the computation is similar to the first-moment calculation that we did above. In brief, to compute $N_{\mu,\nu}$, we have to generate $d$ copies of each literal with shapes (so we generate shaped cards)!

We begin by computing the number of maps $\sigma : V \to \{0, 1, \star\}$ with $|\sigma^{-1}(\star)| = \nu n$. There are $\binom{n}{\nu n}$ possible ways to choose the variables with a star, and once these are fixed, there are $2^{(1-\nu)n}$ ways to give a Boolean value to the other $(1 - \nu)n$ variables. Thus, the number of such maps is $\binom{n}{\nu n} 2^{(1-\nu)n}$. The map $\sigma$ determines which clones are shaped triangle, square or dotted (a clone is dotted $\odot$ if it is either circle or diamond). Thus, to turn $\sigma$ into a shade, we need to determine which of the dotted clones are diamond. So we need to select a total of $\mu m$ dotted clones that are to be shaped diamond, such that at least one clone of each literal $l$ with $\sigma(l) = 1$ is chosen (SD2). Of course, the total number of ways to choose $\mu m$ clones out of the $(1 - \nu)dn$ clones shaped dotted is nothing but $\binom{(1-\nu)dn}{\mu m}$. So, up to now, the total number of colorings is

$$
\underbrace{\binom{n}{\nu n}}_{\text{number of ways to generate } \nu \text{ stars } (\star)} \times \underbrace{2^{(1-\nu)n}}_{\text{fix to 1 or 0}} \times \underbrace{\binom{d(1 - \nu)n}{\mu m}}_{\text{right number of diamond pebbles}}.
$$

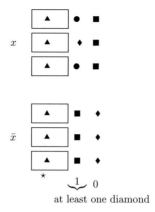

at least one diamond

**Fig. 7.3** A shade for $d = 3$. Note that every literal set to 1 has to have at least one diamond (SD2).

We then have to compute $\mathbb{P}($every literal set to 1 gets a $\blacklozenge)$, i.e., the fraction of these choices that pick at least one clone of each $l \in L$ with $\sigma(l) = 1$. We do this by using an alternative experiment as follows.

**Alternative experiment:** Consider the following experiment: Give each dotted clone (among the total of $(1 - \nu)dn$ clones, $d$ cards for each of the $(1 - \nu)n$ vars) the shape diamond with probability $q$ and circle with probability $1 - q$ independently of the others, for some $q \in (0, 1)$. Let $\mathcal{B}$ be the event that the total number of diamond clones generated in this way equals $\mu m$, and let $\mathcal{R}$ be the event that for each literal $l$ with $\sigma(l) = 1$ at least one clone gets shaped diamond. Then the number of ways of choosing $\mu m$ clones out of the $(1 - \nu)dn$ dotted clones such that for each literal $l$ with $\sigma(l) = 1$ at least one clone gets chosen is $\binom{(1-\nu)dn}{\mu m}\mathbb{P}(\mathcal{R}|\mathcal{B})$. In order to compute $\mathbb{P}(\mathcal{R}|\mathcal{B})$, one can use Bayes' rule. Indeed, it is possible to pick $q$ such that $\mathbb{P}(\mathcal{B}|\mathcal{R}) \sim 1$ (it has exponent 0), and $\mathbb{P}(\mathcal{B}) = \mathbb{P}(\mathrm{Bin}((1 - \nu)nd, q) = \mu m)$, and $\ln(\mathbb{P}(\mathcal{R})) \sim (1 - \nu)n \ln(1 - (1 - q)^d)$. So by plugging all these in and finding the right value for $q$ we get

$$\frac{1}{n} \ln \mathbb{E} N_{\mu,\nu} \sim H(\nu) + (1 - \nu) \ln 2 + (1 - \nu)H\left(\frac{2\mu}{(1 - \nu)k}\right) + (1 - \nu) \ln(1 - (1 - q_\mathrm{r})^d)$$

$$+ d \cdot D_{KL}\left(\frac{2\mu}{(1 - \nu)k}, q_\mathrm{r}\right),$$

where $q_r$ is the solution of

$$\frac{q}{1 - (1 - q)^d} = \frac{2\mu}{(1 - \nu)k}.$$

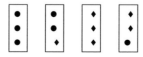

We now intend to use the second-moment method on $\Sigma_{\mu,\nu}$. Similar to what we did before, the idea is to look at pairs of covers, write down the probability that both are SAT in terms of their distance, and then compute the maximum and show that it occurs in the "uncorrelated" case. Let us recall that the main motivation behind the definition of covers was to reduce every cluster of satisfying assignments to a single cover.

Let us recall that $\Sigma_{\nu,\mu} = \#(\nu,\mu)$-covers, and $\Sigma'_{\nu,\mu} = \#$valid $(\nu,\mu)$-covers. To show $\mathbb{E}\Sigma'_{\nu\mu} \sim \mathbb{E}\Sigma_{\nu\mu}$ (if $\nu \sim 2^{-1-k}$), we use the "planted cover model." In this model, we choose a random formula (permutation) conditioned on the event that $\sigma$ is a cover. As a consequence of Bayes' formula and the linearity of expectation, it turns out that it is enough to prove $\mathbb{P}(\sigma \text{ valid} | \sigma \text{ is a cover}) \sim 1$. To do this, we need to replace the $\star$s that the cover corresponding to $\zeta$ assigns to some of the variables by actual truth values, without leaving any clauses unsatisfied. First, we observe that we do not need to worry about the critical clauses of the planted formula. Indeed, any critical clause consists of $k-1$ square and one diamond clone, and thus the underlying variables are assigned either 0 or 1 under $\sigma$ already. So, to summarize, for $k=3$, we can write for example

$$\underbrace{(\blacklozenge,\blacksquare,\blacksquare), (\blacktriangle,\bullet,\blacksquare), (\bullet,\bullet,\blacksquare),}_{\text{do not worry}} \underbrace{(\blacktriangle,\blacktriangle,\blacksquare)}_{\text{worry}} . \tag{7.34}$$

We only need to worry about square/triangle clauses, each containing two or more triangles (and most of them contain precisely two).

The idea is that, out of each square/triangle clause, we pick two triangles randomly. This yields a 2-SAT problem. The objective is then to assign the $\nu n$ triangle variables in a way that this 2-SAT formula is satisfied. The 2-SAT problem is solvable in polynomial time. For instance, one can determine whether there is a satisfying assignment as follows. From the formula, we construct a directed graph whose nodes represent the literals. Each 2-clause is represented by two directed edges: $l_1 \vee l_2$ adds the two edges $(\bar{l}_1 \rightarrow l_2)$ and $(\bar{l}_2 \rightarrow l_1)$). The formula is then satisfiable iff there is no contradictory cycle in the graph (a directed cycle that contains both $x$ and $\bar{x}$). Because the 2-SAT instance that results from the planted cover model is extremely sparse w.h.p. (i.e., it contains far more variables than clauses), standard arguments can be used to show that it does not feature such a cycle w.h.p.

**"Quiet planting":** As an application of the second-moment argument, we extend the "quiet planting" trick from Achlioptas and Coja-Oghlan (2008) to covers. For the present purpose, let us assume that $d, k$ are such that $\mathbb{P}(\ln \Sigma \sim \ln \mathbb{E}\Sigma) \overset{n \to \infty}{\to} 1$; this does not quite follow from the second-moment argument, but a similar statement does (details omitted). We consider the following two sampling processes:

The usual Boltzmann distribution of a random formula and a random cover of it is as follows:

1. Sample a random $\Phi$.
2. Sample a random cover $\sigma$ from the formula, and the output is $(\Phi, \sigma)$.

The "planted" distribution is as follows:

1. Sample a $\sigma : V \to \{0, 1, \star\}$ randomly.
2. Sample a random $\Phi$ such that $\sigma$ is a cover for $\Phi$, and the output is $(\Phi, \sigma)$.

**Lemma 7.8** *Assume that* $\mathbb{P}(\ln \Sigma \sim \ln \mathbb{E}\Sigma) \overset{n \to \infty}{\to} 1$. *Let A be a property of pairs* $(\Phi, \sigma)$. *Then*

$$\mathbb{P}_{\mathrm{Boltz}}[(\Phi, \sigma) \in A | \ln \Sigma \sim \ln \mathbb{E}\Sigma] \le \exp(o(n))\mathbb{P}_{\mathrm{planted}}[(\Phi, \sigma) \in A].$$

This lemma allows us to prove properties of a typical cover of the random formula (i.e., about the Boltzmann distribution) by way of the planted model. For instance, one could use the lemma to determine the cluster size of a typical cover.

## 7.3   Notes and references

For general background on the second-moment method, we refer to the beautiful article of Achlioptas *et al.* (2005). There is a substantial literature on random $k$-SAT; see Coja-Oghlan (2013*a*) and references therein. For results from and references to the physics literature on the problem, we refer Mézard and Montanari (2009) and Krzakala *et al.* (2007).

Theorem 7.1 was first obtained in Rathi *et al.* (2010) via a second-moment argument. However, the details of the proof in Rathi *et al.* (2010) are very different from the approach presented here. More precisely, instead of crafting a probability space in which the clauses can be treated independently, Rathi *et al.* (2010) based their proof on generating functions. The second-moment argument presented in Section 7.1 is actually a special case of the approach developed in Coja-Oghlan and Panagiotou (2013) for $k$-SAT formulas with Poisson degree distributions.

The material of Section 7.2 stems from Coja-Oghlan (2013*b*). The definition of "covers" used here appears in Maneva and Sinclair (2008), although the concept is at least implicit in the physics literature on random communicating sequential processes (CSPs). Furthermore, a similar approach as in 7.2 has been developed independently for the independent set problem and the NAESAT problem (Ding *et al.*, 2013*a*,*b*).

## References

Achlioptas, D. and Coja-Oghlan, A. (2008). Algorithmic barriers from phase transitions. In *Proceedings of IEEE Symposium on Foundations of Computer Science*, Philadelphia, PA, pp. 793–802.

Achlioptas, D., Naor, A., and Peres, Y. (2005). Rigorous location of phase transitions in hard optimization problems. *Nature*, **435**, 759–764.

Coja-Oghlan, A. (2013*a*). The asymptotic $k$-SAT threshold. arXiv:1310.2728v2.

Coja-Oghlan, A. (2013*b*). Random regular $k$-SAT. arXiv:1310.2728v1.

Coja-Oghlan, A. and Panagiotou, K. (2013). Going after the $k$-SAT threshold. In *Proceedings of ACM Symposium on Theory of Computing*, pp. 705–714.

Ding, J., Sly, A., and Sun, N. (2013*a*). Maximum independent sets on random regular graphs. arXiv:1310.4787.

Ding, J., Sly, A., and Sun, N. (2013*b*). Satisfiability threshold for random regular NAE-SAT. arXiv:1310.4784.

Flajolet, F. and Sedgewick, R. (2009). *Analytic Combinatorics*. Cambridge University Press.

Krzakala, F., Montanari, A., Ricci-Tersenghi, F., Semerjian, G., and Zdeborova, L. (2007). Gibbs states and the set of solutions of random constraint satisfaction problems. *Proceedings of the National Academy of Sciences*, **104**, 10318–10323.

Maneva, E. and Sinclair, A. (2008). On the satisfiability threshold and clustering of solutions of random 3-SAT formulas. *Theoretical Computer Science*, **407**, 359–369.

Mézard, M. and Montanari, A. (2009). *Information, Physics, and Computation*. Oxford University Press.

Rathi, V., Aurell, E., Rasmussen, L., and Skoglund, M. (2010). Bounds on threshold of regular random $k$-SAT. In *Theory and Applications of Satisfiability Testing* (ed. O. Strichman and S. Szeider), pp. 264–277. Springer.

# 8

# Local algorithms for graphs

## David GAMARNIK

MIT Sloan School of Management, E62-563
100 Main Street
Cambridge, MA 02139

*Lecture notes taken by*
Mathieu Hemery, Université de Grenoble-Alpes, France
Samuel Hetterich, University of Frankfurt, Germany

*Statistical Physics, Optimization, Inference, and Message-Passing Algorithms.* First Edition.
F. Krzakala et al. © Oxford University Press 2016. Published in 2016 by Oxford University Press.

# Chapter Contents

## 8.1 Introduction

We are going to analyze local algorithms over sparse random graphs. These algorithms are based on local information, where "local" here is with regard to a decision made by exploration of a small neighborhood of a certain vertex plus a belief about the structure of the whole graph and perhaps some added randomness.

### 8.1.1 The independent set problem

Given a graph $G$ with vertex set $V$ and edge set $E$, we say that a set $I \subset V$ is an independent set (i.s.) of $G$ if for all $v, v' \in G$, we have $(v, v') \notin E$. Let $\mathcal{I}_G = \{I \subset V | I$ is an independent set of $G\}$ be the set of all independent sets of $G$. We can think of an algorithm that counts the number of all independent sets of $G$ or we could ask for

$$\max_{I \in \mathcal{I}_G} |I|,$$

the largest independent set of $G$. If $G$ is a weighted graph such that we have a weight $w(v)$ on every vertex $v \in V$, we can ask for

$$\max_{I \in \mathcal{I}_G} w(I),$$

the independent set of $G$ with the largest weight $w(I) = \sum_{v \in I} w(v)$. Later in this chapter, we are going to see an algorithm that counts the number of independent sets $|\mathcal{I}_G|$ for a special class of graphs.

It is still surprising that for a large class of problems, local algorithms provide good prediction and a solution that is nearly optimal, or even the best that we could hope to achieve.

Let us now focus on the problem of counting the total number of independent sets of a graph and denote this number by $Z$. In the case of sparse graphs, this is exponential in the number of vertices. Let us also introduce a parameter $\lambda > 0$ and a partition function

$$Z_\lambda = \sum_{I \in \mathcal{I}_G} \lambda^{|I|}$$

such that $Z_\lambda = Z$ when $\lambda = 1$. In this case, we just write $Z_G$ for the number of all independent sets of $G$. In statistical physics, this is known as the partition function of the *hard-core model* with parameter $\lambda$.

Let us now fix $\lambda = 1$ and let $G$ be a graph with bounded degree such that the degree of every vertex of $G$ has degree less than or equal to $d$. If we choose an independent set $I$ of $G$ uniformly at random out of $\mathcal{I}_G$ (which is algorithmically highly nontrivial), we compute $\Pr[v \notin I]$, the probability that a certain vertex $v \in V$ is not in $I$. It is easily verified that

$$\Pr_G[v \notin I] = \frac{Z_{G-v}}{Z_G} = \frac{\#\text{independent sets without } v}{\#\text{independent sets}}. \tag{8.1}$$

This is equivalent to

$$Z_G = \Pr_G[v \notin I]^{-1} \cdot Z_{G-v}. \tag{8.2}$$

By ordering the vertices in $V$ with $|V| = n$, we have the recursion

$$Z_G = \Pr_G[v_1 \notin I] \cdot Z_{G-v_1} = \Pr_{G-v_1}[v_1 \notin I]^{-1} \cdot \Pr[v_1 \notin I]^{-1} \cdot Z_{G-v_1-v_2} \tag{8.3}$$

$$= \ldots = \prod_i \Pr_{G_{i-1}}[v_i \notin I]^{-1}, \tag{8.4}$$

where $G_i = G - v_1 - v_2 - \ldots - v_i$. Notice that this quantity is invariant under a change in the order of the vertices that are deleted. By estimating this probability, we gain an estimate of the partition function. We will see that this can be done exactly for a certain class of graphs by just exploring a small neighborhood in the thermodynamic limit as $n$ tends to infinity and gives a good approximation for finite $n$.

### 8.1.2 Hosoya index

A second example to which we can adapt local algorithms for estimating the partition function, and which is defined analogously to independent sets, is the problem of counting the number of matchings on a graph, also known as the Hosoya index. A matching is a set $M \subset E$ such that for any two edges $(v_1, v_2), (v_3, v_4) \in M$, all $v_i$ are pairwise-different. The considered counting problem is in $\#P$ for a sparse random graph.

Let us define the girth length $g(G)$ of a graph as the length of the shortest cycle of $G$. We say that a sequence of graphs $G_n$ with bounded maximal degree $d$ is locally tree-like if $g(G_n)$ tends to infinity as $n$ grows. Notice that sparse random regular graphs are only approximately locally tree-like because of the occurrence of a constant number of short cycles.

## 8.2 Power of the local algorithms

### 8.2.1 Theorem

Let $G_n$ be a sequence of a $d$-regular locally tree-like graphs of $n$ nodes. If

$$\lambda \leq \lambda_c(d) = \frac{(d-1)^{d-1}}{(d-2)^d}, \tag{8.5}$$

then we have

$$\lim_{n \to \infty} \frac{\log Z_{G_n}}{n} = -\log\left(p^{d/2}(2-p)^{(d-2)/2}\right) \tag{8.6}$$

for $d = 2, \ldots, 5$ where $p$ is the unique solution of $p = (1 + \lambda p^{d-1})^{-1}$.

## 8.2.2 Remark

- The statement is probably wrong for $d > 5$.
- The statement can be generalized to sparse random regular graphs.
- We cannot hope to get such a general bound for the maximal independent set, since we have an upper bound on it for random $d$-regular graphs that is $0.459n$, and trivially the maximal i.s. of a random $d$-regular bipartite graph is at least $0.5n$. We conclude that the considered class of graphs in Theorem 8.2.1 is too broad to get a similar result on maximal independent sets.
- Theorem 8.2.1 can also be stated for matchings and without the restrictions on $\lambda$ and $d$.

## 8.2.3 Proof

Consider a rooted $d$-regular tree where the root denoted by $v$ has $d$ children and every child has again $d$ children, as depicted in Fig. 8.1. If this tree has depth $t$, it is uniquely defined and we refer to it as $T_{d,t}$. We introduce the shorthand $Z_{T_{d,t}} = Z_t$ if $d$ is specified. Let us denote by $Z_G[B]$ the number of independent sets that satisfy condition $B$. We find that

$$Z_t = Z_t[v \notin I] + Z_t[v \in I] = Z_t[v \notin I] + (Z_{t-1}[v' \notin I])^d, \tag{8.7}$$

where $v'$ is a child of $v$, which is equivalent to

$$\frac{Z_t}{Z_t[v \notin I]} = 1 + \frac{(Z_{t-1}[v' \notin I])^d}{Z_t[v \notin I]} = 1 + \left(\frac{(Z_{t-1}[v' \notin I])}{Z_{t-1}}\right)^d. \tag{8.8}$$

We have already interpreted the quantity on the left-hand side as the probability that $v$ is not contained in an independent set of $T_{d,t}$ chosen uniformly at random, which we denote by $P_t$. Thus, we get

$$P_t = (1 + P_{t-1}^d)^{-1}, \tag{8.9}$$

which is a recursion that hopefully converges to a fixed point.

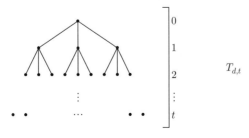

**Fig. 8.1** Sketch of a $d$-regular tree with depth $t$.

We introduce the function $f(x) = (1 - x^d)^{-1}$ defined on $[0, 1]$, which is decreasing. Let us analyze the second iteration of $f$, namely, $f^{(2)} = f \circ f$, which is clearly increasing on $[0, 1]$ and well defined since $f([0, 1]) \subset [0, 1]$. Thus, we get

$$0 \le f^{(2)}(0) \le f^{(4)}(0) \le f^{(6)}(0) \le \ldots \le f^{(2n)}(0) \le \ldots \to x_*, \tag{8.10}$$

$$1 \ge f^{(2)}(1) \ge f^{(4)}(1) \ge f^{(6)}(1) \ge \ldots \ge f^{(2n)}(1) \ge \ldots \to x^*. \tag{8.11}$$

If we are lucky, we get $x_* = x^*$ having a unique fixed point in $[0, 1]$. To check this, we plot the monotonically increasing function $f^{(2)}$ and check how many times it intersects the identity. If this happens only once, then we have a unique fixed point. It turns out that this is the case for $d = 1, \ldots, 4$ and analogously if $\lambda \le \lambda_c(d) = (d-1)^{d-1}/(d-2)^d$. If it exists, let us define the unique fixed point as $\lim_{t \to \infty} P_t = P^*$.

Consider the tree $T_{d,t}$ and define any boundary condition $B$ by fixing every vertex of depth $t$ to be either in an independent set or outside it. Let $B_{\text{in}}$ (respectively, $B_{\text{out}}$) be the event that all vertices of depth $t$ are in (respectively, not in) the independent set. It turns out that we have, for any boundary condition $B$,

$$\Pr[v \notin I | B] \le \max \left\{ \Pr[v \notin I | B_{\text{in}}], \Pr[v \notin I | B_{\text{out}}] \right\}. \tag{8.12}$$

Moreover, we have either

$$\Pr[v \notin I | B_{\text{out}}] \le P_{t-1} \tag{8.13}$$

or

$$\Pr[v \notin I | B_{\text{in}}] \le P_{t-2}, \tag{8.14}$$

and thus $P^* \in [\min \{P_{t-1}, P_{t-2}\}, \max \{P_{t-1}, P_{t-2}\}]$. On the whole graph, we obtain

$$\Pr_G[v \notin I] = \sum_B \Pr_t[v \notin I | B] \cdot \Pr[B|G] \sim P^* \sum_B \Pr(B|G) = P^*, \tag{8.15}$$

which is true as long as we do not have any loops.

We now have two final corrections to complete the proof. The first is a small one. In a regular tree, the root has a degree $d - 1$, while all the other nodes have degree $d$. But this is easily corrected by introducing

$$P_{d-1}^* = \frac{1}{1 + P_{d-1}^{* \, d-1}} \tag{8.16}$$

for the first recursion step.

The other problem is more serious. We want to use the recursive formula $Z_G = p^{-1} Z_{G-v}$ by removing the nodes one by one, but if we do so, $G - v$ is no longer a regular graph. The trick is to remove two vertices $u$ and $v$, and to connect their children in order to preserve the regularity property. However, we must be careful: $u$ and $v$ must be chosen sufficiently far away in order to preserve the local tree structure as well.

We write $G_{n-2} = G_n - u - v + e_1 + \ldots + e_d$, where $\{e_i\}$ denotes the new link between the children of $u$ and $v$, and $H = G_n - u - v$. We want to find the relation between $Z_{G_n}$ and $Z_{G_{n-2}}$ as a function of $p$. We have $Z_{G_n} = (2-p)^2 Z_H$. So let us look at the effect of adding a new link, $e_1$ for example:

$$\frac{Z_{H+e_1}}{Z_H} = \frac{\#\text{indep. sets without } i \text{ and } j}{\#\text{indep. sets in } H} \simeq 1 - (1-p)^2, \tag{8.17}$$

because of the independence of $u$ and $v$, which ensures that $\Pr(u, v \in I) = \Pr(u \in I)\Pr(v \in I) = (1-p)^2$.

When we repeat the operation $d$ times, we have

$$\frac{Z_{G_{n-2}}}{Z_H} = \frac{H + e_1 + \ldots + e_d}{H} = (1 - (1-p)^2)^d = (2p - p^2)^d, \tag{8.18}$$

and, at the end of the day,

$$\frac{Z_{G_n}}{G_{n-2}} = \frac{1}{p^d(2-p)^{d-2}}. \tag{8.19}$$

As this formula is true $n/2 - o(n)$ times, we can write $Z_{G_n} \sim (p^d(2-p)^{d-2})^{-n/2}$; that is,

$$\frac{\log Z_{G_n}}{n} \to -\log\left(p^{d/2}(2-p)^{(d-2)/2}\right). \tag{8.20}$$

## 8.3 Incursion in a loopy land

We are now interested in the case where $G$ is no longer locally tree-like or regular. We just ask that the degree of $G$ be bounded; that is, $\forall i \in V$, $\deg(i) \leq D$, where $\deg(i)$ denotes the degree of the node $i$ and $D$ is an integer.

We still try to compute the partition function $Z_\lambda$, but we will set by default $\lambda = 1$ and use the notation $Z(G) = Z_G(1)$, to emphasize the graph on which we are working. Of course, the formula

$$P_G(u \notin I) = \frac{Z(G-u)}{Z(G)} \tag{8.21}$$

still holds. We denote by $v_0$ the current node and $v_1, \ldots, v_d$ ($d = \deg(v_0)$) its neighbors and we define $G_i = G - v_0 - \ldots - v_i$.

**Exercise:** Check that the formula

$$P_G(v_0 \notin I) = \frac{1}{1 + \displaystyle\prod_{i=0}^{d} P_{G_{i-1}}(v_0 \notin I)} \tag{8.22}$$

is correct.

This defines the computational tree for which $G$ is the root and $G_i$ the children. On this tree, we compute on each node the probability that $v_i$ is outside the independent set. And we can prove that there exist some values of the parameters $\lambda$ and $D$ for which the correlations are decreasing on the computational tree. Indeed, this is a corollary of the contraction property, which states that for a function $f(x_1, \ldots, x_n) = (1 + \lambda x_1 \cdots x_n)^{-1}$, if $\tilde{x}_i$ is an approximation of $x_i$ up to $\delta$, then $f(\tilde{x}_1, \ldots, \tilde{x}_n)$ is also an approximation of $f(x_1, \ldots, x_n)$ up to $\rho\delta$, where $\rho \leq 1$.

## 8.4    The monomer–dimer entropy

Let $B_n^d$ be the regular hypercube of size $n$ and dimension $d$. It can be seen as a graph with $n^d$ vertices. The monomer–dimer entropy is obtained from the number of partial matchings that can be found in such a graph. As usual, $Z(B_n^d)$ is the value of the partition function for $\lambda = 1$. And if we do not care about assigning energies to the different configurations, the entropy is defined as $H = \log\left(Z(B_n^d)\right)$, which grows as $n^d$. It is tempting to define an intensive entropy, and so, by taking the limit $n \to \infty$, we propose

$$h(d) = \lim_{n \to \infty} \frac{\log\left(Z(B_n^d)\right)}{n^d} \tag{8.23}$$

as the entropy of the monomer–dimer model.

How can we compute such a quantity? A physicist will propose the use of a transfer matrix, and this has been done for $d = 3$, giving the bounds

$$0.785 \leq h(3) \leq 0.786. \tag{8.24}$$

But it can be shown that when $h$ is computed with an error $\epsilon$, it scales as $\exp\left(o(1/\epsilon)^{d-1}\right)$, which rapidly becomes intractable.

By using the cavity method on the graph, adding one vertex at a time, the bounds have been increased to

$$0.78595 \leq h(3) \leq 0.785976. \tag{8.25}$$

**Remark:** It is far more difficult to compute the dimer model corresponding to perfect matching than the monomer–dimer model, which by allowing gaps between the edges corresponds to a partial matching only.

## 8.5    Limits of local algorithms

To provide insight into the limits of what a local algorithm can do, we shall focus on the case of random graphs, in particular on the Erdős–Rényi graph $G(n, d/n)$ and the regular graph $G_d(n)$. We remain in the independent set problem, but are now trying to find the maximal independent set $I_n^\star$.

Of course, this quantity scales linearly with $n$, and so, for the Erdős–Rényi graph, we introduce

$$\alpha(d) = \lim_{n \to \infty} \frac{I_n^\star}{n}. \tag{8.26}$$

It can be shown that

$$\alpha(d) \sim \frac{2 \log d}{d}. \tag{8.27}$$

But it seems that the local algorithms always get stuck around the threshold $t_d = (\log d)/d$, which gives us a factor 2 of improvement. Let us describe a very naive algorithm, the so-called "greedy algorithm": choose a random node $v$ in $G$ and put it into the independent set $I$, then erase all of its neighbors and cycle as long as there exists a node to put into $I$. This (stupid) algorithm is always stuck around $t_d$. But no local algorithm can do better!

Let us define more precisely the notion of "local" algorithm.

Let $G$ be a regular random graph of degree $d$. A "local" algorithm can always be represented by a function $f_r : [0, 1]^{|T_{d,r}|} \to \{0, 1\}$ and a realization of the algorithm is given by a set of weight $W = (w_1, \ldots, w_n)$ based on the nodes of the graph, where $T_{d,r}$ is the regular tree of degree $d$ and depth $r$.

$f_r$ should be seen as a decision function based on the neighborhood of each node, which is locally tree-like up to a depth $r$. It then takes a value of 0 or 1 depending on whether the node is in the independent set or not.

Can such a function give a correct realization of the independent set problem? Obviously, the answer is yes. To see this, let $r = 1$ and let $f_r$ be 1 if the weight of the current node is higher than that of all its neighbors and 0 otherwise. This defines a valid independent set, even if it is certainly not the maximal one.

The greedy algorithm can also be defined in term of such a function. For example, $W$ represents the order in which the nodes are chosen by the greedy algorithm. When running $f_r$ on the graph, it is necessary to make sure that no neighbors have been checked before the current node. If this is the case, then everything is satisfactory, but if another node could have been checked before, then this new node must also be checked, and so on and so forth. It is not obvious that this process will come to an end before the whole graph has been explored. However, it should be remarked here that each node increases slightly the value $w$ that the neighbor should beat to be checked. This increasing value ensures that the number of paths goes as $d^r/r!$, where $r = (2 \log n)/(\log \log n)$, and so remains finite. This kind of property is often referred to as influence resistance.

So, why is not possible that such a local algorithm does better than $t_d$?

The analysis shows that above this threshold, independent sets satisfy a clustering property. Roughly speaking, this means that for $I$ and $J$ two different independent sets on $G$, either $|I \cup J|$ is small or it is large.

Now let $f_r$ be the perfect local algorithm such that it produces an independent set and $\alpha(f_r) \sim \alpha(d)$. We can run it twice on the graph $G$ and obtain $f_r(U) = I$ and $f_r(V) = J$, two different independent sets on $G$. As these are two independent realizations, we have $|I \cup J| \sim \alpha(f_r)^2 = o(1/d^2)$, which is a very tiny fraction of the graph, but we can generate a family of such independent sets by running the function successively on the ensemble $W_t = (u_1, \ldots, u_t, v_{t+1}, \ldots, v_n)$. And as $f_r$ depends only on local variables to make its decision, this also means that the successive results $I_t = f_r(W_t)$ differ only locally, which means $|I_t \diamond I_{t+1}| = o(1)$ (where $I \diamond J$ denotes the nodes that change between $I$ and $J$) but also implies that for a value of $t$, we must have $|I_t \cup I|$ falling into the forbidden area, which is impossible according to the clustering property.

# 9
# Expectation propagation

## Manfred OPPER

Department of Artificial Intelligence,
Technische Universität Berlin,
Marchstraße 23, Berlin 10587, Germany

*Lecture notes taken by*
Andre Manoel, Universidade de São Paulo, Brazil
Jack Raymond, La Sapienza Roma, Italy

*Statistical Physics, Optimization, Inference, and Message-Passing Algorithms.* First Edition.
F. Krzakala et al. © Oxford University Press 2016. Published in 2016 by Oxford University Press.

# Chapter Contents

## 9.1 Introduction

This chapter was prepared in October and November 2013. We follow the structure of the original presentation, with the second lecture beginning at Section 9.5. All figures and topics from the original presentation slides are included and the general style of the lectures is followed. For brevity and continuity, we do not attempt to give a detailed description of every application, providing references instead. We have tried to minimize backward references—repeated concepts such as the *tilted distribution* and *Gaussian prior* are redefined locally in most cases. Finally, so as to focus on the important parameter that is the subject of approximation, we frequently omit data argument from the likelihood or the posterior.

Wherever possible, we provide references to the original work discussed. We have also included at our discretion some additional references, in particular the book by Wainwright and Jordan (2008) provides a very elegant introduction to the topics discussed in the first lecture, and the beginning of the second lecture.

## 9.2 Approximate variational inference

We are interested in computing the statistics of hidden variables $\boldsymbol{x} = (x_1, x_2, \ldots, x_n)$ given the observed data $\boldsymbol{y} = (y_1, y_2, \ldots, y_m)$ and a generative model relating $\boldsymbol{x}$ to $\boldsymbol{y}$, specified by the joint distribution $p(\boldsymbol{x}, \boldsymbol{y})$. For this purpose, it is convenient to work with the posterior distribution $p(\boldsymbol{x}|\boldsymbol{y})$, which from Bayes' theorem may be written as

$$p(\boldsymbol{x}|\boldsymbol{y}) = \frac{p(\boldsymbol{x}, \boldsymbol{y})}{p(\boldsymbol{y})}, \tag{9.1}$$

and then focus on one or more of the following tasks (Wainwright and Jordan, 2008):

- determining the mode of the posterior, $\mathrm{argmax}_{\boldsymbol{x}} \, p(\boldsymbol{x}|\boldsymbol{y})$, which gives us the most probable assignment of $\boldsymbol{x}$ for an instance of $\boldsymbol{y}$;
- obtaining the marginals $p(x_i|\boldsymbol{y}) = \int \mathrm{d}\boldsymbol{x}_{\sim i} \, p(\boldsymbol{x}|\boldsymbol{y})$, which allows us to compute $x_i$ moments and estimators ($\boldsymbol{x}_{\sim i}$ denotes the set of variables except $x_i$);
- computing the model evidence $p(\boldsymbol{y}) = \int \mathrm{d}\boldsymbol{x} \, p(\boldsymbol{x}, \boldsymbol{y})$, which may be used for model selection as well for computing the moments of sufficient statistics in exponential families, where $p(\boldsymbol{y})$ is also known as partition function.

Both the marginals and evidence computations involve high-dimensional sums or integrals and are often intractable. A *variational approximation* (Bishop, 2007; Barber, 2011; Murphy, 2012) consists in replacing $p(\boldsymbol{x}|\boldsymbol{y})$ by $q(\boldsymbol{x}) \in \mathcal{F}$, where $\mathcal{F}$ is a tractable[1] family of distributions, so as to minimize any desired measure between

---

[1] For which computing the marginals or the evidence may be performed exactly and in polynomial time in $n$.

$p(\boldsymbol{x}|\boldsymbol{y})$ and $q(\boldsymbol{x})$. In particular, KL variational approximations pick $q(\boldsymbol{x})$ so as to minimize the following Kullback–Leibler divergence:

$$\mathrm{KL}\left[q(\boldsymbol{x})\,\|\,p(\boldsymbol{x}|\boldsymbol{y})\right] = \int \mathrm{d}\boldsymbol{x}\, q(\boldsymbol{x}) \log \frac{q(\boldsymbol{x})}{p(\boldsymbol{x}|\boldsymbol{y})}$$

$$= \underbrace{-\mathbb{E}_q \log p(\boldsymbol{x}, \boldsymbol{y})}_{E[q]} + \underbrace{\int \mathrm{d}\boldsymbol{x}\, q(\boldsymbol{x}) \log q(\boldsymbol{x})}_{-S[q]} + \log p(\boldsymbol{y}) \geq 0. \quad (9.2)$$

Such approximations correspond to the so-called mean-field theories of statistical physics (Opper and Saad, 2001), where the quantities $E[q]$ and $S[q]$ assume the roles of energy and entropy, respectively. By introducing a variational free energy $F[q] = E[q] - S[q]$, we recover the known bound $-\log p(\boldsymbol{y}) \leq F[q]$.

### 9.2.1   Different families/approximations

The next step is to pick a family of distributions. For a fully factorized distribution $q(\boldsymbol{x}) = \mathcal{Z}^{-1} \prod_{i=1}^n q_i(x_i)$, the $\{q_i^*\}$ that minimize $F[q]$ are given by

$$q_j^*(x_j) = \frac{1}{z_j} \exp\left\{\mathbb{E}_{q_{\sim j}^*} \log p(\boldsymbol{x}, \boldsymbol{y})\right\}, \quad (9.3)$$

where $\mathbb{E}_{q_{\sim j}^*}$ denotes the average over all $\{q_i^*\}$ distributions except $q_j^*$. This choice defines the *naive mean-field* approximation. It may be applied to both discrete and continuous models; in particular, for the Ising model

$$\mathbb{E}_{q_{\sim j}} \log p(\boldsymbol{x}, \boldsymbol{y}) = x_j \left( \sum_{k \in \mathcal{N}(j)} J_{jk}\langle x_k \rangle + B_j \right), \quad (9.4)$$

and the familiar expression

$$\langle x_j \rangle = \tanh\left( B_j + \sum_k J_{jk}\langle x_k \rangle \right) \quad (9.5)$$

is recovered. The weak point in this approach is that it neglects dependences between variables. Linear response corrections are possible, and can yield nonzero estimates of correlations correct at leading order, but have themselves some weaknesses.

A Gaussian approximation $q(\boldsymbol{x}) \propto \exp\left\{-\frac{1}{2}(\boldsymbol{x} - \boldsymbol{\mu})^{\mathsf{T}}\Sigma^{-1}(\boldsymbol{x} - \boldsymbol{\mu})\right\}$ leads equivalently to

$$\mathbb{E}_{q^*}\left\{\frac{\partial \log p(\boldsymbol{x}, \boldsymbol{y})}{\partial x_i}\right\} = 0, \qquad \mathbb{E}_{q^*}\left\{\frac{\partial^2 \log p(\boldsymbol{x}, \boldsymbol{y})}{\partial x_i \partial x_j}\right\} = \left(\Sigma^{-1}\right)_{ij}. \quad (9.6)$$

However, discrete models may not be considered in this case, since for the KL divergence (9.2) to be well defined, the support of the distribution $p(x|y)$ must always include that of $q(x)$.

Slightly different approaches lead to other popular approximations. For instance, for $q(\boldsymbol{x}) = \prod_{ij} q_{ij}(x_i, x_j) / \prod q_i^{d_i - 1}(x_i)$, minimizing $F[q]$ while taking into account normalization and consistency constraints leads to the loopy belief propagation algorithm (Yedidia *et al.*, 2000).

## 9.3 Expectation propagation

We could also consider the minimization of the reverse KL divergence $\mathrm{KL}[p(\boldsymbol{x}|\boldsymbol{y}) \,\|\, q(\boldsymbol{y})]$ instead of $\mathrm{KL}[q(\boldsymbol{y}) \,\|\, p(\boldsymbol{x}|\boldsymbol{y})]$; however, this problem is intractable in general, since we would have to compute averages with respect to $p(\boldsymbol{x}|\boldsymbol{y})$. Let us consider $q(\boldsymbol{x})$ distributions belonging to the exponential family

$$q(\boldsymbol{x}) = h(\boldsymbol{x}) \exp\left\{\boldsymbol{\theta}^{\mathsf{T}}\boldsymbol{\phi}(\boldsymbol{x}) + g(\boldsymbol{\theta})\right\} \tag{9.7}$$

of natural parameters $\boldsymbol{\theta}$ and sufficient statistics $\boldsymbol{\phi}(\boldsymbol{x})$; for example, a multivariate normal distribution has $\boldsymbol{\theta} = \left(\Sigma^{-1}\boldsymbol{\mu}, -\frac{1}{2}\Sigma^{-1}\right)$ and $\boldsymbol{\phi}(\boldsymbol{x}) = (\boldsymbol{x}, \boldsymbol{x}\boldsymbol{x}^{\mathsf{T}})$. For this family, the minimization of the reverse KL divergence results in

$$\nabla_{\boldsymbol{\theta}} \mathrm{KL}[p(\boldsymbol{x}|\boldsymbol{y}) \,\|\, q(\boldsymbol{y})] = -\int \mathrm{d}\boldsymbol{x}\, p(\boldsymbol{x}|\boldsymbol{y})\boldsymbol{\phi}(\boldsymbol{x}) - \underbrace{\int \mathrm{d}\boldsymbol{x}\, p(\boldsymbol{x}|\boldsymbol{y})\nabla_{\boldsymbol{\theta}}g(\boldsymbol{\theta})}_{\mathbb{E}_p\nabla_{\boldsymbol{\theta}}g(\boldsymbol{\theta})=-\mathbb{E}_q\boldsymbol{\phi}(\boldsymbol{x})}$$

$$= -\mathbb{E}_p\boldsymbol{\phi}(\boldsymbol{x}) + \mathbb{E}_q\boldsymbol{\phi}(\boldsymbol{x}) = 0, \tag{9.8}$$

that is, in moment matching of the sufficient statistics $\boldsymbol{\phi}(\boldsymbol{x})$; the problem will thus be tractable whenever the moments with respect to $p(\boldsymbol{x}|\boldsymbol{y})$ may be efficiently computed.

### 9.3.1 The assumed density filtering algorithm

If we assume observed data to arrive sequentially, $\mathcal{D}_{t+1} = \{y_1, y_2, \ldots, y_{t+1}\}$, we may incorporate the new measurement to the posterior, at each step, by means of Bayes' rule $p(\boldsymbol{x}|\mathcal{D}_{t+1}) \propto p(y_{t+1}|\boldsymbol{x})\, p(\boldsymbol{x}|\mathcal{D}_t)$.

In the assumed density filtering (ADF) algorithm, the posterior $p(\boldsymbol{x}|\mathcal{D}_{t+1})$ is replaced, at each step, by the distribution $q_\theta(\boldsymbol{x}) \in \mathcal{F}$ that minimizes $\mathrm{KL}[p(\boldsymbol{x}|\mathcal{D}_{t+1}) \,\|\, q_\theta(\boldsymbol{x})]$. An iteration of the algorithm thus consists in the following:

**Initialize** by setting $q_\theta^{(0)}(\boldsymbol{x}) = p_0(\boldsymbol{x})$.

**Update** the posterior

$$p(\boldsymbol{x}|\mathcal{D}_{t+1}) = \frac{p(y_{t+1}|\boldsymbol{x})\, q_\theta^{(t)}(\boldsymbol{x})}{\int \mathrm{d}\boldsymbol{x}\, p(y_{t+1}|\boldsymbol{x})\, q_\theta^{(t)}(\boldsymbol{x})}. \tag{9.9}$$

**Project** it back to $\mathcal{F}$:

$$q_\theta^{(t+1)}(\boldsymbol{x}) = \underset{q_\theta \in \mathcal{F}}{\arg\min}\, \mathrm{KL}[p(\boldsymbol{x}|\mathcal{D}_{t+1}) \,\|\, q_\theta(\boldsymbol{x})]. \tag{9.10}$$

The key premise is that by minimizing the reverse KL divergence while including the contribution of a single factor of the (intractable) likelihood, we ensure that the problem remains tractable. For a simple example, consider the Bayesian classifier given by $y_t = \mathrm{sgn}\left(h_{\boldsymbol{w}}(s_t)\right) \equiv \sum_j w_j \psi_j(s_t)$, leading to a probit likelihood

$$p(y_t|\boldsymbol{w}, s_t) = \frac{1}{2} + \frac{1}{\sqrt{2\pi}} \int_0^{y h_{\boldsymbol{w}}(s_t)} \mathrm{e}^{-u^2/2}\, \mathrm{d}u. \tag{9.11}$$

The update rule (9.9) in this case yields

$$p(\boldsymbol{w}|y_{t+1}, s_{t+1}) \propto p(y_{t+1}|\boldsymbol{w}, s_{t+1}) \, q_\theta(\boldsymbol{w}), \tag{9.12}$$

and for a Gaussian prior over the weights $p_0(\boldsymbol{w}) \propto \exp\left(-\frac{1}{2}\sum_j w_j^2\right)$ and a parametric approximation $q_\theta(\boldsymbol{w}) = \mathcal{N}(\bar{\boldsymbol{w}}, C)$, the minimization procedure (9.10) is easily accomplished, since $p(y_{t+1}|\boldsymbol{w}, s_{t+1})$ depends on a single Gaussian integral. Results obtained by applying the ADF algorithm to this model are presented in Fig. 9.1.

The weakness in ADF lies in the fact that the final result depends on the order of presentation of data. Trying to fix this issue leads us to the expectation propagation (EP) algorithm.

### 9.3.2 The expectation propagation algorithm

Given all the $m$ measurements $y_\mu$, the posterior may be written as

$$p(\boldsymbol{x}|\boldsymbol{y}) = \frac{p_0(\boldsymbol{x}) \displaystyle\prod_{\mu=1}^{m} p(y_\mu|\boldsymbol{x})}{\displaystyle\int \mathrm{d}\boldsymbol{x}\, p_0(\boldsymbol{x}) \prod_{\mu=1}^{m} p(y_\mu|\boldsymbol{x})} = \frac{1}{\mathcal{Z}} f_0(\boldsymbol{x}) \prod_{\mu=1}^{m} f_\mu(\boldsymbol{x}), \tag{9.13}$$

where we assume the prior $f_0(\boldsymbol{x})$ to be a member of the exponential family. The EP algorithm (Minka, 2001) consists in determining a tractable approximation

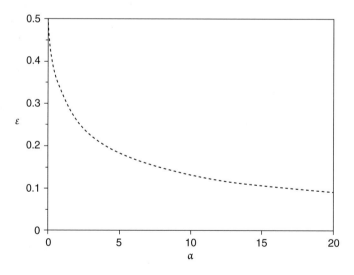

**Fig. 9.1** Generalization error $\varepsilon$ for a Bayesian classifier with $n = 50$ in terms of $\alpha = m/n$. The solid line gives the performance of the ADF algorithm, and the dashed line the Bayes optimal performance obtained via replica calculations. Figure taken from Opper (1989). Reproduced with permission from Cambridge University Press.

$q(\boldsymbol{x}) = \mathcal{Z}^{-1} f_0(\boldsymbol{x}) \prod_{\mu=1}^{m} g_\mu(\boldsymbol{x})$ to $p(\boldsymbol{x}|\boldsymbol{y})$ by repeatedly performing the ADF update (9.9) as if each measurement has never been presented before, effectively by removing its contribution from the current approximation. The whole procedure consists in the following:

**Initialize** by setting $q(\boldsymbol{x}) = f_0(\boldsymbol{x})$ and $g_\mu(\boldsymbol{x}) = 1$, $\mu \in \{1, 2, \ldots, m\}$.

**Repeat** until convergence—pick $\mu \in 1, 2, \ldots, m$ uniformly at random and

    **Remove** $g_\mu(\boldsymbol{x})$ from $q(\boldsymbol{x})$; i.e., build $q_{\sim\mu}(\boldsymbol{x}) \propto q(\boldsymbol{x})/g_\mu(\boldsymbol{x})$.

    **Update** the *tilted* distribution $q_\mu(\boldsymbol{x}) = f_\mu(\boldsymbol{x}) q_{\sim\mu}(\boldsymbol{x})$.

        Note that $q_\mu(\boldsymbol{x}) \propto \big(f_\mu(\boldsymbol{x})/g_\mu(\boldsymbol{x})\big) q(\boldsymbol{x})$ replaces the approximating likelihood $g_\mu(\boldsymbol{x})$ by the exact one $f_\mu(\boldsymbol{x})$, thus taking $q_\mu(\boldsymbol{x})$ outside of the exponential family.

    **Project** $q(\boldsymbol{x})$ back to the exponential family:

$$q^{\text{new}}(\boldsymbol{x}) = \operatorname*{argmin}_{q \in \mathcal{E}} \text{KL}\left[q_\mu(\boldsymbol{x}) \,\|\, q(\boldsymbol{x})\right]. \tag{9.14}$$

    **Refine** the approximated likelihood term

$$g_\mu^{\text{new}}(\boldsymbol{x}) \propto \frac{q^{\text{new}}(\boldsymbol{x})}{q_{\sim\mu}(\boldsymbol{x})} \propto \frac{q^{\text{new}}(\boldsymbol{x})}{q(\boldsymbol{x})} g_\mu(\boldsymbol{x}) \tag{9.15}$$

**At convergence** obtain tractable approximation from $q(\boldsymbol{x}) = f_0(\boldsymbol{x}) \prod_{\mu=1}^{m} f_\mu(\boldsymbol{x})$.

    The approximated posterior $q(\boldsymbol{x})$ and the tilted distributions $q_\mu(\boldsymbol{x}) \propto q(\boldsymbol{x}) \times f_\mu(\boldsymbol{x})/g_\mu(\boldsymbol{x})$ will have matching moments, $\mathbb{E}_q \phi(\boldsymbol{x}) = \mathbb{E}_{q_\mu} \phi(\boldsymbol{x})$ for $\mu \in \{1, 2, \ldots, m\}$.

We assume here that the $\{g_\mu(\boldsymbol{x})\}$ belong to the exponential family; in order to proceed, one needs to choose a distribution from this family, i.e., to specify how $q(\boldsymbol{x})$ is to be factorized. Let us exemplify the procedure by considering a Gaussian latent variable model with $m = n$, $p(\boldsymbol{x}|\boldsymbol{y}) \propto e^{-\frac{1}{2}\boldsymbol{x}^{\mathsf{T}} K \boldsymbol{x}} \prod_{i=1}^{n} p(y_i|\boldsymbol{x})$. We will pick the Gaussian $g_i(x_i) \propto \exp(-\frac{1}{2}\Lambda_i x_i^2 + \gamma_i x_i)$ so that

$$q(\boldsymbol{x}) \propto \exp\left\{-\frac{1}{2}\boldsymbol{x}^{\mathsf{T}} K \boldsymbol{x} - \frac{1}{2}\sum_{i=1}^{n} \Lambda_i x_i^2 + \boldsymbol{\gamma}^{\mathsf{T}} \boldsymbol{x}\right\}. \tag{9.16}$$

The steps described above amount to, at each iteration, removing the $\{\Lambda_i, \gamma_i\}$ terms for a given $i$; computing the marginal $q_i(x_i) = \int d\boldsymbol{x}_{\sim i}\, p(y_i|\boldsymbol{x}) q_{\sim i}(\boldsymbol{x})$ and, from it, the moments $\mathbb{E}_{q_i} x_i$, $\mathbb{E}_{q_i} x_i^2$; and subsequently updating $\{\Lambda_i, \gamma_i\}$ so as to have $\mathbb{E}_q x_i = \mathbb{E}_{q_i} x_i$ and $\mathbb{E}_q x_i^2 = \mathbb{E}_{q_i} x_i^2$.

Empirically, it is verified that EP is a fast algorithm; however, its convergence is not guaranteed. It has the advantage of being applicable to both discrete and continuous variable models, in particular providing remarkable results for Gaussian latent variable models.

### 9.3.3 Relation to BP

It is not hard to show (see, e.g., Murphy, 2012) that by applying the algorithm above to any distribution $p(\boldsymbol{x}|\boldsymbol{y}) \propto f_0(\boldsymbol{x}) \prod_{\mu=1}^{m} f_\mu(\boldsymbol{x})$ with $g_\mu(\boldsymbol{x}) \propto \prod_{i \in \mathcal{N}(\mu)} h_{\mu i}(x_i)$, one recovers the loopy belief propagation algorithm (Minka, 2001), where $\mathcal{N}(\mu)$ are the set of variables coupled through $\mu$. The approximated distribution is given by

$$q(\boldsymbol{x}) \propto f_0(\boldsymbol{x}) \prod_{\mu=1}^{m} \prod_{i \in \mathcal{N}(\mu)} h_{\mu i}(x_i); \tag{9.17}$$

for example, on a pairwise graphical model, one would be replacing likelihoods of the form $f_{ij}(x_i, x_j) = \exp(J_{ij} x_i x_j)$ with $g_{ij}(x_i, x_j) \propto \exp\left(\lambda_{ij}(x_i) + \lambda_{ji}(x_j)\right)$.

By defining $q^{(i)}(x_i) \propto \prod_{\mu \in \mathcal{N}(i)} h_{\mu i}(x_i)$, the removal step for a given $\mu$ leads to

$$q_{\sim\mu}^{(i)}(x_i) \propto \frac{q^{(i)}(x_i)}{h_{\mu i}(x_i)} \propto \prod_{\nu \in \mathcal{N}(i) \setminus \mu} h_{\nu i}(x_i) \qquad \text{for } i \in \mathcal{N}(\mu), \tag{9.18}$$

while the projection step matches the marginals $q_\mu(\boldsymbol{x}) \propto f_\mu(\boldsymbol{x}) \prod_{i \in \mathcal{N}(\mu)} q_{\sim\mu}^{(i)}(x_i)$ and $q(\boldsymbol{x})$, that is,

$$q^{(i)}(x_i) = \sum_{\boldsymbol{x} \sim i} q_\mu(\boldsymbol{x}) \propto \sum_{\boldsymbol{x} \sim i} f_\mu(\boldsymbol{x}) \prod_{i \in \mathcal{N}(\mu)} q_{\sim\mu}^{(i)}(x_i) \qquad \text{for } i \in \mathcal{N}(\mu), \tag{9.19}$$

and at the refine step, finally,

$$h_{\mu i}(x_i) \propto \frac{q^{(i)}(x_i)}{q_{\sim\mu}^{(i)}(x_i)} \propto \sum_{\boldsymbol{x} \sim i} f_\mu(\boldsymbol{x}) \prod_{j \in \mathcal{N}(\mu) \setminus i} q_{\sim\mu}^{(j)}(x_j) \tag{9.20}$$

Comparing these with the BP equations on a factor graph (Wainwright and Jordan, 2008), we can see that (9.17) and (9.20) give the messages from nodes to factors and factors to nodes, respectively, while (9.19) provides approximations to the marginals.

## 9.4 Adaptive TAP

For models with pairwise interactions, the naive mean-field approximation may be improved by means of the linear response correction

$$\frac{\partial \langle x_i \rangle}{\partial B_j} = \langle x_i x_j \rangle - \langle x_i \rangle \langle x_j \rangle, \tag{9.21}$$

where we would compute $\partial \langle x_i \rangle / \partial B_j$ within the current approximation and then set the correlations to $\langle x_i x_j \rangle = \langle x_i \rangle \langle x_j \rangle + \partial \langle x_i \rangle / \partial B_j$. For many interesting families, we can calculate exactly the linear response estimate at leading order, which is sufficient in certain mean-field settings. For the Ising model (9.5) with independent and identically distributed (i.i.d.) couplings that would lead to a new term in the free energy, since now $\mathbb{E}_q J_{ij} x_i x_j = J_{ij} \langle x_i \rangle \langle x_j \rangle + J_{ij}^2 (1 - \langle x_i \rangle^2)(1 - \langle x_j \rangle^2)$, the new mean-field equations are given by

$$\langle x_j \rangle = \tanh \left( B_j + \sum_{k \in \mathcal{N}(j)} J_{jk} \langle x_k \rangle - \langle x_j \rangle \sum_{k \in \mathcal{N}(j)} J_{jk}^2 (1 - \langle x_k \rangle^2) \right), \qquad (9.22)$$

which are known as the TAP equations in statistical physics, while the new term in the equations is called the Onsager reaction term. There are more systematic ways of deriving such corrections: one is the Plefka expansion and another the cavity approach discussed next.

### 9.4.1   Cavity approach

Let us consider a probability distribution over pairwise interactions,

$$p(\boldsymbol{x}) = \frac{1}{Z} \prod_{j=1}^n f_j(x_j) \exp \left( \sum_{jk} J_{jk} x_j x_k + \sum_j B_j x_j \right)$$

$$\propto f_i(x_i) \exp \left\{ x_i \left( B_i + \underbrace{\sum_{j \in \mathcal{N}(i)} J_{ij} x_j}_{h_i} \right) \right\} p(\boldsymbol{x}_{\sim i}), \quad (9.23)$$

whence, introducing the *cavity field* distribution

$$p_{\sim i}(h_i) = \int d\boldsymbol{x}_{\sim i} \, \delta \left( h_i - \sum_{j \in \mathcal{N}(i)} J_{ij} x_j \right) p(\boldsymbol{x}_{\sim i}), \qquad (9.24)$$

the marginal distributions may be written as

$$p_i(x_i) \propto f_i(x_i) \int dh_i \, p_{\sim i}(h_i) \exp\{x_i(B_i + h_i)\}. \qquad (9.25)$$

Repeating the procedure for the $p(\boldsymbol{x}_{\sim i})$ by writing them in terms of $h_{j \in \mathcal{N}(i)}$ would lead to the belief propagation equations; we will proceed, however, by considering the large-connectivity limit, and apply the central limit theorem, according to which $p_{\sim i}(h_i)$ should be well approximated by a Gaussian. By setting

$$p_{\sim i}(h_i) \propto \exp \left\{ -(h_i - a_i)^2 / 2 V_i \right\}, \qquad (9.26)$$

the marginals may be computed from

$$p_i(x_i) = \frac{1}{z_i} f_i(x_i) \exp \left\{ (B_i + a_i) x_i + \frac{V_i}{2} x_i^2 \right\}. \qquad (9.27)$$

The $\{a_i\}$ can be determined using the identity

$$\langle h_i \rangle \propto \int \mathrm{d}x_i \, f_i(x_i) \int h_i \, \mathrm{d}h_i \, p_{\sim i}(h_i) \exp(x_i h_i) = a_i + V_i \langle x_i \rangle = \sum_{j \in \mathcal{N}(i)} J_{ij} \langle x_j \rangle, \quad (9.28)$$

so that

$$a_i = \sum_{j \in \mathcal{N}(i)} J_{ij} \langle x_j \rangle - V_i \langle x_i \rangle. \quad (9.29)$$

The $\{V_i\}$ are. by definition, $V_i = \sum_{jk} J_{ij} J_{jk} (\mathbb{E}_{p_{\sim i}} x_j x_k - \mathbb{E}_{p_{\sim i}} x_j \, \mathbb{E}_{p_{\sim i}} x_k)$; for independently sampled couplings and assuming that $\mathbb{E}_{p_{\sim i}} \sim \mathbb{E}_p$, one may write

$$V_i = \sum_{j \in \mathcal{N}(i)} J_{ij}^2 (1 - \langle x_j \rangle^2), \quad (9.30)$$

and, by substituting (9.30) into (9.29), we get

$$a_i = \sum_{j \in \mathcal{N}(i)} J_{ij} \langle x_j \rangle - \langle x_i \rangle \sum_{j \in N_i} J_{ij}^2 (1 - \langle x_j \rangle^2). \quad (9.31)$$

For $x_i \in \{\pm 1\}$, $\langle x_i \rangle = \tanh(B_i + a_i)$, thus recovering the TAP equations (9.22).

### 9.4.2  Adaptive correction for general $J$ ensembles

In deriving (9.30), we have assumed $J_{ij}$ and $J_{ik}$ to be statistically independent, so that in the thermodynamical limit $n \to \infty$, the off-diagonal terms $j \neq k$ vanish. While this assumption is true for the SK model where $J_{ij} \sim \mathcal{N}(0, 1/n)$, it breaks down when there are higher-order correlations between the $\{J_{ij}\}$. For instance, in Hopfield-like models, $J_{ij} = \sum_{p=1}^{\alpha n} \xi_i^{(p)} \xi_j^{(p)}$, these off-diagonal contributions do not vanish (Fig. 9.2). Also, when working with real data (Fig. 9.3), the distribution of the $\{J_{ij}\}$ is unknown, and it is important to have a scheme that works independently of any such assumptions.

While noting that $\langle x_i \rangle = (\partial / \partial B_i) \log z_i (B_i, a_i, V_i)$, let us consider again the linear response relation (9.21), from which a matrix $\chi$ of susceptibilities can be defined:

$$\chi_{ij} = \frac{\partial \langle x_i \rangle}{\partial B_j} = \frac{\partial \langle x_i \rangle}{\partial B_i} \frac{\partial B_i}{\partial B_j} + \frac{\partial \langle x_i \rangle}{\partial a_i} \frac{\partial a_i}{\partial B_j} = \frac{\partial \langle x_i \rangle}{\partial B_i} \left( \delta_{ij} + \frac{\partial a_i}{\partial B_j} \right) \quad (9.32)$$

and an approximation that the $\{V_i\}$ are kept fixed has been made. By further making use of (9.29), we get

$$\chi_{ij} = \frac{\partial \langle x_i \rangle}{\partial B_i} \left\{ \delta_{ij} + \sum_{k \in \mathcal{N}(i)} (J_{ik} - V_k \delta_{ik}) \chi_{ik} \right\}, \quad (9.33)$$

which can be solved for $\chi$ to yield

$$\chi_{ij} = \left[ (\Lambda - J)^{-1} \right]_{ij}, \quad (9.34)$$

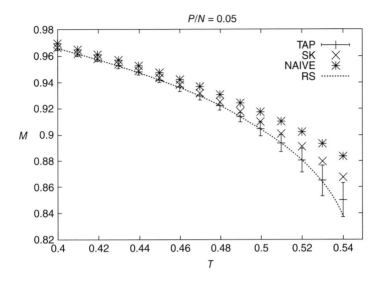

**Fig. 9.2** Comparison of the average overlap ($M$) in the Hopfield model obtained by means of iterating the naive mean-field equations (NAIVE), the TAP equations with i.i.d. assumptions (SK), and the correct TAP equations (TAP) derived (see Kabashima and Saad, 2001) for the statistics of the Hopfield couplings in a system with $N = 10\,000$. The Hopfield TAP equations successfully reproduce the replica-symmetric (RS) results, exactly for $N \to \infty$. Figure taken from Kabashima and Saad (2001).

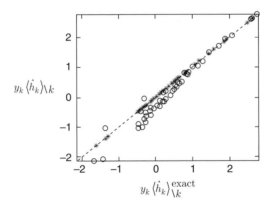

**Fig. 9.3** Testing TAP self-consistency in real data: the vertical axis gives the cavity field computed from the solution of the TAP equations, whereas the horizontal axis gives the *exact* cavity field obtained by removing each sample from the training data and iterating the $m - 1$ remaining TAP equations. Stars and circles give the results for adaptive and conventional TAP, respectively. Figure taken from Opper and Winther (2001).

where

$$\Lambda \equiv \text{diag}\left(V_1 + \left(\frac{\partial\langle x_1\rangle}{\partial B_1}\right)^{-1}, \ldots, V_n + \left(\frac{\partial\langle x_n\rangle}{\partial B_n}\right)^{-1}\right) \tag{9.35}$$

has been introduced. The diagonal elements $\chi_{ii}$ may be computed from this relation, but also from

$$\chi_{ii} = \frac{\partial^2}{\partial B_i^2}\log z_i = \frac{1}{\Lambda_i - V_i}; \tag{9.36}$$

by enforcing both solutions to be consistent, we obtain the following set of equations:

$$\frac{1}{\Lambda_i - V_i} = \left[(\Lambda - J)^{-1}\right]_{ii}, \tag{9.37}$$

which should be solved for the $V_i$, effectively replacing (9.30). Means and variances are then computed from

$$\langle x_i\rangle = \frac{\partial}{\partial B_i}\log z_i(B_i, a_i^*, V_i^*), \qquad \langle x_i^2\rangle - \langle x_i\rangle^2 = \frac{\partial^2}{\partial B_i^2}\log z_i(B_i, a_i^*, V_i^*), \tag{9.38}$$

given the updated values $\{a_i^*, V_i^*\}$.

### 9.4.3 Relation to EP

Another way of deriving the adaptive correction comes from replacing the $f_i(x_i)$ in (9.27) by a Gaussian $g_i(x_i) \propto \exp(-\frac{1}{2}\Lambda_i x_i^2 + \gamma_i x_i)$, with $\Lambda_i$ and $\gamma_i$ chosen as to be consistent with $p_i(x_i)$ in the first two moments $\langle x_i\rangle$ and $\langle x_i^2\rangle$. Let

$$z_i = \int dx_i\, f_i(x_i)\exp\left\{a_i x_i + \frac{V_i}{2}x_i^2\right\}, \qquad \tilde{z}_i = \int dx_i\, g_i(x_i)\exp\left\{a_i x_i + \frac{V_i}{2}x_i^2\right\}; \tag{9.39}$$

then, by the moment matching requirement, we have

$$\langle x_i\rangle = \frac{d}{da_i}\log z_i = \frac{d}{da_i}\log\tilde{z}_i = \frac{\gamma_i + a_i}{\Lambda_i - V_i},$$

$$\langle x_i^2\rangle - \langle x_i\rangle^2 = \frac{d^2}{da_i^2}\log z_i = \frac{d^2}{da_i^2}\log\tilde{z}_i = \frac{1}{\Lambda_i - V_i}. \tag{9.40}$$

On the other hand, by direct computation of the moments of

$$p(\boldsymbol{x}|\boldsymbol{y}) \propto \exp\left(\frac{1}{2}\boldsymbol{x}^\mathsf{T} J\boldsymbol{x}\right)\prod_{i=1}^{n} g_i(x_i), \tag{9.41}$$

we find

$$\langle x_i\rangle = \left[(\Lambda - J)^{-1}\boldsymbol{\gamma}\right]_i, \qquad \langle x_i^2\rangle - \langle x_i\rangle^2 = \left[(\Lambda - J)^{-1}\right]_{ii}, \tag{9.42}$$

and combining (9.40) and (9.42) leads us to the relations obtained previously. This approach, which generalizes an idea by Parisi and Potters, better illustrates the similarity between adaptive TAP and EP. In fact, running EP on a Gaussian latent model while using a Gaussian approximation gives us the same fixed points as iterating the equations above (Csató *et al.*, 2001).

### 9.4.4 TAP approximation to free energy

In order to write the TAP approximation to the free energy (Opper and Winther, 2001), we will introduce the $t$ variable, which mediates the strength between pairwise interactions:

$$p(\boldsymbol{x}) = \frac{1}{\mathcal{Z}_t} \exp\left(\frac{t}{2}\sum_{ij} J_{ij} x_i x_j\right) \prod_{i=1}^{n} f_i(x_i). \tag{9.43}$$

Since the TAP equations provide us with the first and second moments of $\boldsymbol{x}$, $m_i = \langle x_i \rangle$ and $M_i = \langle x_i^2 \rangle$, we work with the Legendre transform of the free energy, which keeps these quantities fixed:

$$G_t(\boldsymbol{m}, \boldsymbol{M}) = \min_q \left\{ \mathrm{KL}\left[q(\boldsymbol{x}) \| p(\boldsymbol{x})\right] \mid \langle x_i \rangle_q = m_i, \langle x_i^2 \rangle_q = M_i \right\} - \log \mathcal{Z}_t. \tag{9.44}$$

We want to compute $G$ for $t = 1$, which may be done by using

$$G_1 = G_0 + \int_0^1 \mathrm{d}t\, \frac{\partial G_t}{\partial t}. \tag{9.45}$$

Taking the derivative of $G_t$ with respect to $t$ gives

$$\frac{\partial G_t}{\partial t}(\boldsymbol{m}, \boldsymbol{M}) = -\frac{1}{2}\sum_{ij} J_{ij}\langle x_i x_j \rangle = -\frac{1}{2}\sum_{ij} J_{ij} m_i m_j - \frac{1}{2}\,\mathrm{Tr}\,\chi_t J; \tag{9.46}$$

then using the TAP approximation to $\chi$, $\chi_t = (\Lambda - tJ)^{-1}$, yields, after integration,

$$G_1 = G_0 - \frac{1}{2}\sum_{ij} J_{ij} m_i m_j + \frac{1}{2}\log\det(\Lambda - J)$$

$$- \sum_i V_i(M_i - m_i^2) + \frac{1}{2}\sum_i \log(M_i - m_i^2). \tag{9.47}$$

## 9.5 The Gibbs free energy

It has been shown (Opper and Winther, 2001; Csató *et al.*, 2001) that the Gibbs TAP free energy (9.47) has an interpretation in terms of a weighted sum of free energies for solvable models:

$$G_{\mathrm{TAP}}(\mu) = G^{\mathrm{Gauss}}(\mu) + G_0(\mu) - G_0^{\mathrm{Gauss}}(\mu). \tag{9.48}$$

These are the free energies for a global multivariate Gaussian approximation $G^{\mathrm{Gauss}}(\mu)$, a model that is factorized but contains non-Gaussian marginal distributions $G_0(\mu)$ and a part $G_0^{\mathrm{Gauss}}(\mu)$ that is Gaussian and factorized.

Each Gibbs free energy is defined by the moments $(m, M)$ corresponding to the single-site statistics $\phi(x) = (x, -\frac{1}{2}x^2)$, which are

$$\mu = (\langle \phi(x_1) \rangle, \dots, \langle \phi(x_N) \rangle). \tag{9.49}$$

The Gibbs free energy is obtained from the Legendre transform of the partition function for each of these models: given a set of fields $\lambda = \{(\gamma_i, \Lambda_i) : i = 1, \dots, N\}$, we can write the partition functions as

$$Z^{\text{Gauss}}(\lambda^1) = \int \mathrm{d}x \, f_0(x) \exp\left(\sum_{i=1}^{N}(\lambda_i^1)^{\mathsf{T}}\phi(x_i)\right), \tag{9.50}$$

$$Z_0(\lambda^2) = \int \mathrm{d}x \prod_{i=1}^{N} f_i(x) \exp\left(\sum_{i=1}^{N}(\lambda_i^2)^{\mathsf{T}}\phi(x_i)\right), \tag{9.51}$$

$$Z_0^{\text{Gauss}}(\lambda^3) = \int \mathrm{d}x \, \exp\left(\sum_{i=1}^{N}(\lambda_i^3)^{\mathsf{T}}\phi(x_i)\right). \tag{9.52}$$

where $f_0(x) = \exp\left(\sum_{ij} x_i J_{ij} x_j\right)$.

The Legendre transform relates the representation in terms of fields to the representation in terms of moments. The Gibbs free energy is defined as

$$G(\mu) = \max_{\lambda}\{-\ln Z(\lambda) + \lambda^{\mathsf{T}}\mu\}. \tag{9.53}$$

Stationarity of the TAP free energy $\nabla_\mu G_{\text{TAP}}(\mu) = 0$ implies that only two of the three fields $\lambda^*$ are independent:

$$-\log Z \approx -\log(Z_{\text{TAP}}) = -\log Z^{\text{Gauss}}(\lambda^1) - \log Z_0(\lambda^2) + \log Z_0^{\text{Gauss}}(\lambda^1 + \lambda^2). \tag{9.54}$$

We find the TAP free energy by requiring stationarity of the right-hand side with respect to the fields.

### 9.5.1   Double-loop algorithms (minimization of nonconvex Gibbs free energies)

Typically, one can formulate iterative algorithms as minimizers of some free energy, which are convergent under some sufficient set of criteria; one pervasive criterion required for the success of these algorithms (in general) is convexity of the free energy. Unfortunately, unlike the exact Gibbs free energy, approximate Gibbs free energies $G_{\text{approx}}$ are often not convex—this is true of adaptive TAP for example. However, in many cases, such as (9.48), there is a decomposition of the form

$$G_{\text{approx}}(\mu) = G_A(\mu) - G_B(\mu), \tag{9.55}$$

where both $G_A$ and $G_B$ are convex. There exists a type of algorithm, called a double-loop algorithm, that is guaranteed to find local minima of approximate free energies with this form (Heskes *et al.*, 2003). This is achieved by noting that there is an upper bound $L(\mu)$ to the concave part $-G_B(\mu)$, defined by

$$L(\mu) = -G_B(\mu_{\text{old}}) - (\mu - \mu_{\text{old}})\nabla G_B(\mu_{\text{old}}). \tag{9.56}$$

The function $G_{\text{vex}}(\mu) = G_A(\mu) + L(\mu)$ is now convex and we can apply standard methods. It can then be shown that by iteratively updating $\mu_{\text{old}}$, we can converge to a minimum; a standard implementation is as follows:

**Initialize** From random initial conditions $\mu_{\text{old}}$,

**Repeat** until convergence
    **Update** $\mu_{\text{new}} = \text{argmin}_\mu(G_{\text{vex}}(\mu))$,
    **Update** $\mu_{\text{old}} = \mu_{\text{new}}$.
    **At convergence** $\mu = \mu_{\text{new}}$ minimizes the approximate free energy $G_{\text{approx}}(\mu)$.

Alternatives to double-loop algorithms nevertheless continue to be popular, despite problematic behavior in some regimes. These include standard message-passing procedures such as loopy belief propagation and expectation propagation (using the recursive assumed density filtering procedure), and other naive gradient descent methods. It is found that in many interesting application domains, these methods do indeed converge. These algorithms are often preferred because of their significantly faster convergence.

## 9.6   Improving and quantifying the accuracy of EP

We have so far considered very simple approximations to $q$, involving approximations to the marginal distributions by Gaussians. There are a number of ways in which the approximations obtained might be improved. Two ways are to consider more structured approximations (on trees rather than single variables) and to consider expansions about the obtained solutions.

1. For discrete random variables, one can consider extension to moment matching approximations, where consistency of diagonal statistics $\phi(x) = \{x, x^2\}$ is extended to pair statistics $\phi(x_i, x_j) = x_i x_j$ (Opper and Winther, 2005; Minka and Qi, 2003).
2. We can expand about the approximation $q(x)$ to account at leading order for the structure ignored in the approximation (Opper *et al.*, 2009; Cseke and Heskes, 2011). We will consider two approximations based on expansions of the tilted distributions $q_n(x)$ about the EP approximation $q(x)$ (Opper *et al.*, 2009). An expansion can proceed in either the difference of the two distributions or in the higher-order cumulants (first and second cumulants agree by definition).

### 9.6.1   The tree approximation

In this case, and particularly for sparse graphs, we note that it is possible to consider a more substantial part of the interaction structure exactly; for example, we can include pair statistics in the approximations and require consistency of these moments.

The choice of additional pair interactions has an important effect on the quality of the approximation that will be obtained: we will want to include in the edge set the most important interactions, and as many as possible such that the approximation remains tractable. A practical extension is to include all the pair and vertex statistics defined by a tree within a Gaussian approximation (Opper and Winther, 2005). The tree can be chosen to cover the most important correlations (by some practical computable criteria). For example, it can be chosen as a maximum weight spanning tree, with weights given by the absolute values of the couplings.

Consider a spanning tree $\mathcal{T}$ that includes all the variables $\{n\}$ and a set of edges $\{(m, n)\}$. With respect to the tree,[2] each variable $n$ is said to have connectivity $d_n$. Assuming the probability distribution is described by a Gaussian part and a product of single variable distributions, it can be rewritten as

$$
p(x) = \frac{1}{Z} \exp\left(-\frac{1}{2}x^{\mathsf{T}} K^{-1} x\right) \frac{\displaystyle\prod_{(m,n)\in\mathcal{T}} f_m(x_m) f_n(x_n)}{\displaystyle\prod_{n\in\mathcal{T}} f_n(x_n)^{d_n-1}}. \tag{9.57}
$$

We will then approximate the latter part by a Gaussian restricted to the tree:

$$
p(x) = \frac{1}{Z} \exp\left(-\frac{1}{2}x^{\mathsf{T}} K^{-1} x\right) \frac{\displaystyle\prod_{(m,n)\in\mathcal{T}} g_{m,n}(x_m, x_n)}{\displaystyle\prod_{n\in\mathcal{T}} g_n(x_n)^{d_n-1}}. \tag{9.58}
$$

The moment matching algorithm can then be developed requiring matching of the moments $\langle x_m x_n \rangle \in \mathcal{T}$, in addition to the single-variable moments $\langle x_n \rangle$ and $\langle x_n^2 \rangle$.

For discrete random variables, Minka and Qi (2003) have proposed a method, tree-EP, that is in a similar spirit. It involves a non-Gaussian approximation with consistency of the moments on both edges and vertices required.

### 9.6.2   Expansion methods

The aim will be to demonstrate the connection between the approximate distribution $q$ and the true distribution $p$ as an expansion in some small terms. The expansion can indicate sufficient conditions for EP to succeed, and can be used in practice to improve estimation. In the selection of expansion methods, we do not assume that the interactions reveal any innate structure—if this were so, we might select or refine the approximation tailored to account for this clustering.

Consider the difference between the exact and approximate probability distributions for the standard approximation on marginal statistics, as described by

$$
p(x) = \frac{1}{Z} \prod_n f_n(x), \qquad q(x) = \frac{1}{Z_q} \prod_n g_n(x). \tag{9.59}
$$

---

[2] For simplicity, we assume a single connected component; otherwise the notation applies for a collection of trees.

A Gaussian interacting part can be included, since $f_0 = g_0$. The tilted distribution we recall as

$$q_n(x) = \frac{1}{Z_n}\left(\frac{q(x)f_n(x)}{g_n(x)}\right). \tag{9.60}$$

Solving for $f_n$ yields

$$\prod_n f_n(x) = \prod_n \frac{Z_n q_n(x) g_n(x)}{q(x)} = Z_{EP} q(x) \prod_n \left(\frac{q_n(x)}{q(x)}\right), \tag{9.61}$$

where we introduce the definition of the EP partition function

$$Z_{EP} = Z_q \prod_n Z_n. \tag{9.62}$$

In terms of the function

$$F(x) = \prod_n \left(\frac{q_n(x)}{q(x)}\right), \tag{9.63}$$

we can write

$$p(x) = \frac{1}{Z}\prod_n f_n(x) = \frac{Z_{EP}}{Z} q(x)F(x). \tag{9.64}$$

The ratio of the true to approximate partition functions defines the normalization constant

$$\frac{Z}{Z_{EP}} = \int q(x)F(x)\,\mathrm{d}x. \tag{9.65}$$

Note that the term $F(x)$ should be close to 1, and $Z \approx Z_{EP}$, when the approximation method works well.

### An expansion in $q_n(x)/q(x) - 1$

So long as the approximation is meaningful, the tilted and fully approximated distributions should be close, and hence we can take the quantity

$$\epsilon(x) = \frac{q_n(x)}{q(x)} - 1 \tag{9.66}$$

to be typically small. The exact probability is then

$$p(x) = q(x)\frac{1 + \sum_n \epsilon_n(x) + \sum_{n_1 < n_2} \epsilon_{n_1}(x)\epsilon_{n_2}(x) + \dots}{1 + \sum_{n_1 < n_2} \langle \epsilon_{n_1}(x)\epsilon_{n_2}(x)\rangle_q + \dots}. \tag{9.67}$$

The ratio of the partition functions (the denominator) is expanded as

$$R = 1 + \sum_{n_1 < n_2} \langle \epsilon_{n_1}(x) \epsilon_{n_2}(x) \rangle_q + \sum_{n_1 < n_2 < n_3} \langle \epsilon_{n_1}(x) \epsilon_{n_2}(x) \epsilon_{n_3}(x) \rangle + \dots. \tag{9.68}$$

At first order, $\sum_n \langle \epsilon_n(x) \rangle_q = 0$, by normalization. Thus, the first-order correction to the probability in $\epsilon_n$ is particularly simple in that it *does not require the computation of expectations*:

$$p(x) \approx \sum_n q_n(x) - (N-1)q(x). \tag{9.69}$$

A related calculation by Cseke and Heskes (2011) leads to a similar correction identity for marginal probabilities:

$$p(x_i) \approx q_i(x_i) \prod_{j(\neq i)} \int \mathrm{d}x_j \, q(x_j | x_i) \frac{f_j(x)}{g_j(x)}. \tag{9.70}$$

For an illustration of this expansion, we consider a class of models called Bayesian mixture of Gaussians, which is used to fit a mixture of $K$ Gaussians to data points $\zeta_n$. The latent variables of the model are $x = \{\pi_\kappa, \mu_\kappa, \Gamma_\kappa\}_{\kappa=1}^K$; each component $x_\kappa$ gives the weight, the mean, and the covariance matrix of a Gaussian in the mixture. Given a set of data $\zeta_n$, $n = 1, \dots, N$, and a prior, we will be interested in inferring these parameters. The likelihood given a data point $\zeta_n$ we will describe as

$$f_n(x) = \sum_\kappa \pi_\kappa \mathcal{N}(\zeta_n; \mu_\kappa, \Gamma_\kappa^{-1}), \tag{9.71}$$

where $\mathcal{N}(a; b, c)$ denotes a Gaussian distribution of the random variable $a$ with mean $b$ and covariance $c$. A convenient choice for the prior is a product of Dirichlet and Wishart distributions:

$$f_0 = \mathcal{D}(\pi) \prod_k \mathcal{W}(\mu_k, \Gamma_k). \tag{9.72}$$

From the product of the prior and likelihoods for different data, we obtain the posterior

$$p(x | \zeta_1, \dots, \zeta_N) = \frac{1}{Z} \prod_n f_n(x). \tag{9.73}$$

The approximation we make assumes the same form as the prior:

$$q(x) = \mathcal{D}(\pi) \prod_k \mathcal{W}(\mu_k, \Gamma_k). \tag{9.74}$$

A second class of models is given by the so-called Gaussian process popular in the area of machine learning (Rasmussen and Williams, 2006). Here one assumes latent variables with a joint prior distribution given by a Gaussian.

The Gaussian process classification problem involves a likelihood that leads to a non-Gaussian posterior probability $p(x)$. The classification can, however, be modeled by a Gaussian, yielding the approximation $q(x)$; thus, the structure of the problem is suitable for EP.

Within the context of a mixture of Gaussians and Gaussian process classification, the expansion method (9.67) can be very effective, as shown in Figs. 9.4–9.6.

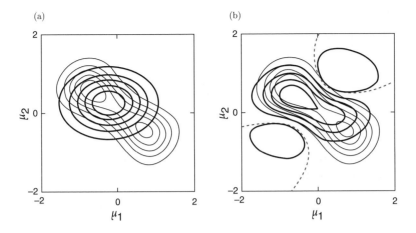

**Fig. 9.4** We generate a toy mixture model of two Gaussians, where only the means $\mu_{1,2}$ of the Gaussians are assumed to be unknown latent variables. The thin curves show the posterior probability distribution contour plots alongside the approximation. The thick curves show the EP estimate in (a) and the leading-order correction in (b). The dashed curves indicate a region in which the corrected probability becomes unphysical (negative).

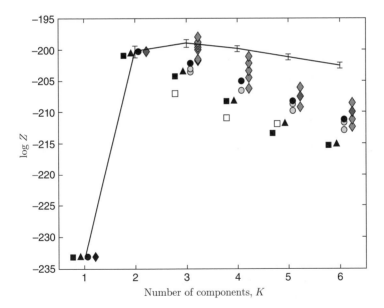

**Fig. 9.5** Approximations to a $K$-component Gaussian mixture model parametrized by a standard data set (the acidity data set) which gives data on acidity levels across 155 US lakes. The line with error bars indicates the Monte Carlo estimate to the log partition function. Significantly faster approximations are variational Bayes (squares) (Attias, 2000), Minka's $\alpha = \frac{1}{2}$ divergence method (triangles) (Minka, 2005), standard EP (circles), and EP with the second-order corrections (9.67) (diamonds). Symbol intensity indicates the frequency of the outcome based on 20 independent runs. Figure taken from Opper *et al.* (2009).

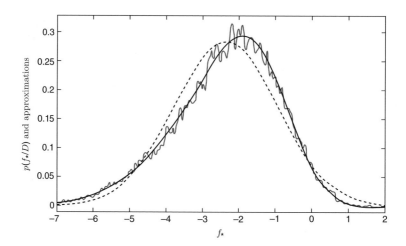

**Fig. 9.6** Marginal posterior for a toy Gaussian process classification. An MCMC approximation (jagged curve) compared with faster approximations: standard EP (dashed curve), and EP with the first-order correction (full curve) (9.67).

### *An expansion in cumulants*

The cumulants of a distribution are an alternative description, they represent a convenient framework for expansions about Gaussians.

Consider the latent Gaussian model, where

$$q(x) \propto \exp\left[-\frac{1}{2}x^{\mathsf{T}}K^{-1}x\right] \prod_n \exp\left(\gamma_n x_n - \frac{1}{2}\lambda_n x_n^2\right) \tag{9.75}$$

is used to approximate

$$p(x) = \frac{1}{Z}\exp\left[-\frac{1}{2}x^{\mathsf{T}}K^{-1}x\right] \prod_n f_n(x_n). \tag{9.76}$$

The tilted distribution (9.60) is required to match $q(x)$ in the first and second cumulant. The error can be quantified by the ratio of the partition functions ($R = Z/Z_{\text{EP}}$), which can be expressed as

$$R = \int q(x) \prod_n \left(\frac{q_n(x)}{q(x)}\right)$$

$$= \int q(x) \prod_n \left(\frac{q(x_{\sim n}|x_n)q_n(x_n)}{q(x_{\sim n}|x_n)q(x_n)}\right)$$

$$= \int \mathrm{d}x\, q(x) \prod_n \left(\frac{q_n(x_n)}{q(x_n)}\right), \tag{9.77}$$

where $x_{\sim n}$ denotes the set of all variables excluding $n$, and so $q(x_{\sim n}|x_n)$ is the conditional probability. So long as $q_n$ is Gaussian, the ratio is one, the errors can be accounted for by the higher-order cumulants of the tilted distribution, so these are good candidates for an expansion.

In terms of the Fourier transform $\chi_n(k)$ of the marginal tilted distribution, the original distribution is described by

$$q_n(x_n) = \int_{-\infty}^{\infty} \frac{dk}{2\pi} \exp(-ikx_n)\chi_n(k) \,. \tag{9.78}$$

In the Fourier basis, we have a simple expression for all the neglected cumulants $r_n(k)$ (of third order and higher) as

$$\log \chi_n(k) = \sum_l (i)^l \frac{c_{nl}}{l!} k^l = im_n k - S_n \frac{k^2}{2} + r_n(k) \,. \tag{9.79}$$

The first two moments are equal in the tilted and untilted distributions by the moment matching conditions. The ratio of the two distributions can be found by resubstitution, in terms of an integral

$$\frac{q_n(x_n)}{q(x_n)} = \sqrt{\frac{S_{nn}}{2\pi}} \exp\left(\frac{(x_n - m_n)^2}{2S_{nn}}\right) \int_{-\infty}^{\infty} \frac{dk}{2\pi} \exp(-ikx_n)\chi_n(k) \,. \tag{9.80}$$

Using the linear change of integration variable

$$\nu_n = k + i\frac{x_n - m_n}{S_{nn}} \,, \tag{9.81}$$

we can abbreviate

$$\frac{q_n(x_n)}{q(x_n)} = \int_{-\infty}^{\infty} d\nu_n \sqrt{\frac{S_{nn}}{2\pi}} \exp\left[-\sum_n \frac{S_{nn}\nu_n^2}{2}\right] \exp\left[r_n\left(\nu_n - i\frac{x_n - m_n}{S_{nn}}\right)\right] \,. \tag{9.82}$$

The ratio of partition functions

$$R = E_q \left[\prod_n \left(\frac{q_n(x_n)}{q(x_n)}\right)\right] \tag{9.83}$$

requires an integral over the multivariate Gaussian $q(x) = \mathcal{N}(x; m, S)$, in addition to $\nu_n$. We have a double integration over an expression that depends on a weighted sums of the parameters $\nu_n$ and $x_n$. Since these are Gaussian variables, and the sum of two Gaussian variables is itself Gaussian, we can replace the weighted sums by a new complex Gaussian random variable

$$z_n = \nu_n - i\frac{x_n - m_n}{S_{nn}} \tag{9.84}$$

of zero mean and with a distribution described by covariances

$$\langle z_n^2 \rangle_z = 0\,, \qquad \langle z_m z_n \rangle_z = -\frac{S_{mn}}{S_{nn} S_{mm}}\,. \tag{9.85}$$

The double integral is thereby replaced by a single integral

$$\frac{Z}{Z_{\text{EP}}} = \left\langle \exp\left[\sum_n r_n(z_n)\right]\right\rangle_z \,. \tag{9.86}$$

Assuming that the cumulants $c_{ln} > 2$ are small, we can make a power series expansion in $r_n(z_n)$, to obtain

$$\log(R) = \frac{1}{2} \sum_{m \neq n} \langle r_m r_n \rangle_z + \dots$$

$$= \sum_{m \neq n} \sum_{l \geq 3} \frac{c_{ln} c_{lm}}{l!} \left(\frac{S_{mn}}{S_{nn} S_{mm}}\right)^l + \dots . \tag{9.87}$$

The self-interaction (diagonal) term is absent, which indicates that corrections may not scale with $N$, a desirable property—at least insofar as theory is concerned.

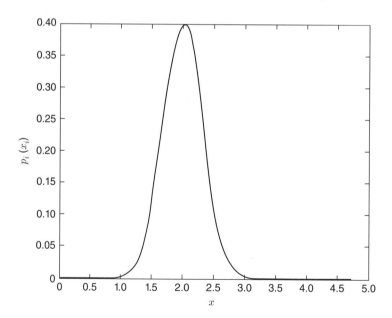

**Fig. 9.7**  A toy problem considering classification likelihoods $f_i(x_i) = \Theta(y_i x_i)$. Provided posterior variance is small compared with the mean, the portion of the Gaussian extending across the threshold at zero is small. Truncation of the Gaussian in the tail results in only a small modification of the cumulants in the tilted function $q_i(x_i)$.

(a)                                              (b)

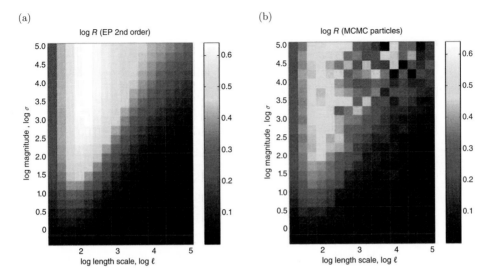

**Fig. 9.8** Analysis of Gaussian process classification on US postal service data. The data set consists of $16 \times 16$ grayscale images (real-valued vectors with components on $[0, 1]$) of handwritten digits ($[0, 9]$). The free energy is already well approximated by EP. The remaining difference in the free energy ($\log R$) is mostly accounted for by including $l \leq 4$ corrections (a), by comparison with the exact Monte Carlo evaluation (b). The exact evaluation is slower by orders of magnitude. The free parameters are a length scale $l$ and the width of the Gaussian prior, $\sigma$. Figure taken from Opper *et al.* (2013); see also Rasmussen and Williams (2006).

(a)                                              (b)

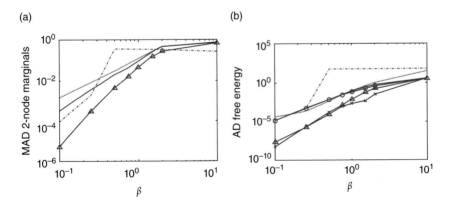

**Fig. 9.9** An Ising spin model of $N$ spins, with zero field and random couplings of variance $\beta^2$. The absolute difference (AD) between exactly calculated values and approximation values is shown for the two-node marginals (a) and true free energy ($-\log Z$) (b). The approximate methods used are loopy belief propagation (LBP) (dotted line), generalized LBP (with shortest loops included in outer region) (dashed line), and standard EP (full line without symbols). The second-order cumulant corrections (9.87) with $l \leq 3$ (circle), 4 (triangle), and 5 (cross) ameliorate the error. Figure taken from Opper *et al.* (2009).

| Graph | Coupling | $d_{\mathrm{coup}}$ | EP | EP c | EP t | E P tc |
|-------|----------|---------------------|------|------|------|--------|
| Full  | Repulsive | 0.25 | 0.0310 | 0.0018 | 0 .0104 | 0 .0010 |
|       | Repulsive | 0.50 | 0.3358 | 0.0639 | 0.1412 | 0.0440 |
|       | Mixed | 0.25 | 0.0235 | 0.0013 | 0.0129 | 0.0009 |
|       | Mixed | 0.50 | 0.3362 | 0.0655 | 0.1798 | 0.0620 |
|       | Attractive | 0.06 | 0.0236 | 0.0028 | 0.0166 | 0.0006 |
|       | Attractive | 0.12 | 0.8297 | 0.1882 | 0.2672 | 0.2094 |
| Grid  | Repulsive | 1.0 | 1.7776 | 0.8461 | 0.0279 | 0.0115 |
|       | Repulsive | 2.0 | 4.3555 | 2.9239 | 0.0086 | 0.0077 |
|       | Mixed | 1.0 | 0.3539 | 0.1443 | 0.0133 | 0.0039 |
|       | Mixed | 2.0 | 1.2960 | 0.7057 | 0.0566 | 0.0179 |
|       | Attractive | 1.0 | 1.6114 | 0.7916 | 0.0282 | 0.0111 |
|       | Attractive | 2.0 | 4.2861 | 2.9350 | 0.0441 | 0.0433 |

**Fig. 9.10** In the Wainwright–Jordan set-up, there are $N = 16$ Ising spins, which are either (upper table) fully connected or (lower table) connected to nearest neighbors in a $4 \times 4$ grid. Fields are sampled uniformly at random in $[-0.25, 0.25]$ and couplings are sampled from distributions of width $[a - d_{\mathrm{coup}}, a + d_{\mathrm{coup}}]$, where $a = d_{\mathrm{coup}}$ is the attractive case, $a = 0$ the mixed case, and $a = -d_{\mathrm{coup}}$ the repulsive case. Average absolute deviations of $\log Z$ are shown, comparing standard EP (EP), EP with tree corrections (EP-t), and the same approximations with $l = 4$ second-order cumulant corrections (EP-c and EP-tc, respectively). Figure taken from Opper *et al.* (2009).

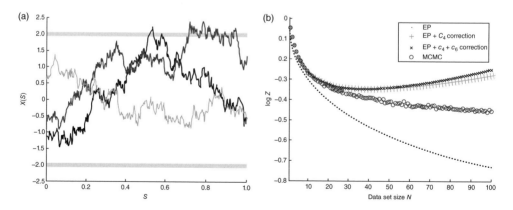

**Fig. 9.11** Gaussian process in a box. We are interested in the probability distribution over a Gaussian process constrained to remain in the interval $[-a, a]$ at all times: $p(x) = Z^{-1} \prod_n \theta(a - |x_n|) \mathcal{N}(x; 0, K)$, where $\theta$ is the Heaviside function. Realizations of $x$ are shown in (a) (of which one is invalid). $Z$ is the fraction of processes that remain in the box; as the number of points $n$ increases, we anticipate convergence. EP strongly underestimates the number of solutions, while the corrections, after initial improvements, significantly overestimate it.

The cumulant expansion allows us to consider the scenarios in which EP may be accurate. From the calculation, we note that the correction is small if either the cumulants $c_{ln}$ are small, which is often true in classification problems (see, e.g., Fig. 9.7), or the posterior covariances $S_{mn}$ are small for $m \neq n$.

Applications of the cumulant expansions are shown in Figs. 9.8–9.11. In the first four cases, we see significant improvements in the estimation. In the final example of a Gaussian process in a box, the EP estimate becomes increasingly inaccurate as the number of observations, $N$, increases, and the corrections about this result do not greatly improve the estimation. Standard EP is clearly not universally applicable, and a better approximation to capture the problem structure is required before expansion methods can be useful.

## 9.7   Some advanced topics and applications

It is worth noting that already in a form closely related to that presented here there are many applications of EP, examples being TrueSkill (Dangauthier *et al.*, 2008) and recommender systems at Microsoft. In the TrueSkill application, one has a set of data that is the outcome of games between players matched online. One can imagine that each player has a certain positive skill level (which might be time-dependent) modeled by a latent variable $x_i$ (the true skill). The discrete (binary) outcome of a game is the data: victory or defeat might be modeled by the sign of $x_i - x_j + \nu$, where $\nu$ is some noise. This defines a simple likelihood suitable for the methods outlined in previous sections, and EP is being used in practice to determine player skill levels. The true model is more complicated, since it involves a dynamic element, but it is still simple enough to process an estimated $O(10^6)$ games for $O(10^5)$ players every day in real time.

In the remainder of this section, we outline some other advanced applications and generalizations of the methods presented in earlier sections. These examples show how EP can be used out of the box in combination with other analytical techniques such as the replica method, and for continuous-time and other frameworks outside the standard machine learning context.

### 9.7.1   Bootstrap estimation for Gaussian regression models, an application of the replica method

A standard problem in inference is to determine the mean square error of a statistical estimator $\hat{E}[x_i|D]$, such as a method for predicting the mean or variance of a distribution. We now outline a method to combine EP with the replica method to reduce the complexity of bootstrap estimation (Malzahn and Opper, 2002).

In practice, we do not have access to the distribution, only $N$ data points $D_0$. To get around this problem, *we pull ourselves up by our bootstraps*—using a subset of the data for training, while the remaining set constructs an empirical distribution $D$. The *pseudo-data* that builds the empirical distribution is determined via resampling with replacement.

Consider a vector $m_i$ to represent the sampling process: each data point can be included zero, one, or several times in forming the empirical distribution $D$. Those elements not occupied $m_i = 0$ are independent of the empirical distribution, and can be used as the test set. The test error (the Efrons estimator) can then be constructed as

$$\epsilon(m) = \frac{1}{N} \sum_i \frac{E_D[\delta_{m_i,0} (E[x_i|D] - y_i)^2]}{E_D[\delta_{m_i,0}]}. \tag{9.88}$$

One problem with the scheme is that for each sample we must reevaluate the error, and many samples are required for a robust estimate. One can consider an approximation to this quantity using statistical physics insight as discussed in other chapters in this volume. Asymptotically, occupation numbers $m_i$ become independently and identically distributed (i.i.d.) Poisson random variables of mean $m/N$, so the denominator in (9.88) is $\exp(-m/N)$. Still, for a particular sample, we can write

$$\epsilon_n(m) =$$

$$\frac{1}{\exp(-m/N)N} \sum_{i=1}^{N} E_D\left[\delta_{m_i,0} Z^{n-2} \int \prod_{j=1,2} [\mathrm{d}x^{(j)} \, p_0(x^{(j)}) P(D|x^{(j)})(x_i^{(j)} - y_i)]\right]. \tag{9.89}$$

The idea of the replica method is to compute $\epsilon_n(m)$ (approximately) analytically for integer $n > 2$ and then to take the limit $n \to 0$. $Z$ is the partition function for a single replica; the full calculation is found in Malzahn and Opper (2002). We use EP and take the limit $n \to 0$ to approximate the Efrons estimator (9.88). The analytical approximation to $E_D$ saves significantly on the computation. Comparisons between analytical and empirical test errors obtained for different problems and datasets are given in Fig. 9.12.

### 9.7.2   Gaussian approximations for generalized models

For Gaussian latent models, we have shown the power of the methods outlined. We have, however, focused on pairwise models, and we now identify one significant generalization (Opper and Winther, 2001). We can generalize to a wider class of models defined by

$$p(x) \propto \prod_{i=1}^{N} f_i(x_i) \exp\left[\sum_{i<j}^{N} x_i J_{ij} x_j\right] \prod_{k=1}^{m} F\left(\sum_i \hat{J}_{ik} x_k\right). \tag{9.90}$$

Thus, we are able to extend straightforwardly our consideration of latent Gaussian models to classifiers as one example (e.g., a perceptron classifier can take the form $F_k(x) = \Theta\left(\sum_i \hat{J}_{ik} x_k - b_k\right)$) and other interesting models. Defining augmented random variables $\sigma = (x, \hat{x})$, we can cast this model in the form

$$p(\sigma) \propto \prod_i \rho_i(\sigma) \exp\left[\sum_{i<j} \sigma_i A_{ij} \sigma_j\right], \tag{9.91}$$

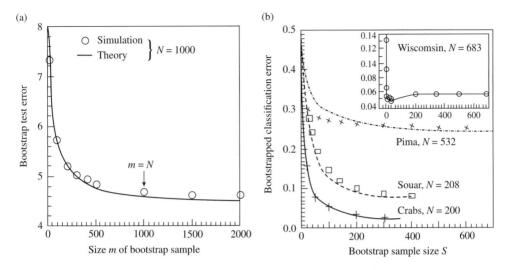

**Fig. 9.12** Example of results for regression and support vector machine classification. (a) Averaged bootstrapped generalization error on the Abalone data (the simulation results are shown by the symbols and the theoretical results by the curve, each using the same $D_0$ of size $N = 1000$). The Abalone data set relates a set of physical characteristics of abalones to their age, the latter being the object of classification. (b) Average bootstrapped generalization error for hard margin support vector classification on different standard data sets (again symbols represent simulation results and curves theoretical results). Figures are taken respectively from Malzahn and Opper (2002, 2003).

where the augmented coupling matrix is

$$A = \begin{pmatrix} J & \hat{J} \\ \hat{J}^{\mathsf{T}} & 0 \end{pmatrix} \tag{9.92}$$

and $\rho_i(\sigma_i) = f_i(x_i)\hat{f}_i(\hat{x}_i)$, where

$$f_i(\hat{x}_i) = \int \frac{\mathrm{d}h}{2\pi i} \exp\left(-\hat{x}h\right) F_i(h). \tag{9.93}$$

### 9.7.3 Inference in continuous-time stochastic dynamics

The application of EP to continuous processes introduces new challenges, but these can be overcome—we give one example (Cseke *et al.*, 2013). Assume a prior process (Ornstein–Uhlenbeck), with variables $x$ having a continuous time index $t$ rather than the discrete $n$:

$$\mathrm{d}x_t = (A_t x_t + c_t)\,\mathrm{d}t + B_t^{1/2}\,\mathrm{d}W_t. \tag{9.94}$$

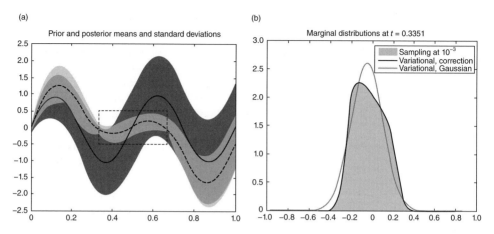

**Fig. 9.13** (a) A Gaussian prior with a periodic drift (dark shading, mean plus error bars), and the posterior process with a hard box constraint added (light shading, mean plus error bars). The continuous-time potential is defined as $(2x_t)^8 I_{[1/2,2/3]}(t)$, which describes soft box constraints, and we assume two hard box discrete likelihood terms $I_{[-0.25,0.25]}(x_{t_1})$ and $I_{[-0.25,0.25]}(x_{t_2})$ placed at $t_1 = 1/3$ and $t_2 = 2/3$. The prior is defined by the parameters $a_t = -1$, $c_t = 4\pi \cos(4\pi t)$ and $b_t = 4$. (b) The purely Gaussian approximation of EP fails to anticipate the skew of the distribution, which a correction to EP in the style of (9.70) is able to capture. Figure taken from Cseke *et al.* (2013).

We can build the likelihood given *continuous-* and discrete-time observations $y$. The simplest case might be one in which we measure spike counts, and a statistical log-Cox process would be described by the likelihood

$$p(\{y_{t_i}^d\}, \{y_t^c\}|\{x_t\}) \propto \prod_{t_i \in T_d} p(y_{t_i}^d|x_{t_i}) \exp\left\{-\int_0^1 dt\, V(t, y_t^c, x^t)\right\}, \tag{9.95}$$

where $T_d$ is the set of spike observation times. We wish to estimate the rate process. We can apply our EP approximation if we discretize in time:

$$p(\{y_{t_i}^d\}, \{y_t^c\}|\{x_t\}) \propto \prod_{t_i \in T_d} p(y_{t_i}^d|x_{t_i}) \prod_k \exp\{-\Delta_{t_k} V(t_k, y_t^c, x^{t_k})\}. \tag{9.96}$$

An important question is whether EP remains meaningful in the limit $\Delta t \to 0$. It can be shown that EP continues to be applicable in this limit for smooth approximations. In Fig. 9.13, we have a process with a drift, with a hard box constraint that is softened in order to apply the approximation.

## 9.8 Conclusion: open problems with EP

We have seen how EP applies across a broad range of theoretical and practical inference problems. We now conclude with some open problems.

A major open problem with EP is scaling for structured approximation. Applications like TrueSkill can work with independent variable approximations, but structured approximations should be more powerful. Another challenge is creating versions of the algorithm that parallelize.

Convergence properties of EP-related approximations are not well understood. In certain approximation algorithms, or in special applications, there are non-rigorous arguments that connect non-convergence of fast methods with weakness of the underlying model assumptions, but this needs to be generalized.

The free energies are not known to be bounds except for a few special cases. Bounds are important since they would allow proofs for convergence of various schemes, as well as allowing certain systematic extensions of approximations.

It would be interesting to understand performance bounds based on very general criteria: one such framework with which a connection might be made is that of PAC–Bayes bounds.

# References

Attias, H. (2000). A variational Bayesian framework for graphical models. In *Advances in Neural Information Processing Systems 12*, pp. 209–215. MIT Press.

Barber, D. (2011). *Bayesian Reasoning and Machine Learning*. Cambridge University Press.

Bishop, C. (2007). *Pattern Recognition and Machine Learning*. Springer.

Csató, L., Opper, M., and Winther, O. (2001). TAP Gibbs free energy, belief propagation and sparsity. In *Advances in Neural Information Processing Systems*, pp. 657–663.

Cseke, B. and Heskes, T. (2011). Approximate marginals in latent Gaussian models. *Journal of Machine Learning Research*, **12**, 417–454.

Cseke, B., Opper, M., and Sanguinetti, G. (2013). Approximate inference in latent diffusion processes from continuous time observations. In *Advances in Neural Information Processing Systems 26*.

Dangauthier, P., Herbrich, R., Tom, M., and Graepel, T. (2008). Trueskill through time: Revisiting the history of Chess. In *Advances in Neural Information Processing Systems 20* (ed. J.C. Platt, D. Koller, Y. Singer, and S.T. Roweis), pp. 337–344. Curran Associates, Inc. http://papers.nips.cc/paper/3331-trueskill-through-time-revisiting-the-history-of-chess.pdf

Heskes, T., Albers, K., and Kappen, B. (2003). Approximate inference and constrained optimization. In *Proceedings of the 19th Conference on Uncertainty in Artificial Intelligence*, UAI'03, San Francisco, CA, pp. 313–320. Morgan Kaufmann Publishers.

Kabashima, Y. and Saad, D. (2001). The TAP approach to intensive and extensive connectivity systems. *Advanced Mean Field Methods—Theory and Practice*, **6**, 65–84.

Malzahn, D. and Opper, M. (2002). A statistical mechanics approach to approximate analytical bootstrap averages. In *Advances in Neural Information Processing Systems 15*, pp. 327–334.

Malzahn, D. and Opper, M. (2003). Approximate analytical bootstrap averages for support vector classifiers. In *Advances in Neural Information Processing Systems 16*.

Minka, T. (2001). Expectation propagation for approximate bayesian inference. In *Proceedings of the 17th Conference on Uncertainty in Artificial Intelligence*, UAI'01, San Francisco, CA, USA, pp. 362–369. Morgan Kaufmann.

Minka, T. and Qi, Y. (2003). Tree-structured approximations by expectation propagation. In *Advances in Neural Information Processing Systems 16*.

Minka, T. P. (2005). Divergence measures and message passing. Technical Report MSR-TR-2005-173, Microsoft Research, Cambridge, UK.

Murphy, K. P. (2012). *Machine Learning: A Probabilistic Perspective*. MIT Press.

Opper, Manfred (1989). A Bayesian approach to online learning. In *On-Line Learning in Neural Networks* (ed. D. Saad), pp. 363–378. Cambridge University Press.

Opper, M., Paquet, U., and Winther, O. (2009). Improving on expectation propagation. In *Advances in Neural Information Processing Systems 21*.

Opper, M., Paquet, U., and Winther, O. (2013). Perturbative corrections for approximate inference in Gaussian latent variable models. *Journal of Machine Learning Research*, **14**, 2857–2898.

Opper, M. and Saad, D. (2001). *Advanced Mean Field Methods: Theory and Practice*. MIT Press.

Opper, M. and Winther, O. (2001). Adaptive and self-averaging Thouless–Anderson–Palmer mean-field theory for probabilistic modeling. *Physical Review E*, **64**(5), 056131.

Opper, M. and Winther, O. (2005). Expectation consistent free energies for approximate inference. In *Advances in Neural Information Processing Systems 17*.

Rasmussen, C.E. and Williams, C.K.I. (2006). *Gaussian Processes for Machine Learning*. MIT Press.

Wainwright, M. and Jordan, M. (2008). Graphical models, exponential families, and variational inference. *Foundations and Trends in Machine Learning*, **1**(1-2), 1–305.

Yedidia, J., Freeman, W., and Weiss, Y. (2000). Bethe free energy, Kikuchi approximations and belief propagation algorithms. Technical report, Mitsubishi Electric Research Laboratories, Inc.